C000241633

PRAISE FOR

THE PRIVILEGED PLANET

"In a book of magnificent sweep and daring, Guillermo Gonzalez and Jay Richards drive home the argument that the old cliché of no place like home is eerily true of Earth. Not only that, but if the scientific method were to emerge anywhere, Earth is about as suitable as you can get. Gonzalez and Richards have flung down the gauntlet. Let the debate begin; it is a question that involves us all."
— Simon Conway Morris, author of *Life's Solution:
Inevitable Humans in a Lonely Universe*

"This thoughtful, delightfully contrarian book will rile up those who believe the 'Copernican Principle' is an essential philosophical component of modern science. Is our universe designedly congenial to intelligent, observing life? Passionate advocates of the search for extraterrestrial intelligence (SETI) will find much to ponder in this carefully documented analysis."
— Owen Gingerich, Harvard-Smithsonian Center
for Astrophysics, author of *The Book Nobody Read:
Chasing the Revolution of Nicolaus Copernicus*

"Not only have Guillermo Gonzalez and Jay Richards written a book with a remarkable thesis, they have constructed their argument on an abundance of evidence and with a cautiousness of statement that make their volume even more remarkable. In my opinion, *The Privileged Planet* deserves very careful attention."
— Michael J. Crowe, Cavanaugh Professor Emeritus
at the University of Notre Dame and author of
The Extraterrestrial Life Debate, 1750–1900

"The Privileged Planet will surely rattle, if not finally dislodge, a pet assumption held by many interpreters of modern science: the so-called Copernican Principle. Gonzalez's and Richards's argument is so carefully and moderately presented that any reasonable critique of it must itself address the astonishing evidence. I expect this book to renew the whole scientific and philosophic debate about Earth's cosmic significance. It is a high-class piece of work that deserves the widest possible audience."
— Dennis Danielson, Professor of English at the
University of British Columbia and editor of
The Book of the Cosmos

The

PRIVILEGED
PLANET

HOW OUR PLACE IN THE COSMOS IS
DESIGNED FOR DISCOVERY

GUILLERMO GONZALEZ AND JAY W. RICHARDS

REGNERY
GATEWAY

Regnery books may be purchased in bulk at special discounts for sales promotion, corporate gifts, fund-raising, or educational purposes. Special editions can also be created to specifications. For details, contact the Special Sales Department, Regnery, 307 West 36th Street, 11th Floor, New York, NY 10018 or info@skyhorsepublishing.com.

Regnery® is an imprint of Skyhorse Publishing, Inc.®, a Delaware corporation.

Visit our website at www.regnery.com.
Please follow our publisher Tony Lyons on Instagram @tonylyonsisuncertain.

10 9 8 7 6 5 4 3 2 1

Library of Congress Cataloging-in-Publication Data is available on file.

Cover design by Brian Gage and David Ter-Avanesyan
Cover image by Guillermo Gonzalez and Jay W. Richards

Print ISBN: 978-1-68451-702-2
Ebook ISBN: 978-1-59698-707-4

Printed in the United States of America

CONTENTS

FOREWORD TO THE 2024 EDITION

The gratuitous beauty of the starry heavens above. The chimerical appearance of rainbows. The austere splendor of solar eclipses. These inspire people in every time and place. For most of human history, they were also mysteries.

In a sense, we've now removed the mysteries. Scientists routinely measure the distances to the stars. We know how sunlight passing through millions of suspended water droplets in the atmosphere renders a rainbow. We can predict to within one second the time and place of solar eclipses years in advance anywhere on Earth.

Yet, these discoveries point to even deeper mysteries. Why is our world, from the local and galactic environments to the constants of physics to the remnant echo of the big bang, set up so that we can see the stars, rainbows, and solar eclipses? Being able to see these things per se is not a prerequisite for our existence. Surely, the universe could have been otherwise.

In 2004, when this book was first published, we connected the dots. We argued not just that the universe is vaguely habitable and open to science, but that those rare places in the universe best suited for complex life and observers are also the best places, overall, for scientific discovery.

Moreover, we argued that this is evidence of a conspiracy, rather than a mere coincidence. We set out to test this hypothesis against the best evidence from natural science. We argued that, together, the evidence forms a cumulative case that the universe is designed for discovery.

Though we stated the argument precisely, we didn't try to immunize it against future discoveries. On the contrary, we put it at risk. That's why we predicted that relevant new observations (or new analyses of existing observations) would confirm our argument.

So, how has our argument fared in the intervening years? Here's a brief survey of what nature has revealed.

NATURE STILL SPEAKS

We opened the book by pondering why we can observe *perfect* solar eclipses. We enjoy them because the Sun and Moon appear to be the same size from the Earth's surface. While this coincidence has been long known, it still befuddles scholars. Seven years after *The Privileged Planet* was published, John Gribbin observed in *Alone in the Universe: Why Our Planet Is Unique*:

> Just now the Moon is about 400 times smaller than the Sun, but the Sun is 400 times farther away than the Moon, so that they look the same size on the sky. At the present moment of cosmic time, during an eclipse, the disc of the Moon almost exactly covers the disc of the Sun. In the past the Moon would have looked much bigger and would have completely obscured the Sun during eclipses; in the future, the Moon will look much smaller from Earth and a ring of sunlight will be visible even during an eclipse. Nobody has been able to think of a reason why intelligent beings capable of noticing this oddity should have evolved on Earth just at the time that the coincidence was there to be noticed. It worries me, but most people seem to accept it as just one of those things.[1]

Either Gribbin didn't read *The Privileged Planet* or he didn't want to admit that he had.

Although we explored this topic in detail, we left two issues unresolved. First, while we showed that Earth enjoys solar eclipses far better than the other planets with moons, we didn't consider solar eclipses viewed *from* other moons. So-called mutual eclipses occur when one moon eclipses another moon in orbit around the same planet—as they do on some outer planets such as Jupiter, which have many moons. One of us (Guillermo) later studied mutual eclipses in the solar system and found that Earth's solar eclipses are indeed the best. The results of this research were published in 2009.[2]

Second, while we explained why the apparent sizes of the Sun and Moon closely match, we didn't explain why they are such a good match. The same year that *The Privileged Planet* came out, Dave Waltham of the University of London published a paper arguing that the Moon's size is almost too large to stabilize Earth's rotation axis well—within 22 kilometers (about 14 miles)![3] In a later paper, Waltham buttressed the case for the Moon's role in stabilizing Earth's climate.[4] This strengthens our argument.

In chapters 2–4 we discussed certain ways that Earth is a good platform for scientific discovery. In chapter 3, we argued that Earth's accessible, abundant, and diverse minerals and fossil fuels prepared the way for technology. Robert Hazen, a geologist at Carnegie Institution's Geophysical Laboratory and George Mason University, has quantified just how special Earth's mineral resources are. In his 2012 book, *The Story of Earth: The First 4.5 Billion Years, from Stardust to Living Planet*, Hazen notes that Earth has the greatest diversity of mineral species of any body in the solar system.[5] Over 4,600 mineral species are known on Earth. Mars only has about 500 and Venus about 1,000. What's more, Hazen discovered that life formed about two-thirds of Earth's mineral species. His work reinforces our argument linking life and scientific discovery.

Because our book covered so many topics, it was, in effect, an hors d'oeuvres reception with a dozen small appetizers. Michael Denton's recent books in his Privileged Species series, *Fire-Maker: How Humans Were Designed to Harness Fire and Transform our Planet* (2016),[6] *The Wonder of Water: Water's Profound Fitness for Life on Earth and Mankind* (2017),[7] and *Children of the Light: Astonishing Properties of Sunlight That Make Us Possible* (2018),[8] provide a three-course meal. If our treatment whets your appetite, then we encourage you to check out Denton's trilogy.

In chapter 5 we discussed exoplanets, but only briefly, because, at the time, just over 100 such planets were known.[9] The number of confirmed exoplanets has doubled about every 27 months (about two and a half years) since then. As we write this, over 5,500 confirmed exoplanets have been discovered.

Some critics, who hadn't read our book, assumed that every new exoplanet discovery would somehow weaken our argument. Our claim, they seemed to think, was that our planet was unique, and therefore, designed. That is a clunker of an argument! But it's not the one we made. We didn't know in 2004 if Earth was uniquely habitable. We still don't know that in 2024. For many people, the sheer number of new planet discoveries seems to guarantee many other earthlike planets in the Milky Way galaxy. But other research in astrobiology rejects this impression. While astrobiologists have been finding exoplanets, they have also been discovering new ways that a planetary system must be fine-tuned to allow for habitable planets. Taken as a whole, astrobiology research in the last twenty years has not changed our view that earthlike planets are very rare.[10] If anything, that claim is less risky now than it was at the time. We can't yet assign numbers, but we're quite confident that the number of earthlike planets will be a tiny fraction of the exoplanets we discover, and that any habitable planets we do find will be Earth-twins.

Turning to the broader cosmos, the theory of general relativity received new confirmations from the first direct measurement of gravity waves in 2015, the first (synthetic) image of the region around a supermassive black hole in 2019, and the gravitational wave background in 2023. Why do these discoveries matter? Because general relativity is the foundational theory for big bang cosmology, which points to a beginning. In chapter 9, where we covered cosmology, we concluded that we live, not just during what we dubbed the cosmic habitable age—which is no surprise— but also during the best time to study cosmology. In a much later, less habitable era, evidence of a beginning and of cosmic expansion would be undetectable.

In 2007, we received support for this idea from an unlikely source. Atheist cosmologist Lawrence Krauss with Vanderbilt cosmologist Robert Scherrer published an award-winning paper on the loss of key information about the cosmos in the distant future. Their concluding paragraphs are worth quoting at length:

> The remarkable cosmic coincidence that we happen to live at the only time in the history of the universe when the magnitude of dark energy and dark matter densities are comparable has been a source of great current speculation, leading to a resurgence of interest in possible anthropic arguments limiting the value of the vacuum energy... But this coincidence endows our current epoch with another special feature, namely that we can actually infer both the existence of the cosmological expansion, and the existence of dark

energy. Thus, we live in a very special time in the evolution of the universe: the time at which we can observationally verify that we live in a very special time in the evolution of the universe!

Observers when the universe was an order of magnitude younger would not have been able to discern any effects of dark energy on the expansion, and observers when the universe is more than an order of magnitude older will be hard pressed to know that they live in an expanding universe at all, or that the expansion is dominated by dark energy. By the time the longest lived main sequence stars are nearing the end of their lives, for all intents and purposes, the universe will appear static, and all evidence that now forms the basis of our current understanding of cosmology will have disappeared.[11]

While it had escaped our notice in 2004, Tony Rothman and George F. R. Ellis had thought in 1987 about alternate worlds where cosmology would go wrong. "It is even possible," they wrote in the final sentence of their paper, "that universes could exist in which life arises only at times when observations lead to a deceptive understanding of cosmology."[12]

These all, in one way or another, support elements of our original argument which we extended five years ago: The Earth (and the broader solar system) is an excellent platform for space travel.[13] Many factors have converged to permit us to put men on the Moon and send probes to all the planets in the solar system. Earth contains many crucial chemical rocket propellants (such as hydrogen plus oxygen in water). In contrast, inhabitants on planets only a bit more massive than Earth would have a much harder time getting rockets with large payloads into space. People on planets within the habitable zones of less massive host stars would find it much harder to launch interstellar missions. As it happens, our solar system is in the best place in its 225-million-year orbit around the center of the galaxy for interstellar travel to nearby stars. This is just at the moment when such travel has become conceivable. (Elon Musk: Take note.)

A NEEDLESSLY HOSTILE REACTION

When we were working on this book at the dawn of the new millennium, we wanted to propose a new hypothesis. We also wanted to add to the growing case for purpose and design in the universe. Our argument hinges on the eerie overlap between the conditions for life and for scientific discovery. Those rare places where observers can exist are, as it happens, the best overall places for discovery. This pattern, we argue, makes much more sense if the universe is designed for discovery than if it is not. So, far from being anti-science, our design argument implies that the world is made for scientific discovery!

We were, perhaps, a bit naive about the hostility and metaphysical panic this book would provoke among self-appointed defenders of science and atheist religion professors. These hostile attacks have had serious repercussions for us over the years, but especially for Guillermo, (formerly) a research and teaching astronomer.

While we won't recount these trials and tribulations here, we should note that none of these bad-faith attacks have touched our argument or its supporting evidence. On the contrary, confirming evidence has continued to emerge. We're happy to trust the fate of our argument to future discoveries and are grateful that this book continues to be read and debated.

INTRODUCTION

*Discovery is seeing what everyone else saw and
thinking what no one thought.*

Albert von Szent-Györgyi[1]

On Christmas Eve, 1968, the Apollo 8 astronauts Frank Borman, James Lovell, and William Anders became the first men to see the far side of the Moon.[2] They had been wrested from Earth's gravity and hurled into space by the massive, barely tested Saturn V rocket. One of their main tasks was to take pictures of the Moon in search of future landing sites. The first lunar landing would take place just seven months later. But a different photograph etched their mission in our collective memories. It's known as *Earthrise*. (See plate 1).

Emerging from the Moon's far side during their fourth orbit, the astronauts were transfixed by their vision of Earth—a delicate, gleaming swirl of blue and white, set against the monochromatic, barren lunar horizon. Earth had never looked so small to human eyes, and yet it was never more the center of attention.

To mark the timing of the event on Christmas Eve, the crew had decided to read the opening words of Genesis: "In the beginning, God created the heavens and the Earth . . ." The reading, and the reverent silence that followed, were telecast to an estimated one billion viewers, the largest audience in television history.

In his book about the Apollo 8 mission, Robert Zimmerman notes that the spacemen chose those words "to include the feelings and beliefs of as many people as possible."[3] Indeed, when most Earthlings look out at the wonders of nature or Apollo 8's awe-inspiring photo, they see a grand design.

But a contrary opinion holds that such a response is emotional rather than rational, and that our existence is not only mundane, but without purpose. In his book *Pale Blue Dot*, the late astronomer Carl Sagan typifies this view. While reflecting on another image of Earth, this one taken by Voyager 1 in 1990 from some four billion miles away, Sagan writes (See plate 2):

> Because of the reflection of sunlight. . . . Earth seems to be sitting in a beam of light, as if there were some special significance to this small world. But it's just an accident of geometry and optics. . . . Our posturings, our imagined self-importance, the delusion that we have some privileged position in the Universe, are challenged by this point of pale light. Our planet is a lonely speck in the great enveloping cosmic dark. In our obscurity, in all this vastness, there is no hint that help will come from elsewhere to save us from ourselves.[4]

But what if this melancholy belief, despite its heroic pretense, is wrong? What if the scientific knowledge gained in the last century, when rightly interpreted, bespeaks a purpose behind our place in the cosmos? What if, when viewing *Earthrise*, an overwhelming sense of awe and gratitude turns out to be the reasonable response?

In the following pages we hope to justify that response by means of a striking feature of the natural world, one as widely grounded in the evidence of nature as it is wide-ranging in its meaning. Simply stated, the conditions that allow for intelligent life on Earth also make our planet strangely well-suited for discovering the universe.

The fact that our atmosphere is clear; that our Moon is just the right size and distance from Earth, and that its gravity stabilizes Earth's rotation; that Earth's geology records and preserves events about its past; that our place in our galaxy is just so; that our sun is its precise mass and composition: all of these and many more features are needed to make Earth habitable. But they also have been crucial for scientists to measure and discover the universe — at both the smallest and largest scales.

Mankind is oddly well-placed to decipher the cosmos. Scrutinize our world with the best tools of modern science, and you'll find that a place with the right ingredients for intelligent life will also afford that life a clear view of the universe. Such so-called habitable zones are rare in the universe, and most of these may be devoid of life. But if there is another civilization out there, it will also enjoy a clear vantage point for searching the cosmos, and maybe even for finding us.

To put our argument both more generally and more technically, *discoverability*—including what we call measurability—*correlates with habitability*.[5] Is this correlation just a coincidence? Are we just lucky? We think not, not least because this evidence contradicts a popular idea called the Copernican principle or principle of mediocrity. This principle is far more than the observation that the cosmos doesn't literally revolve around Earth. It's a metaphysical claim. The argument goes like this: Modern science since Copernicus has moved human beings ever farther from the "center" of the cosmos. As a result, scientists should assume that our place, literally and metaphysically, is unexceptional. This principle is often paired with what philosophers call materialism—the view that the material world is "all that is, or ever was, or ever will be," as Carl Sagan famously put it.[6]

Following this principle, most scientists have supposed that since the universe is vast and ancient, our planetary system must be typical and the emergence of life quite likely. Thus, most have assumed that the universe is teeming with life. In the early 1960s, for instance, astronomer Frank Drake proposed what became known as the Drake equation. With it, he sought to list the factors needed to detect chatty ETs. The factors range from the rate at which stars form to the likely age of civilizations prone to sending signals to strangers on other planets.[7]

Ten years after he proposed it, Drake's colleague Carl Sagan used the equation to conjecture that our Milky Way galaxy alone might contain as many as one million candidates. Probability theorist Amir Aczel is similarly sanguine about the possibility of ETIs, as his book title reveals: *Probability One: Why There Must Be Intelligent Life in the Universe.*[8]

This optimism finds its practical form in the Search for Extraterrestrial Intelligence, or SETI, a project that scans the skies for radio signatures of ETIs. SETI seeks evidence that would prove that ETIs exist. In contrast, some advocates of ETIs are content to rely on math and best-guess estimates of the various factors in the Drake equation.

We were weaned on *Star Trek* and other science fiction and wish it were that simple. But such certainty is misplaced. Mounting evidence suggests that animal life needs just the right setting, and that the chance of getting all the right details at the same place and time is remote.

A few scientists have begun to face these facts. For instance, in 1998, Australian planetary scientist Stuart Ross Taylor stressed the rare, chance events that formed our planetary system, with Earth nestled in its narrow habitable zone.[9] He argued that we should not assume that other planetary systems are like ours.

Similarly, in *Rare Earth: Why Complex Life Is Uncommon in the Universe*, paleontologist Peter Ward and astronomer Donald Brownlee focused on the many happenstances that joined forces to give complex life a chance on Earth.[10]

These views challenge the Copernican principle. But while violating its letter, Taylor, Ward, and Brownlee followed its spirit. They assumed, for instance, that the origin of life is more or less a matter of getting liquid water in one place for a few million years. As a result, they still expected "simple" microbial life to be common in the universe.

And they all keep faith with the philosophy behind the Copernican principle. "We are just incredibly lucky," Peter Ward remarked after the release of *Rare Earth*. "Somebody had to win the big lottery, and we were it."[11]

But there's a better explanation. To see this, we must connect these recent insights about habitability—the prerequisites for complex life— with *discoverability*. Discoverability refers to those features of the universe, and our location in it—both in space and time—that allow us to detect, observe, uncover, and determine the size, age, history, laws, and other properties of the physical universe. Discoverability is what makes natural science possible.

Crucial to discoverability is *measurability*. Although scientists don't often discuss it, the degree to which we can measure the wider universe— and not just our local neighborhood—is surprising. Most scientists assume that the world can be measured. Read any book on the history of science, and you'll find plucky tales of ingenuity, grit, and dumb luck. What you likely won't see is any mention of the countless fortuitous features of nature and our place in it that were needed for such feats.

We think this eerie overlap between the needs of complex life and scientific discovery suggests not coincidence but conspiracy. However, our argument is subtle. First, we don't argue that every aspect of discovery is *individually* optimized on Earth's surface. Nor are we saying that it's always easy to measure and make discoveries. Our claim is that our Earthly perch allows a stunning diversity of discoveries, including a diversity of measurements across the sciences, across the universe, and at every relevant size scale.

Yes, there are places that could be better for measuring this or that feature of nature. But repeatedly, we find that if we were in one of those places, the cost for doing other kinds of science would greatly exceed the benefit. For instance, intergalactic space might be a better spot to study certain distant objects than the surface of any planet with an atmosphere and a host star. But intergalactic space would be a terrible place for learning about the recent past, the structure of stars or how they form, and the laws of celestial mechanics.

Likewise, a planet in a giant molecular cloud in a spiral arm might be a great perch to study how stars form, but it would hide the distant universe. Earth, in contrast, strikes a winning balance. It offers great views of the distant and nearby universe while providing a special platform for discovering the laws of physics.

When we say that habitable places are optimal for science, we're referring to an optimal balance of competing conditions. Engineer and historian Henry Petroski calls this *constrained optimization*. "All design involves conflicting objectives and hence compromise," he explains in his book *Invention by Design*, "and the best designs will always be those that come up with the best compromise."[12] Think of the laptop computer. Engineers design laptops that strike the best compromise among various competing features. Users might prefer large screens and keyboards, all things being equal, to small ones. But in a laptop, all things aren't equal. The engineer must account for trade-offs among such matters as CPU speed, hard drive size, peripherals, weight, screen resolution, cost, aesthetics, durability, ease of production, and the like. The best design will be the best compromise. Similarly, to make discoveries in fields from geology to cosmology, an environment must balance trade-offs. The best place for science will be one that meets or exceeds a host of thresholds for discovery.

For instance, there's a threshold for detecting the cosmic background radiation that was left over from the big bang. (To discover and measure something, you must be able to detect it.) If our atmosphere or solar system blocked this radiation, or if we lived in the future when it has disappeared, we could not discover and measure this stunning confirmation of the big bang.

Now, the space between galaxies might give us a slightly better view of the cosmic background radiation, but we would be cut off from everything that can't be measured from deep space, such as the information-rich layering processes on a terrestrial planet.

The best setting for scientific discovery, then, will be one that meets many diverse thresholds needed for fruitful discovery. In this sense, our local environment is optimal for science.[13] The cosmos, our solar system, and our exceptional planet are themselves a laboratory, and Earth is the best bench in the lab.

Even more mysterious: the features of our home that provide the best overall setting to discover the world around us are the same rare properties that allow for our existence. This is the correlation between life and discovery. It's a surprise we had no right to expect. We don't believe this is mere coincidence. It cries out for another explanation, one that suggests there's more to the world than many have been willing to entertain.

SECTION 1

OUR LOCAL ENVIRONMENT

CHAPTER 1

WONDERFUL ECLIPSES

Perhaps that was the necessary condition of planetary life:
Your Sun must fit your Moon.

Martin Amis[1]

I t was a cool clear morning on October 24, 1995, in Neem Ka Thana, a small town in the dry region of Rajasthan, India — perfect for an eclipse. I (Guillermo) bounced up at 5 a.m., along with the other astronomers in our group. By 6 a.m. I had staked my claim within a roped-off compound in a schoolyard and was setting up my instruments. Half a dozen other setups were scattered around me in the compound, each with its own team of astronomers. Some had mounted their experiments on stable concrete piers built weeks before. Around the compound were TV and radio news crews plus hundreds of curious onlookers. They were staring at us as if we were rare zoo exhibits. The Indian Institute of Astrophysics in Bangalore had invited me to join the expedition. The eclipse was not even the main purpose of my trip, but I couldn't pass up this chance.

Solar eclipses are like snowflakes: no two are exactly alike. Still, astronomers sort them into three types: partial, annular, and total. In a partial eclipse, the Moon does not quite cover the Sun's bright photosphere.[2] In an annular eclipse, their centers may pass very close to each other, but the Moon's disk is too small to cover the Sun's photosphere. To qualify as a total eclipse, the Moon's disk must cover the bright solar disk as seen from Earth.

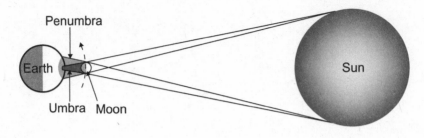

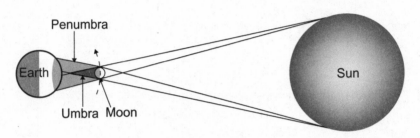

Figure 1.1: A total solar eclipse (above) compared to an annular eclipse (below). As the name implies, viewers of a total eclipse see the Moon block the Sun's entire photosphere. An annular eclipse leaves a bright ring of the Sun's photosphere visible even for the best-placed viewers. Sizes and separations are not drawn to scale.

Those far from the centerline see only a partial eclipse. Only total eclipses really darken the sky and reveal the eerie pink chromosphere and silvery-white corona. During totality, when the Moon fully eclipses the Sun, the chromosphere looks like a fragile, jagged crown, with pink flames protruding around it like a ring of fire. The corona—the outer-most part of the Sun's atmosphere—extends out into the darkness like a ghostly halo.

Given the choice, you want to see a total eclipse—not a mere partial or annular one. I had seen several partial ones, including two annular ones in 1984 and 1994, but this was my first chance to see the real deal.[3] I performed a simple experiment to measure changes in the air temperature, pressure, and humidity. And I captured the event with a 35 mm camera and a telephoto lens.

It worked. The weather was perfect that day and also the day before. This allowed me to compare the changes during the eclipse.[4] I managed to shoot thirty frames during the fifty-one seconds of totality. I could see the long coronal streamers with my naked eyes (See plate 3). Alas, I was so busy snapping photos that my best view of totality was through the camera's viewfinder.

To experience a total solar eclipse is far more than simply to see it. The sudden chill in the air, the confused chatter of birds and insects, and the gasps, oohs, and aahs from the crowd mingle with the darkness of the blotted Sun. In India, just after the total phase ended, the crowd burst into applause, as if praising a ballerina for a well-executed finale.

This was only the fourth total eclipse seen in India in the twentieth century. National television had crews set up at three or four spots spread across the eclipse path. One of them shared our site. Reporters asked scholars to explain the science of eclipses. They had others discuss Indian eclipse mythology. The media, it seemed, were trying to show the world that India had discarded superstitious rituals and entered the era of scientific enlightenment. But the practices in evidence during this eclipse didn't quite fit that storyline.[5]

Completing the cast were the amateur astronomers and eclipse chasers. These are folks who try to see as many total solar eclipses as they can fit into a lifetime. Serge Brunier is one of them. In *Glorious Eclipses: Their Past, Present, and Future*, he writes that

> there is the same astonishment and, each time, the feeling grows stronger that eclipses are not just astronomical events, that they are more than that, and that the emotion, the real internal upheaval, that they produce—a mixture of respect and also empathy with nature—far exceeds the purely aesthetic shock to one's system.[6]

After his first total eclipse, Brunier was hooked. He writes that

> the sight is so staggering, so ethereal, and so enchanting that tears come to everyone's eyes. It is not really night. A soft twilight bathes the Mauna Kea volcano. Along the ridge, the silvery domes, like ghostly silhouettes of a temple to the heavens, stand rigidly beneath the Moon. The solar corona, which spreads its diaphanous silken veil around the dark pit that is the Moon, glows with an otherworldly light. It is a perfect moment.[7]

Amateur astronomers who have traveled abroad to watch solar eclipses give the same reports. They and the locals are equally in awe and often in tears. Scientists can predict when total solar eclipses will appear within a second of time anywhere on Earth, but this has not dispelled their mystery. Nor has it stopped a modern astronomer like Brunier from describing them as ethereal, as spiritual.

Is there more to such eclipses than just the mechanics of the Earth, Moon, and Sun? Do they connect, somehow, to our very existence on Earth?

Yes, there is. And yes, they do.

THE PHYSICS OF THE MOON

First, consider this. Our Moon keeps the tilt of Earth's axis, or obliquity — the angle between the axis around which it rotates and an imaginary axis perpendicular to the plane in which it orbits the Sun — from varying over a large range.[8] If Earth's tilt wobbled more, our climate would fluctuate far more than it does.[9] At present, our planet tilts 23.5 degrees but varies from 22.1 to 24.5 degrees over several thousand years.

The Moon's large mass compared to Earth — about 1/82 — makes all the difference. Contrast this with the two tiny potato-shaped moons of Mars, Phobos, and Deimos. If our Moon were that small, Earth's tilt would vary not 3 degrees but over 30 degrees. That might not sound like anything to fuss over. But tell that to someone trying to eke out a life on an Earth with a 60-degree tilt. If the North Pole leaned sunward through the summer for half of the year, most of the northern hemisphere would swelter in months of perpetual daylight. The high north would be hot enough to make Death Valley in July feel like a shady spring picnic. Any survivors would then suffer months of frigid night during the other half of the year.

It's not just a large tilt that would make for a hard life. On Earth, a smaller tilt might lead to very mild seasons, but it would also prevent the wide distribution of the rain so friendly to surface life. Tilted at 23.5 degrees, however, Earth's wind patterns change throughout the year, bringing seasonal monsoons to areas that would otherwise remain parched. A planet with little or no tilt would likely be mostly covered with arid land.

The Moon also raises Earth's ocean tides. The tides mix nutrients from the land to the oceans, creating the fecund intertidal zone, where seawater regularly bathes the land. (Without the Moon, Earth's tides would be only about one-third as strong. Only the solar tides would be regular.) About one-third of the tidal energy is spent on the rugged areas of the deep ocean

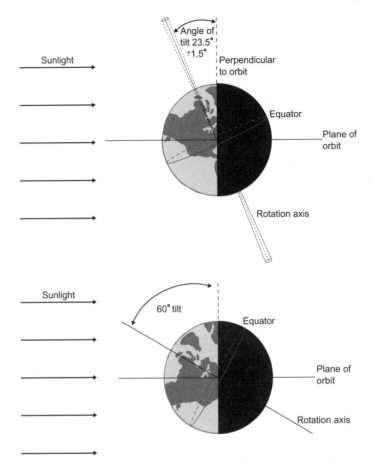

Figure 1.2: Earth's spin axis currently tilts 23.5 degrees relative to the perpendicular to the plane formed by the Earth's orbit around the Sun. It varies a modest 2.5 degrees over thousands of years. Such stability is thanks to the Moon's gravity. Without a large Moon, Earth's tilt could vary by 30 degrees or more, even 60 degrees, which would render Earth less habitable for complex life.

floor, contrary to the previous idea that it was all dissipated in the ocean shallows. This may be a key driver of ocean currents,[10] which circulate vast amounts of heat.[11] If Earth lacked such lunar tides, Seattle would look more like northern Siberia than the lush, temperate Emerald City.

The story of life includes an early chapter on the origin of the Moon. The most popular version of this chapter describes a glancing blow to the proto-Earth by a body a few times more massive than Mars.[12] That violent collision would have helped form Earth's iron core by melting the planet and allowing liquid iron to sink more fully to the center.[13] This, in turn, could have led to a strong planetary magnetic field—a protector of life that we'll discuss later.

Oxygen would have taken longer to build up in our atmosphere if more iron had remained in the crust, because exposed iron would consume the free oxygen. The collision is also thought to have removed some of Earth's original crust. If this hadn't happened, the thick crust might have prevented plate tectonics, still another key ingredient for building a habitable planet.

In short: no Moon, no us.[14]

A smaller impact would not have done the trick. The Moon would have quickly receded from the Earth—slowing the Earth's spin and orbit around the Sun in the process.[15] If the Moon had been even 10 percent more massive—a mere 70 miles greater in diameter—it would no longer stabilize the Earth's tilt.[16] We have about the largest moon that can still do this job after 4 1/2 billion years. And that's just what we should want, since there are other reasons to prefer a large Moon for building a life-friendly planet.

Of course, with eclipses, it takes three to tango: a star, a planet, and the planet's moon. As long as they're the right relative sizes and distances apart, a total eclipse can happen with a larger or smaller moon or star. But planets vary a lot when it comes to hosting life and producing eclipses that are useful for science. Let's start with the former.

For life, the sizes of a planet and its host star and distance between them make all the difference. The type of star also matters for complex life. More on that in chapter 7. Like a high-risk driver, a more massive star lives in the fast lane. Water is the key ingredient here. A smaller, cooler star is dimmer and gives off less radiant heat, so a planet must orbit closer in to keep liquid water on its surface.

The distance between the star and planet is crucial. The band around a star wherein a terrestrial planet must orbit to keep liquid water on its surface is called the *circumstellar habitable zone*. Think of it as the Goldilocks zone—where it's not hot or too cold, but just right.

Orbiting too close to the host star leads to rapid tidal locking or "rotational synchronization," in which one side of the planet perpetually faces its host star. (The Moon is so synchronized in its orbit around Earth.) Its day side would be brutally hot and its night side brutally cold. Even if the thin boundary between day and night on such a planet, called the terminator, were habitable, life would be nasty and brutish around a less massive star.

So you don't get lost in the details, let's start with our conclusion: The host star for any planet with observers will look much the same size in the sky as our Sun does from Earth—*because it's the result of key requirements for complex life in this universe.* Planets with host stars looming much larger or smaller in their sky will be hostile to complex life, and so won't have any (native) observers peering out from their surface.

If a planet's moon is farther away, it would need to be bigger than our Moon to generate similar tidal energy and secure the planet's tilt.[17] The Moon is already weirdly large compared to Earth. A smaller moon would have to be closer, but then it would probably be less round. That would create other problems.

Assume we're trying to build a habitable system. We'd want the host planet to be about Earth's size to keep the tectonic plates moving, to keep some land above the oceans, and to retain an atmosphere (more on these in chapter 3). To maintain a stable tilt, a planet needs a certain tidal force from a moon. A larger planet would need a larger moon. So even Earth's size is relevant. In short, the very conditions that allow for complex life on a terrestrial planet also provide that life with total solar eclipses.

SUPER ECLIPSES AND PERFECT ECLIPSES

What if the Moon were closer to Earth? About 2.5 billion years ago, the Moon was, on average, about 16 percent closer than it is now.[18] "Super-eclipses" would have been more common and visible over more of Earth's surface. During a super-eclipse, we can see the pink chromosphere and parts of the innermost corona briefly only near the start and end of totality. Today, we can see the whole chromosphere throughout most of totality.

Take that eclipse on October 24, 1995, when the Moon's black disc just barely covered the Sun's bright photosphere.[19] We could see the Sun's extended atmosphere for almost a minute. We'll refer to this as a perfect eclipse, since totality lasts long enough for an observer to take it in. The Moon is just large enough to block the bright photosphere, but not so large that it obscures the colorful chromosphere. A briefer total eclipse leaves a brighter sky, providing less time for our eyes to adapt to the darkness and to see the faint outer corona. A slightly larger moon would give us longer eclipses but block more of the chromosphere so valuable for science.

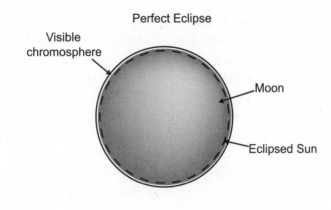

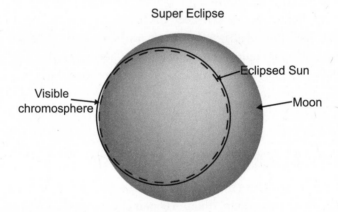

Figure 1.3: A perfect solar eclipse compared to a super eclipse. For scientific discovery, perfect eclipses are better than super eclipses. In a perfect eclipse, the Moon just covers the Sun's bright photosphere, revealing the Sun's thin chromosphere. In contrast, a super eclipse would reveal only a small sickle of the scientifically valuable chromosphere, and then only at the beginning and end of totality. The thickness of the chromosphere has been exaggerated for clarity; in reality, it is about one three-hundredth the radius of the Sun.

If the Moon were less round, we could still enjoy solar eclipses as long as its minor axis appeared larger than the Sun. But in that case, the chromosphere would be obscured along the Moon's major axis during mid-totality. The Moon and the Sun, as it happens, are two of the roundest known bodies in the solar system.[20]

The moons in the outskirts of our system are a mixture of rock and ice, which leads to a rounder shape as ice is less resistant to stress than rock. Our Moon has almost no ice, and yet—surprisingly—it's quite round. This is probably due to the way it formed, in contrast to the moons in the outskirts. If the Moon had formed from a giant impact with the proto-Earth, then ejected material would have quickly coalesced while some of it was still partly molten. The remaining material would have accreted onto the Moon soon thereafter.[21]

The Moon's scant atmosphere also makes for better eclipses. Total lunar eclipses provide clues for what solar eclipses would look like if our lunar neighbor had a thicker atmosphere. The Moon turns deep sunset red during the central phase of a total lunar eclipse, as sunlight refracts and scatters through Earth's atmosphere on its way to the Moon. Deep red light would bathe an astronaut on the Moon, and he would see a bright red ring encircling Earth. And we would see such a ring around the Moon during a total solar eclipse. Sounds nice, but it would obscure the pink chromosphere and much, if not all, of the corona.[22]

Finally, what if we were on another planet in the solar system? Figure 1.4 shows how big a moon looks to an observer on its host planet compared to the Sun.[23] The apparent size of a moon is what an observer at the equator of the parent planet would observe; for the gas giants, imagine the observer floating above the cloud tops in a research balloon. The results are astonishing. Of the more than 200 moons in our solar system, ours yields the best match to the Sun as viewed from a planet's surface. The Sun is some 400 times farther than the Moon, but it's also 400 times larger. What's more, this perfect match happens during a narrow window of Earth's history. As a result, both bodies appear the same size in our sky just when we're here to observe it.

Jupiter's "Galilean" moons cast large shadows on the Jovian cloud tops. In general, the Sun looks smaller and total eclipses become more common as one goes outward from the Sun. Total solar eclipses are much harder to pull off when the Sun looms close and large.

In fact, if your only goal were mere total eclipses, you might wish to relocate to a planet farther from the Sun. But, for science, Earth's eclipses are the best on offer, since the farther a planet is from the Sun, the briefer its eclipses. Because the Sun looks smaller on those outer planets, their average moon passes over the Sun's disk more quickly. What's more, the heft of the outer planets causes their moons to orbit much faster than our Moon orbits Earth. Finally, only four moons in the solar system are larger than our Moon. As a result, the typical eclipse seen on the outermost planets lasts mere seconds.

Of the sixty-four moons plotted in Figure 1.4, only two mark the same average size on their planet's sky as the Sun. Our Moon is one;

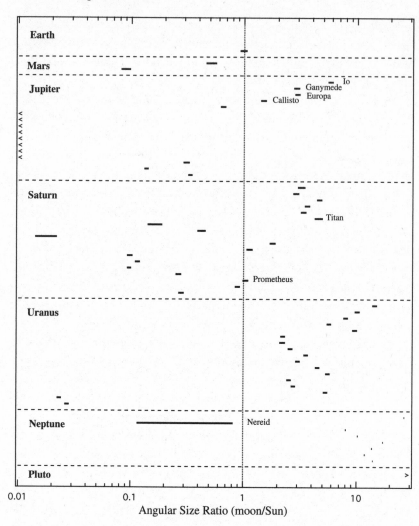

Figure 1.4: Comparison of the average angular size ratios of sixty-four moons to the Sun from the surfaces of their host planets. (Many smaller, recently discovered moons around the giant planets are not included.) The ratios are plotted on a logarithmic scale on the horizontal axis, so the main tick marks represent multiples of ten. If a moon is nonspherical, then its smallest dimension is used for calculating its apparent size. If a moon has a ratio of one, it's a perfect match for the Sun from the surface of its host planet. There are only two such matches in our solar system: Earth's Moon, and Prometheus—a small potato-shaped moon of Saturn. Unlike our Moon, however, Prometheus's eclipses last less than one second. Moons have a range of angular size ratios because the orbits of the planets and moons are not perfectly circular. As a result, the angular sizes of the Sun and the moons vary from the respective planetary surfaces. Hence, in this figure, a line rather than a point represents them. Nereid, one of Neptune's moons, has a quite eccentric orbit. "<" represents moons too small to appear on the chart. ">" represents the one moon too large for the chart—Pluto's Charon.

Prometheus—a small moon of Saturn—is the other. But Prometheus's eclipses last less than one second as it whips around Saturn. And its shape—like a fingerling potato—spoils the view of the chromosphere. As the figure shows, a typical moon appears larger than the Sun in the outer solar system. The average ratio is close to one at Saturn, so it's not so surprising that one of its moons most closely matches the Sun compared to the other planets.

But the Moon bucks this trend. In fact, it gives us "better than perfect" eclipses, since the Sun appears larger from Earth than from any other planet with a moon. An Earthbound eclipse observer can discern finer details in the Sun's chromosphere and corona than from any other planet.

Okay. But what about eclipses on the moons themselves? In a system with more than one moon, a moon can produce an eclipse on one of its siblings. These "mutual eclipses" are pretty common in the solar system. But none come close to the eclipses we enjoy—either in detail or duration.[24] In one of the best four cases around Saturn, for instance, the eclipses last less than one second!

What about eclipses in systems around other stars? We have yet to detect a moon around an exoplanet. But we can consider eclipses produced by one planet casting its shadow on another. The planets in our system are too far apart to produce total eclipses in this way. In contrast, the systems with the most detected planets tend to be highly compact. Still, none seem to be good candidates for producing eclipses as ideal as our own.[25] In a 2017 study of solar eclipses in exoplanetary systems, astronomers Dimitri Veras and Elmé Breedt state, "The Sun–Earth–Moon cases emphasize how unusual the situation is in which humans find themselves."[26]

REVEALING ECLIPSES

Now, we come to the point. Eclipses aren't just beautiful. They have played a key role in scientific discovery. They have helped solve the mystery of stars, for instance. Eclipses also provided a natural experiment for testing Einstein's general theory of relativity and allowed us to measure the slowdown of Earth's rotation. Let's consider each of these.

SPECTRA AND THE SUN'S ATMOSPHERE

Observers on the ground can see the Sun's full corona only during a total eclipse.[27] It's why people want to see more than one; the corona never looks the same twice. Even today, astronomers conduct experiments during eclipses.

Figure 1.5: In 1811, Joseph von Franhaufer (1787–1826) first described the dark gaps that cross the solar spectrum.

More important: total eclipses helped us unlock the mystery of stars. Astronomers use instruments called spectroscopes to separate light into its constituent colors. The different colors in the light spectrum, we now know, correspond to different wavelengths of electromagnetic radiation. The traditional colors of the visible spectrum are the stripes of a rainbow: red, orange, yellow, green, blue, indigo, and violet. Wavelengths get longer as we go from the blue to the red end of the spectrum. (Visible light is a minute part of the electromagnetic spectrum, which extends from radio waves on the long end to X-rays and gamma rays on the short end.) Scientists since just before Isaac Newton (1666) had known that sunlight splits into all the colors of the spectrum when passed through a prism. And if you've seen Pink Floyd's album cover *Dark Side of the Moon*, you know that too. But it was only in 1811 that Joseph Franhaufer first described the dark gaps that intersperse the smooth continuum of the solar spectrum, often called Franhaufer lines.

Over the following decades, lab tests revealed that atoms and molecules both emit and absorb light at characteristic points on the spectrum, called *emission* and *absorption* lines. At a certain temperature, a gas emits a unique, telltale light. Such a gas also absorbs light when lit from behind. This produces absorption lines in the spectrum like a bar code superimposed on a rainbow. Each element impresses its own unique fingerprint on the spectrum. From these experiments, astronomers could identify many of the Franhaufer lines in the Sun's spectrum with emission lines produced by specific elements.

But at the time, astronomers did not know where the Sun's Franhaufer lines formed or understand how gas absorbed the light. Two notable eclipses in the latter half of the nineteenth century provided the answer. During the few minutes of totality on August 18, 1868, French astronomer Pierre Jules César Janssen pointed his spectroscope at prominences — plumes of gas that surge out from the photosphere into the corona. It revealed a spectrum of bright emission lines.

Most of these were quickly identified as hydrogen by comparing them with laboratory spectra.[28] But the bright emission lines inspired Janssen to search for — and find — prominences the following day when there was no eclipse. Soon thereafter, he invented the spectrohelioscope, which produces an image of the Sun in the light of one spectral line. This allows astronomers to study the gas motions in the Sun's atmosphere in great detail.

These two discoveries made during a perfect eclipse — the prominences and the chromosphere against the backdrop of dark space — revealed that stars are made of hot, low-density gas, like gas-filled glass tubes excited by an electric current. In fact, the color of such a hydrogen-filled tube is much like that of the chromosphere and the prominences. (See plate 4.) This confirmed the 1864 conjecture of Jesuit priest Angelo Secchi and John Herschel that the Sun is a ball of hot gas.

More insights soon followed. George Airy was the first astronomer to describe what we now know as the chromosphere, during an 1851 eclipse. He had called it the sierra, mistaking it for a range of mountains on the Sun.

The English astronomer Joseph Norman Lockyer recorded spectra of prominences without an eclipse, though he certainly knew of the results from eclipse expeditions. Janssen and Lockyer independently discovered a bright emission line in the yellow part of the Sun's emission line spectrum. Lockyer identified this with an unknown element he named helium, after the Greek word for the Sun god Helios. Helium was not isolated in the lab until 1895. The second element in the periodic table doesn't reveal itself in the absorption spectrum of the Sun. So, who knows how long it would have taken astronomers to find it if they had nowhere else to look? Today, we know helium makes up about 28 percent of the Sun's mass; it's the second most common element in the universe. And yet, it's very unlikely that either Janssen or Lockyer would have found it if eclipse observers had not led the way.

This, however, wasn't the last solar secret eclipses would reveal. During the eclipse of December 22, 1870, American astronomer and onetime missionary Charles A. Young noticed that the Sun's spectrum changed from its usual sharp, dark lines superimposed on a bright continuum to emission lines just as totality began. In Young's own words:

> As the Moon advances, making narrower and narrower the remain-
> ing sickle of the solar disk, the dark lines of the spectrum for the
> most part remain sensibly unchanged, though becoming somewhat
> more intense. A few, however, begin to fade out, and some even
> begin to turn palely bright a minute or two before totality begins.
> But the moment the Sun is hidden, through the whole length of

the spectrum—in the red, the green, the violet—the bright lines
flash out by the hundreds and thousands almost startlingly, for the
whole thing is over in two or three seconds. The layer seems to
be only something under a thousand miles in thickness, and the
Moon's motion covers it very quickly.[29]

Hence, this thin region came to be called the *reversing layer*. We now
know it's part of the chromosphere. Young was the first to show the location
and state of the gas producing the absorption lines in the out-of-eclipse
solar spectrum.[30] By applying Gustav Kirchhoff's laws of spectroscopy (See
plate 5),[31] Young realized that the reversing layer is made of cooler gas
than the underlying photosphere. It's only during a total eclipse that the
bright photosphere is blocked. Had the Moon loomed larger, Young could
only have run his experiment over a short segment of the Sun's limb—the
apparent edge of the photosphere.

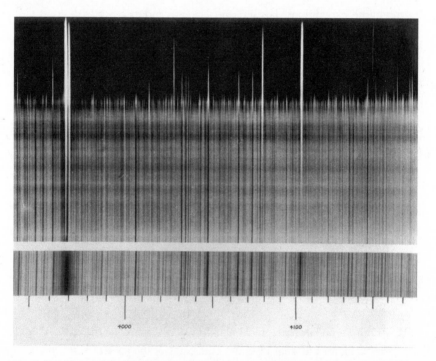

Figure 1.6: Ultraviolet region of the solar spectrum obtained by W. W. Campbell during
the total eclipse of August 30, 1905. Campbell used a clever moving plate method to
record the changing solar spectrum as the edge of the Moon covered the last bit of the
Sun's bright disk. Wavelength runs horizontally and time vertically on the photo. Note
how the spectrum changes from absorption to emission. And for comparison, note the
spectrum of the Sun's photosphere shown on the bottom panel.

Since Young's historic observations of the 1870 eclipse, the so-called flash spectrum of the chromosphere has been photographed several times. Think of the Moon's edge as a giant slit, allowing only a thin sliver of light from the chromosphere to reach the observer during the first and last few seconds of totality. If there were a rainbow during an eclipse, you could, with good video tools, see it change from a continuous spectrum to an emission line spectrum for a few brief seconds.[32] In effect, Earth, the Moon, and the Sun supply astronomers with a giant spectroscope. They just have to hold the prism to their eyes.

The insights afforded by the 1868 and 1870 eclipses cleared the path for stellar astrophysics. It's only because we know how absorption lines form in the Sun's atmosphere that we can interpret the spectra of distant stars—and so figure out what they're made of—without leaving our tiny planet. Such knowledge is the gateway to modern astrophysics and cosmology.

EDDINGTON AND EINSTEIN'S THEORY OF GENERAL RELATIVITY

Arthur Eddington was a famous astrophysicist of the early twentieth century. Today, most know him as the man who confirmed a prediction of Einstein's general theory of relativity—that gravity bends light. On May 29, 1919, two teams, one led by Eddington and Edwin Cottingham on Principe Island off the coast of West Africa and the other led by Andrew Crommelin and Charles Davidson in Brazil, used a total eclipse to test Einstein's 1915 theory. They hoped to measure changes in the positions of stars near the Sun compared to where they were months later or after. Both teams confirmed Einstein's predictions and won him instant acclaim.[33]

Astronomers have repeated the 1919 experiment many times since. The best one was conducted during the eclipse on June 30, 1973. The results of these experiments confirmed general relativity.[34] Other tests involving radio transmissions from space probes have also confirmed related aspects of general relativity. Despite these later refinements, solar eclipse experiments played a vital role in confirming Einstein's theory.

DISCERNING THE PAST RATE OF EARTH'S ROTATION

Okay, one final example: Historical records of total solar eclipses are the best-known way to measure how the length of Earth's day has changed over the last few thousand years.[35] Careful studies of stars using telescopes show that Earth's spin around its axis is slowing at a rate of two

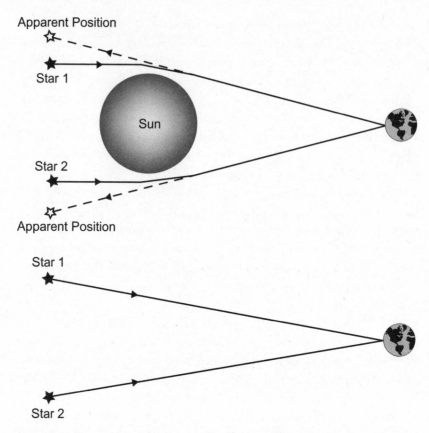

Figure 1.7: According to Einstein's general theory of relativity, gravity should cause starlight passing near the edge of the Sun's disk to follow the path of the curved space around it. A perfect total solar eclipse creates the best natural experiment for testing this prediction. Sizes and separations of bodies are not drawn to scale; the amount of bending has been exaggerated for clarity.

milliseconds per day per century, due mostly to the tug of the tides on Earth by the Sun and Moon.[36] Our days are getting slightly longer. However, such precise observations have only been possible for the last couple of centuries.

Since an eclipse casts a narrow shadow across Earth's surface, only the lucky or ardent few see one in all its glory. This makes historical records quite narrow and precise. Astronomers can predict when and where eclipses should have occurred in the past and then compare their predictions to ancient accounts of eclipses. If there's a discrepancy, they can translate it into an error in solar time. Changes in Earth's rotation period translate into errors in the placement of the predicted shadow track. This kind of information has several uses. For example, knowing changes in the length of our days helps us to discern subtle changes in Earth's shape over centuries and millennia, such as changes due to the retreat of the glaciers in the northern hemisphere.

Eclipses also allow historians to translate ancient calendar systems into our modern Gregorian system. This allows us to place events from different civilizations on a common timeline. We can then establish the configuration of the Sun, Moon, and planets at any place and any date on that calendar. Other events, such as lunar eclipses and conjunctions of planets, are not as useful as total solar eclipses for historical studies since they're visible over a much broader area and/or last much longer.

Perfect eclipses are ideal for all three of these purposes—discovering the nature of the Sun's atmosphere, testing general relativity, and timing Earth's spin. If, instead, we could just see super-eclipses, we would see the chromosphere over only a small fraction of the solar limb.[37] We could not measure the deflection of starlight as closely to the solar limb.[38] And the eclipse shadow on Earth would be larger and much less useful for studying Earth's spin on its axis.

We would see far less if the Moon's disk did not cover the Sun's bright face, yielding only annular eclipses. The difference between an annular and a total eclipse is like the difference between day and night. To a casual observer, an annular solar eclipse looks much like a partial one. Since the chromosphere is wafer thin, we would have learned far less about stellar atmospheres if the Moon's disk were even slightly smaller in our sky.

The sheer beauty of total eclipses no doubt motivated astronomers in the last two centuries to travel great distances to observe them. This may seem trivial, but it's clear from their diaries and accounts that the experience of an eclipse made all the difference. Several key discoveries about the Sun were unplanned. If eclipses had not drawn astronomers to their shadow's narrow tracks, who knows how long such discoveries would have been postponed?

Today, observatories in space can image the Sun's outer corona, and mountaintop telescopes with coronagraphs can image the inner corona. But the occulting disks in the space-based coronagraphs cover far more than the bright photosphere. And they lack the high resolution of observations from the ground during total eclipses.[39] Since we can see the full corona only during total eclipses, these events still provide useful and inexpensive data about this elusive part of the Sun's atmosphere.

There's a final, even more bizarre twist. The Moon is receding from Earth at 3.82 centimeters per year.[40] In ten million years, the Moon will look much smaller. At the same time, the Sun appears to swell by some six centimeters per year—which is a normal part of aging for stars of its type, as it is for modern Americans. As a result, total eclipses will occur for only about 250 million years, a mere 5 percent of the age of Earth. This small window of opportunity happens to coincide with the existence of intelligent life.[41] In other words, the most habitable place in the solar system yields the best view of solar eclipses just when observers are around to see them.[42]

FROM AN ECLIPSE TO AN INSIGHT

Perfect solar eclipses may be the most exquisite example of the correlation between habitability and observability—between the needs for life and for doing a certain kind of natural science. But scientists need more than a brief chance to see this or that event. They need time, resources, and opportunity to analyze and measure the details precisely—at vastly different scales of the very large and very small. This often-tedious work, in turn, must not return void, but must provide keys to unlock natural mysteries that would otherwise remain hidden. Scientific discovery is not about finding one gold nugget buried in a field. It's like finding a hidden treasure map in one place that provides clues about treasures and treasure maps buried elsewhere.

Earth is ideal for observing eclipses. Our eclipses, as it happens, are optimal for measuring the solar flash spectrum, prominences, the deflection of starlight passing near the otherwise bright disk of the Sun, and long-term changes in the Earth's rotation. These eclipses allowed us to discover the Sun's atmosphere and helium—neither of which scientists expected. And our Sun was a patient tutor for introducing us to the vast sea of stars and galaxies beyond our local neighborhood. We had no reason to expect the world to be set up just so.

Perfect eclipses also played a key role in our own story, and in the argument we lay out in these pages. They were the occasion for discovering the correlation between habitability and discoverability itself. The dark beauty of eclipses enticed human beings to seek them out. But this darkness cast a bright light—for both science itself and the argument that follows. Once we noticed, in the late 1990s, the eclipses were connected to both the requirements for complex life and for science, we wondered if this might be one instance of a larger, more pervasive pattern. The answer, we learned, was *yes*. And the evidence has grown even more compelling in the intervening years.

In the pages that follow, we'll describe that evidence, working our way out in concentric spheres. We begin where we all begin—on the surface of this pale blue dot. Then we will move on to the sky, the planets, the starry heavens, and finally, the cosmos itself.

CHAPTER 2

AT HOME ON A DATA
RECORDER

But now ask the beasts, and let them teach you;
And the birds of the heavens, and let them tell you.
Or speak to the earth, and let it teach you;
And let the fish of the sea declare to you.

Job 12:7–8[1]

In the warmth of the summer of 1975, a young male fur seal waddled across the steep rocks of Signy Island, off the coast of continental Antarctica. He had pressed his way through the ice-filled waters of the Stygian Cove, where the ocean meets the tiny island. Upon cresting the rocky shore, he found what he hoped for—a freshwater lake. Although Sombre Lake was still mostly frozen, nonstop sunlight had melted enough of its icy crust to give him all the water he could wish to drink. After a few minutes in the fresh slushy ice, he returned to the cove to rejoin the scores of breeding fur seals. He left a trace of his visit: a few thick, black hairs.

In 1992, Dominic Hodgson and other members of the British Antarctic Survey lowered sediment traps deep into Sombre Lake, extracting cores from the dark, chilly bottom. The variation of seal hairs in the layers of extracted sediment revealed startling details. The researchers discerned the annual fur seal populations directly. But with careful analysis, the hairs gave them detailed clues about the effects of seal and whale hunting on Antarctic flora and fauna over the past two hundred years.[2]

Hodgson's team had tapped into a tiny sliver of the information stored on our planet by regular layering processes. Their ingenuity in deciphering such information is apparent. Less apparent but equally important are the rare conditions that preserve it. Indeed, this layering process is just one of dozens of natural devices that record the past.

NATURE'S DATA LOGGERS

As children, we learn to estimate the age of a tree by counting the concentric rings in a cross section of its trunk. But scientists can discern more than age from such rings. Comparing their thickness, for instance, reveals changes in temperature and precipitation in a tree's vicinity while it was alive.

Tree rings, as well as many other layering and growth processes, are like scientists' data recording devices. Take the dreaded lie detector—which most of us have seen on TV dramas—with its black spindly pens scribbling across a scroll of paper. The heart of these lie detectors is a strip chart recorder, which is a lot like many natural records. Consider the deposits of snow and ice at the poles. The ice is the paper. The diurnal and annual ice layers are the pens that mark time intervals. And any other substances in the ice sensitive to environmental changes are the other pens that record signals from the transducers, which convert environmental data to electrical signals.

Of course, to be useful, the paper roll must be shielded from damage so that new information can be added without spoiling previously recorded data. If the motor on a chart recorder breaks down and the paper fails to advance, the pens will keep writing over the same small area of paper. The result will be an indecipherable glob of ink. Likewise, in a snow deposit, if several feet of snow accumulate over the course of a few years, and the topmost layers erode before the next year's snow falls, then the process won't be a good data recorder. A proper recorder preserves the sequence of events as layers grow one at a time, new over old, accreting like memories not easily forgotten.

Of course, to interpret a record objectively, it must be calibrated. To calibrate a recorder strip, researchers relate the ink markings on the paper to some property in the environment that they know independently, such as temperature, pressure, or humidity. As we'll show, there are several ways to calibrate natural recorders as well.[3] Our planet contains vast libraries of time-stamped information that we're only now learning how to read.

Let's open a few books in that library.

IN COLD STORAGE

Arguably, Earth's best data recorders are the snow and ice deposits near the north and south poles.[4] Ice contains many things scientists employ as proxies, which, as the name suggests, stand in for facts about Earth's past. Such proxies include carbon dioxide, oxygen, and methane gas trapped as bubbles; dust; marine aerosols; ash and sulfuric acid from large volcanic eruptions; soot from forest fires; pollen; micrometeorites; and the ratios of isotopes.

Isotopes of common elements, such as hydrogen and oxygen, have the same number of protons but different numbers of neutrons in their atoms' nuclei. Though isotopes of the same element behave in the same way in most chemical reactions, changes in the environment can subtly alter their ratios in deposited layers. In a warmer climate, for instance, more water evaporates, leaving behind more of the heavier isotopes. This changes the ratio of isotopes in both the water that remains and in the water that precipitates elsewhere. This water, if preserved in layered deposits, can leave clues about changes in the climate going back hundreds of thousands of years.[5]

Today, scientists travel to the frigid polar regions to drill for "white gold." They drill deep into the ice sheets, carefully extracting and storing the ice cores for later study. In the late 1990s, scientists working in Antarctica extracted and measured the Vostok station ice core, a 2.25-mile-deep record of snowfall on East Antarctica going back about 420,000 years. Although it comes from just one place, an ice core contains data that are both local, such as snowfall amount and temperature; regional, such as windblown dust, sea salt, volcanic dust, and other aerosols; and global, such as atmospheric levels of methane and carbon dioxide and total ice volume.

This feature of ice cores impressed Richard Alley, professor of geosciences at Pennsylvania State University, and Michael Bender, professor of Geosciences at Princeton University. "The ice brings all this information together," they noted, "providing a surprisingly complete history of climate changes over much of Earth's surface."[6]

From the Vostok cores, we've learned that the levels of carbon dioxide and methane have changed with temperature and ice volume. We've also

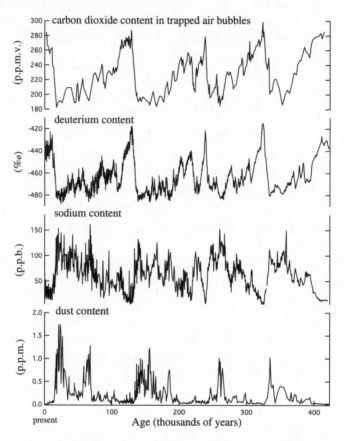

Figure 2.1: Environmental information derived from four of the many diverse proxies that can be traced back some 400,000 years in the Vostok ice core from Antarctica. Trapped air bubbles tell us about the atmosphere, including its carbon dioxide and methane content. Deuterium in the ice tells us about the local temperature. Sodium tells us about wind speeds over the oceans surrounding Antarctica. And dust tells us about global winds and the extent of deserts.

learned that the cold glacial periods were dusty and windy. (Since the sea level was low enough to expose the continental shelves during a glacial period, strong winds swept what was once fine underwater sediment into the atmosphere.) Perhaps most importantly, we now know that Earth's climate has been colder and less stable than the present, with brief warmings, for most of the last 740,000 years, the length of the longest continuous ice core.[7]

Researchers have also obtained ice cores in Greenland, the forbidding island near the North Pole that was named by Norsemen when it was still green. Since snow accumulates there more rapidly than in Antarctica, these ice cores go back only about 100,000 years. But while shorter in total time span, the Greenland cores have a higher resolution than the Antarctic cores. As a result, we can assign fairly accurate dates to short-lived events like volcanic eruptions and sudden climate shifts.

Large volcanic eruptions appear in ice cores as spikes in the acidity of the ice or even as ash layers. By comparing these icy archives with historical records, such as the eruption of Vesuvius in 79 A.D., we can confirm that eruptions leave a signal in the ice. These serve as independent checks for the methods used in dating ice cores. Once scientists can identify known eruptions in a core, they can catalog other eruptions not well documented elsewhere. For example, this method has been used to fix the date of the Mount Mazama eruption (which formed Crater Lake in Oregon) at 7627 ± 150 years ago.[8]

The Greenland cores also show us that the global climate has changed widely over the past 100,000 years. The most recent one was the Younger Dryas, a glacial period lasting about a thousand years, which ended abruptly some 12,000 years ago.

Because of their differences, the two sets of ice cores—Greenland in the north and Antarctica in the south—complement each other. As we noted above, the Greenland cores are quite high resolution but preserve records only about a hundred thousand years into the past; the Antarctica cores, in contrast, reach some three-quarters of a million years back but with lower resolution. This allows one core to be a check on the other, giving us a concise picture of the past climate. What's more, once one ice core is carefully dated, scientists can use it to date other cores by matching patterns of change in carbon dioxide and methane. Since these gases are well mixed in the atmosphere, measuring the trapped gas at one spot gives us the global value.

Ice cores have also revealed changes in the Earth's magnetic field. When high-energy cosmic ray particles—mostly protons—strike atoms in the upper atmosphere, they produce the unstable isotopes carbon 14, beryllium 10, and chlorine 36. Beryllium 10 dissolves in water and is transported to the surface via precipitation. The presence of beryllium 10 in ice (and marine sediment) cores recorded the Laschamp event some forty thousand years ago, when Earth's magnetic field temporarily weakened.[9]

There isn't space here to describe everything that ice cores have taught us about Earth's past, but here's a final example. Ice cores, as well as tree rings and marine sediments, even record extraterrestrial events.[10] Fluctuations in the Sun—for example, the eleven-year sunspot cycle—affect the amount of these isotopes made in the atmosphere in any given year. Those fluctuations are recorded as changes in the concentration of beryllium 10 in the Greenland cores and carbon 14 in tree rings.[11] Evidence of other, longer-term changes in the Sun's energy output over the last twelve thousand years show up as well.

Finally, deep-sea archives hold evidence of much more distant events. In these archives, scientists have found spikes in the concentration of interstellar iron 60 from supernovae 1.5–3.2 million years ago and, earlier still, 6.5–8.7 million years ago.[12] One or a series of these in the more recent period coincide with the end-Pliocene extinction; some researchers have attributed both it and the changed climate to the supernovae.[13]

LIFE'S DATA RECORDERS

Like trees, other life-forms provide some of nature's most sensitive records about Earth's past. They may do this directly by producing growth layers that record sequential information about the environment, or indirectly by enhancing inorganic deposits.[14] Marine sediments would record data in a lifeless world, for instance, but the skeletal shells of single-celled marine organisms embedded in these sediments make them far more informative.[15]

Many organisms show a slight preference for certain isotopes. For example, in 1946, Harold Urey discovered that the oxygen isotope ratios in the skeletons of tiny planktonic foraminifera, or foram, are sensitive to water temperature. This discovery was later confirmed in the lab. As a result, researchers are now able to convert the isotope ratios they measure in fauna-enriched ocean sediments to changes in temperature. As we noted earlier, however, the ratio of oxygen isotopes in ocean water also depends on the volume of ice on the planet. Thus, the ratios measured in the planktonic foram skeletons can reveal both water temperature and global ice volume.

There are at least two ways of telling these effects apart. One method uses another species of foram. Benthonic forams live near the ocean floor where temperatures hover near freezing. Changes in the ratios of oxygen isotope in these forams reflect changes in the volume of ice on the globe. By measuring the sediment-trapped skeletons of both species in the same region, scientists can discern changes in both temperature and global ice volume.[16]

See an important example of reconstructed ocean temperatures during the Holocene in Figure 2.3. It's based on measurements of foram remains in the ocean sediments. Overplotted in the figure is a reconstruction of the surface temperature of Greenland from ice cores.[17] Notice how closely the two curves agree on millennial timescales.

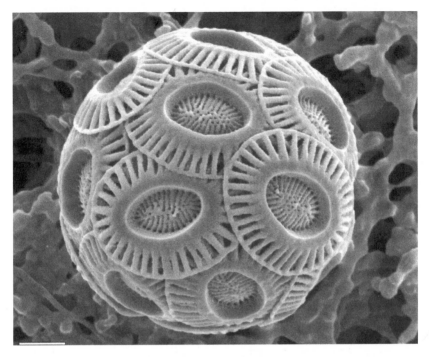

Figure 2.2: *Emiliania huxleyi* (type A), the most common species of phytoplankton. The platelets, called coccoliths, are made of calcium carbonate. These tiny organisms play two vital roles in regulating the climate. First, when deposited on the ocean floor, their skeletons play an important role in the carbon cycle. Second, by emitting dimethyl sulfide into the atmosphere, they help form marine clouds. The scale bar on the lower left is equal to one micron.

Freshwater deposits are also good data recorders. Lake Baikal in Siberia provides the longest continuous sediment record in Asia's interior. These provide detailed temperature data going back about 1.8 million years.[18] While not as data rich as ice cores, marine and lake sediment cores give us a pretty detailed look at even more ancient climates.[19]

Other processes allow us to peer much further back in time. We can measure how quickly the Moon is receding from Earth today, for instance, by timing the round trip of laser beams reflected off the mirrors left on its surface by the Apollo astronauts, and doing so several years apart and then measuring the difference.[20] The method is remarkably precise, but as it turns out, we don't have to deliver a mirror to the Moon to learn a lot about the Moon's recession rate over its long history. Geologists, working with astronomers, have estimated the Moon's orbital period and Earth's rotation period back to about 3 billion years ago.[21] They can do this only

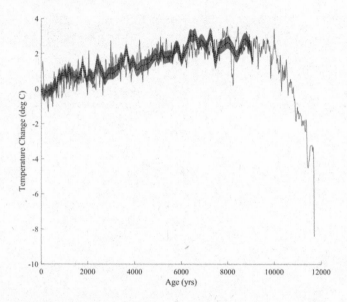

Figure 2.3: Temperature reconstructions of the Holocene from two very different paleoclimate records. The thin solid curve is the average Greenland surface temperature derived from ice cores. The dashed curve with the error bands is the temperature at five hundred meters' depth at Indonesia's Makassar Strait. This region samples water from the northern and southern hemispheres. The temperatures are relative to the average of recent decades.

because the tides—as established by the cycles of Earth's orbit, Earth's rotation, and the Moon's orbit—leave repeating patterns in the growth layers of corals, mollusk shells, and tidal flats.[22] To reconstruct these cycles, they needed a good measure of the patterns in the preserved layers and, happily, Earth delivered.[23]

Earth provides other such records. For example, we learn from ancient tidalites—fossilized remains of sediments deposited by periodic coastal floods—that 500 million years ago, days were twenty hours long and months were 27.5 modern Earth days.[24]

Taken together, these data lines tell us that the Moon has been receding constantly over the last few hundred million years.

By providing more precise and recent information, tree rings complement more ancient climate data. Thicker tree rings imply warmer and wetter conditions during the growing season, while thinner rings suggest the opposite. They can also reveal past droughts and fires that low-res global records might miss.[25] Those trees most sensitive to changes in local climate grow in mountain areas without direct access to steady groundwater. For this reason, the western United States has been a prime field laboratory for those scientists, called dendrochro-nologists, who specialize in this work.[26] The trees you see in a forest are all tightly wound scrolls recording data—for those who know how to read it.

Even long-dead trees hold secrets. The leaves of fossilized trees can serve as "palaeobarometers" of carbon dioxide in the atmosphere.[27] Trees exchange gas with the atmosphere through small openings on leaf sur-faces called stomata. From laboratory experiments and historical her-barium data, botanists have found that the concentration of stomata on a leaf surface is sensitive to the concentration of carbon dioxide in the atmosphere. (Leaf fossils often reveal details fine enough to enable counting of the tiny stomata.) These studies, when compared to tem-perature reconstructions from oxygen isotopes in marine fossils, give us precious clues about the long-term relationship between temperature and carbon dioxide.

Some tiny organisms record data almost everywhere, even under Earth's surface. Mites, for instance, occupy almost every nook and cranny, like the clutter that fills the house of a hoarder. We find mites entombed in dark caves, preserved in stalagmites and stalactites by mineral-rich water. Because each species is adapted to a specific climate, the domi-nance of certain species of mites can teach us about the environment in which they lived. One study cataloged twelve types of mites in stalagmites in a cave near Carlsbad Cavern dating back 3,200 years,[28] revealing that the southwestern United States was wetter and cooler from 3,200 to 800 years ago.

EARTH'S CLOCKS

The most basic way to date deposited or grown records is just to count layers. Tree ring counting has been especially useful for calibrating the carbon 14 dating technique, since dendrochronologists can measure the carbon 14 in individual rings. This is necessary because carbon 14 isn't produced in the atmosphere at a constant rate.

The organic excretions of living things contain carbon 14. So, any growth layers or organic materials in layered deposits, like pollens and tree rings, are potential tools for absolute dating. Pollens are especially valu-able. Trees and plants proliferate pollens specific to their species. Charles Darwin complained that the process was extravagant and even wasteful; but that very extravagance allows us to use pollen granules to date layered

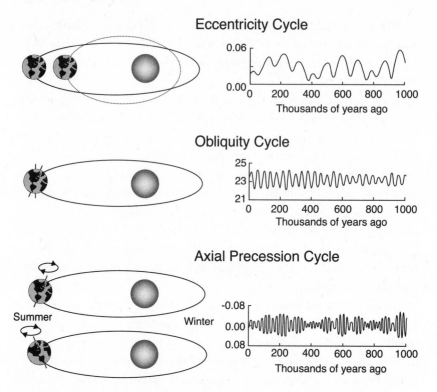

Figure 2.4: Milankovitch astronomical cycles for the Earth going back one million years. The small variations in Earth's orbit change the climate in ways that leave marks on its surface. All three types of variation—eccentricity, obliquity (tilt), and precession—are observed in the geological record. The dominant cycles are eccentricity, 100,000 and 400,000; obliquity, 41,000 years; precession, 23,000 and 19,000 years. The actual precession period is 26,000 years, but modulation by the eccentricity cycles results in the two periods seen in the geological record. Earth and Sun are not drawn to scale.

deposits. Pollens, cast far and wide by the wind, have been dated in lake sediments with carbon 14 to 60,000 years ago and in bogs to about 12,000 years ago.

There are other fairly short-lived radioactive isotopes in the oceans, including a number of intermediate decay products from long-lived radioactive isotopes. For example, uranium 234 and thorium 230 hide in corals and ocean sediments, allowing us to date them back a few tens of thousands of years.[29]

A very different sort of clock uses the polarity reversals in Earth's magnetic field (more on this in chapter 3). Magnetic minerals align themselves with Earth's magnetic field when they're deposited as sediments and left undisturbed. Oceanographers have mapped magnetic variations back nearly 200 million years, providing another way to check other records.[30]

Finally, there are the Milankovitch cycles, probably the single most useful type of clock for layered deposits.[31] These are the long-term dynamic rhythms such as the changes in the angle and direction of Earth's tilt and the subtle changes in its orbit. These cycles affect anything that involves water condensing or evaporating, or that is sensitive to changes in temperature or sunshine. Many such cycles occur on Earth, with ranges from 19,000 to 400,000 years, and are cataloged in ice, marine, and deep lake sediment cores.[32]

Once calibrated, researchers can use one record as an independent check on others.[33] (Such layered records are much more useful when more than one kind is available.) With Earth's many distinct records, therefore, scientists can reconstruct much of the ancient global climate confidently, as these records preserve events from Earth's deep interior to the distant stars.

There are many more examples, but we won't try your patience. We hope this brief survey is enough to establish the point: Earth's surface is peppered with millions of natural data recorders.

WHAT'S ALL THIS HAVE TO DO WITH HABITABILITY?

This may come as a surprise, but these recording devices are closely related to the life-friendly properties of our environment. To see this, we must bear in mind life's basic requirements.

Astrobiologists tend to assume that extraterrestrial life, if it exists, will resemble Earthly life. In seeking life on other worlds, it's much easier if they can use what they know of Earthly life. Critics claim this assumption betrays a certain Earth-centered provincialism. But this complaint overlooks a second reason for expecting life elsewhere to resemble life on Earth: namely, complex biological life of any kind has certain needs that can only be delivered by specific chemicals with specific virtues. Moreover, only a very specific planetary setting will meet the chemical requirements for complex life. Earth meets those requirements better than any place we've observed.

No life-form is actually simple. Still, we can distinguish three fuzzy categories of life relative to each other: simple, complex, and technological — each fuzzier than the one that precedes it.[34] Technological life involves, at minimum, a macroscopic aerobic metazoan—that is, a largish, oxygen-breathing, multicellular animal. Large mobile animals, especially those with large brains, need oxygen. This is not a limit imposed by a blinkered imagination, but by basic chemistry and physiology.

LIFE'S CHEMISTRY

Science fiction stories often describe alien life based on exotic chemistry.[35] The most popular option is silicon-based life—no doubt, because silicon is close to carbon on the periodic table. But as the old saying goes, close only counts in horseshoes and hand grenades.

CARBON

At bottom, chemical life must carry the instructions for building its progeny using basic atomic units.[36] These blueprints require many things, chief among them a complex molecule as the carrier. This molecule must be stable enough to withstand distress from heat and chemicals, but not be so stable that it won't react with other molecules at low temperatures. In other words, it must be *metastable*.[37] It must also have an affinity for many other kinds of atoms similar to the affinity it has for itself. Carbon excels at this task, but silicon does not. Other elements aren't even in the conversation.

Carbon's fitness for life goes far beyond the fact that it's metastable.[38] For instance, it forms gases when combined with oxygen (carbon dioxide) and hydrogen (methane), which allow free exchange with the atmosphere and oceans.[39] And most importantly, when combined with other key atoms—hydrogen, nitrogen, oxygen, and phosphorous—carbon can form the informational backbones of life, DNA and RNA, as well as the structural molecules—amino acids and proteins. Together, these make up a vast menagerie of molecules, which perform countless specialized jobs in living things. These molecules can store *far* more information than can the alternatives.[40] In fact, carbon has the half dozen or so key chemical requirements for life that are rare or absent in other elements.[41]

WATER

Life also needs a solvent as a medium for chemical reactions. The best solvent should dissolve all sorts of molecules and transport them to reaction sites, while preserving their integrity. It should be in the liquid state, since the solid state doesn't allow for movement, and the gaseous one doesn't allow for frequent reactions. Further, the solvent should be liquid over the same range of temperatures where the molecules of life remain intact in either the liquid or gaseous state. At low temperatures, reactions become too slow, while at high temperatures, organic compounds become unstable.[42] Water, the most abundant compound in the universe, exquisitely qualifies, and no other substance comes close to matching it.[43]

In fact, water far exceeds these needs for the chemistry of life. In 1913, Harvard chemist Lawrence J. Henderson described the unique fitness of water and carbon for life in his classic work, *The Fitness of the Environment*.[44] Everything we've learned over the past century has reinforced his arguments.[45]

The topic is too vast to cover here, but let's note a few good examples. First, and bizarrely, water is denser as a liquid than as a solid. As a result, ice floats on water, insulating the water underneath from further loss of heat.[46] If ice sank, lakes and oceans would freeze from the bottom up. It would likely remain there, frozen in the deep, estranged from the warmth of the Sun. Surface ice also helps regulate the climate by altering the amount of sunlight absorbed by, or reflected from, Earth's surface, as we'll further explore in chapter 4.[47]

Second, water has very high latent heats when changing from solid to liquid to gas. That is, it takes more heat to vaporize water than is required for the same amount of *any* other known substance at ambient surface temperature. (In fact, it takes more heat to vaporize water than almost any other substance at any temperature.[48]) Similarly, water vapor releases the same amount of heat when it condenses back to liquid. As a result, water moderates both Earth's climate and the temperatures of larger organisms. This feature of water also allows small bodies of water to exist on land; otherwise, ponds and lakes would easily evaporate.

Third, water's surface tension, which is higher than almost all other liquids, gives it better capillary action in soils, trees, and circulatory systems; a greater ability to form discrete structures with membranes; and the power to speed up chemical reactions at its surface. Finally, water is probably essential for Earth's plate tectonics and distributing the minerals of life.[49]

No other substance comes close to water's suite of life-essential properties.[50] Frank H. Stillinger, an expert on water, observed that "it is striking that so many eccentricities should occur together in one substance."[51]

ALL TOGETHER NOW

John S. Lewis, a planetary scientist at the University of Arizona, agrees that carbon and water have no equals. After pondering the alternatives, he concludes: "Despite our best efforts to step aside from terrestrial chauvinism and to seek out other solvents and structural chemistries for life, we are forced to conclude that water is the best of all possible solvents, and carbon compounds are apparently the best of all possible carriers of complex information."[52]

Earth's ability to regulate its climate hinges on both water and carbon, not least because carbon dioxide (which is highly soluble in water), water vapor, and, to a lesser extent, methane are key greenhouse gases. Our planet's living creatures, atmosphere, oceans, and solid interior exchange these life-giving vapors. Together, they create a unified climate feedback system, keeping Earth a lush planet for the past 500 million years.[53]

Life is not dross clinging to an inert surface. Life, rocks, and the atmosphere interact in a complex web of feedback loops. This fact poses a chicken-and-egg dilemma: Life needs a habitable planet to exist, but planets need simple organisms to make them habitable. James Lovelock even goes so far as to speak of a planetary physiology, because, like an animal's metabolism, our environment remains relatively stable despite changing external conditions. But the analogy goes deeper since both systems use the special properties of carbon and water.

We suspect carbon and water are uniquely well suited for life at the scales of molecules, cells, organisms, and planets.[54] If so, then environments without enough carbon and water will be lifeless. Once we recognize the exquisite fitness of carbon and water for life, we must also accept that this constrains the types of planets that can host life. A planet less flexible than Earth at regulating its climate with water and carbon will surely be less habitable.

Of course, even simple life requires far more elements than carbon, hydrogen, and oxygen. A tiny bacterium needs seventeen elements, and humans need twenty-seven.[55] In general, the larger and more complex the organism, the more diverse the proteins and enzymes it requires. Most but not all the vital elements for life are concentrated in seawater. But the atmosphere is the main source of nitrogen, and the continents are the main source of several mineral nutrients, including molybdenum.[56] So, planets without a nitrogen-rich atmosphere and continents may be unable to support much, if any, life.[57]

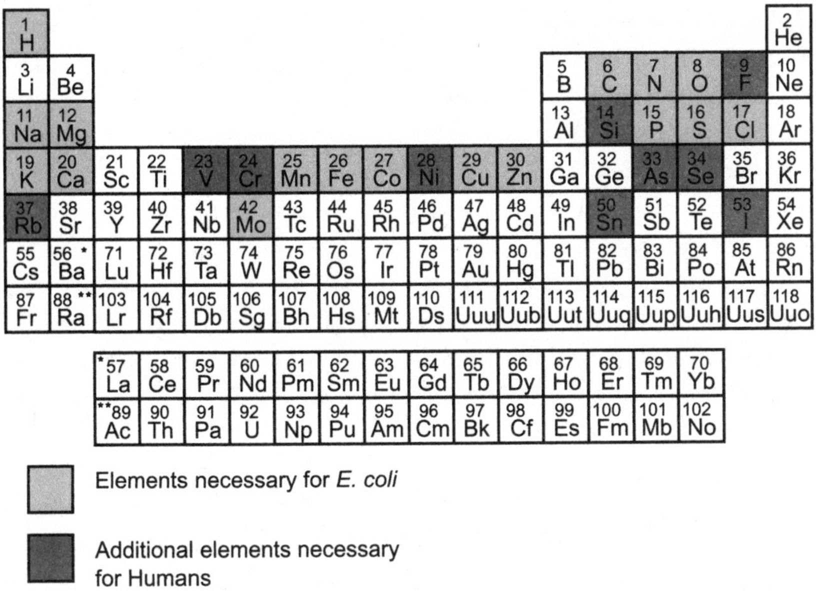

Elements necessary for *E. coli*

Additional elements necessary
for Humans

Figure 2.5: The periodic table of elements. Hydrogen (H), the lightest element, is number one on the table. Helium (He) is two. Complex life requires more elements than does "simple," single-celled life. A bacterium needs seventeen elements, while a human being needs twenty-seven. Additionally, some of these essential elements, such as iron, along with radioactive elements located near the end of the table, are required for geological processes that produce a habitable planet.

Apart from these essential elements, life needs a stable, long-term energy source. The basic sources are stars, geothermal heat, and chemicals. An environment needs enough energy to maintain liquid water, but even with water, a lush biosphere requires widely available energy.

Complex life also needs a support system of autotrophs, creatures that synthesize organic molecules from simple, inorganic matter.[58] For example, photosynthetic algae and some bacteria synthesize food from such inorganic materials as carbon dioxide, nitrogen, methane, hydrogen, and minerals. These algae and bacteria, and their organic products, then become food for other organisms that require organic food—heterotrophs like us. Some environments might support low-level microbial life. But if they can't sustain a vast swarm of autotrophs, there's no hope for larger, more complex creatures.

Life relies on chemical energy for its immediate metabolic needs, and such energy is all about the exchange of electrons. The most energy is released when elements located on opposite ends of the periodic table exchange electrons. Oxygen is second only to fluorine for releasing energy when combined with other elements.[59] Hydrogen, and carbon combined with hydrogen, or hydrocarbons, are the best elements to combine with oxygen. All complex life-forms use such oxidation reactions. Other common reactions yield far less chemical energy. And not incidentally, the products of oxidation are, once again, our friends water and carbon dioxide. So, hydrogen, carbon, and oxygen, together, offer the best source of chemical energy. As Henderson noted: "The very chemical changes, which for so many other reasons seem to be best fitted to become the process of physiology, turn out to be the very ones which can divert the greatest flood of energy into the stream of life."[60]

EXTREME ENVIRONMENTS

Studies of so-called extremophiles have made many astrobiologists more sanguine about finding such life on other planets.[61] We find these hardy critters in the severely salty Dead Sea and Great Salt Lake, the superheated water in deep-sea thermal vents, the stinky, gurgling Yellowstone springs, the frigid ice fields of the Arctic, and the dry valleys of the Antarctic. But these life-forms are not nearly as independent of other life as it appears. For example, many of the creatures found around some deep-sea thermal vents require oxygen. Surface-dwelling organisms produce oxygen by photosynthesis, which is then mixed into the deep ocean waters. What's more, these animals may even have migrated to the deep from shallower waters. Such extremophiles may not be able to prosper on an isolated planet with no other life.

Extremophiles also make a go of it off rock and hydrogen about a mile deep in the Columbia River basalt in eastern Washington and other subsurface spots around the globe. They typically have very slow metabolisms compared to life at the surface, and the concentration of the cells is typically very dilute. These subsurface life-forms also probably depend on surface life, and not just for oxygen. There's mounting evidence that deep subsurface microbial communities feed off dissolved organic matter, either from fossil soils or from fresher material brought down from the surface.[62] As a result, thermal vent and deep subsurface life may not be able to exist on a world that has never had abundant life on its surface or is far from a source of light energy.

Research by Abel Méndez, an astrobiologist at the University of Puerto Rico at Arecibo, suggests that most prokaryotes—organisms without a nucleus—grow best between 70 to 126 degrees Fahrenheit, with optimum growth at 96.8 degrees.[63] This is important because the biodiversity in the tropical regions depends mostly on the growth rate of such organisms.[64] Temperatures quite different from this optimum would provide much less support for a complex biosphere.

Méndez also notes that life cannot survive at arbitrarily high pressures. The limit is nearly one thousand times the pressure at Earth's surface. Several environments in the solar system where liquid water may exist exceed this limit. Moreover, while some species of extremophiles can tolerate extremes in temperature, salt content, moisture, and pH, few can tolerate a very broad *range* of conditions.[65] In fact, they're tough to maintain even in the laboratory.[66] So, while we can learn something of the range of settings in which life can exist by studying extremophiles, we should not assume we'll find them on planets far different from ours.

Studying the other planets in the solar system also helps us figure out the range of conditions required for life. And the more we learn, the more we realize that Earth is an exceptional host for both simple and complex life.

SPREADING LIFE?

We may still find life elsewhere in the solar system, because the rocky planets, moons, and asteroids in our solar system have been exchanging material, especially during its early history. Even today, we can find pieces of Mars and the Moon on Earth. In the same way, Earth might have contaminated other bodies in the solar system with its microbes (see Appendix B). On most of those bodies—such as Mercury, the Moon, Jupiter, and asteroids—life can't flourish. Mars, however, was probably wet in its early history, and might have supported life. But today, we find no evidence of life on its surface.

Even with an early helping of our microbes, though, the harshness of conditions on the other bodies prevented those organisms from surviving on or making their host planets more habitable. One or two hardy species clinging to a few oases in a mostly barren world have no chance to regulate the climate: they're at the mercy of runaway climatic shifts. If so, simple life may be less widespread in the universe than astrobiologists believe, even given what we know about extremophiles on Earth.

THE FEEDBACK FROM LIFE TO DISCOVERY

In taking basic minerals and energy sources to produce organic compounds, autotrophs improve the environment for all life. For example, marine organisms deposit carbonates on the ocean floor, which are a key part of the carbon cycle. (More on this in chapter 3.) In addition, marine phytoplankton produce most of the oxygen in the atmosphere. We and our animal neighbors also depend on simple life as food sources, digestive aids, and decomposers.

As mentioned above, life also makes Earth a more measurable place—through tree rings, stomata on leaves, mites in cave stalagmites, foram skeletons in deep ocean sediments, and pollen in lake sediments, to name a few. All the layering processes depend on the water cycle. This cycle includes snowfall on Antarctica and Greenland; rainfall that replenishes rivers, lakes, and springs and nourishes trees and other living things; and mountain erosion that provides the life-essential minerals for lakes and oceans. But not every water cycle will provide the high-quality natural recorders we find on Earth. Too little water would result in the erosion of deposited sediments; too much would leave scant land surface for stable ice sheets, trees, or corals. Both extremes would be less hospitable for life and discovery. Since this is easier to appreciate by comparing Earth to the other planets, we'll wait until chapter 5 to explore this.

If you took high school or college chemistry, you may recall how often water entered into the lab experiments. This isn't just because it's plentiful and cheap. The true value of water for chemists is that it's an amazing solvent. If it were not, chemistry would have developed at a snail's pace, perhaps never blossoming into the robust science we recognize today.

FORETELLING THE FUTURE

So, Earth's processes encode high-grade, accessible information—despite the fact that this conferred no survival advantage on living things throughout Earth's long past. After all, we've only recently noticed this and are still perfecting the technology required to recover and read data. At the same time, Earth's data-recording could help us survive far in the future.[67] For instance, it could help us keep Earth habitable long into the future by teaching us how temperature and atmospheric carbon dioxide are related.

Here's what we mean. Ice cores have revealed that the climate can swing wildly over just a few years—at least in the past. The last time was the Younger Dryas, which occurred about 12,000 years ago. Cooling our climate to the glacial zone over just a few years would disrupt global food production and render cities far from the equator uninhabitable. The ice core record from central Greenland shows that events like the Younger Dryas were the norm for most of the last 100,000 years. Recorded human history has been, by comparison, like the early summer in the French Riviera.[68] Extending the record back in time with the less detailed Antarctic ice cores, it looks like our present warm spell is the longest-lived one of the past 420,000 years.

These records allow us to test computer simulations against the real world. Climatologists can simulate the long-term global climate by adjusting their models to the present climate and testing them on the ancient data derived from the diverse Earthly archives. With this growing database, they should get much better than they are now at predicting future climate changes. Accurate long-term forecasting once seemed a pipe dream—and it's still far too speculative and imprecise. But the ice-filled pipes of cores alongside other records may one day make that dream a reality.

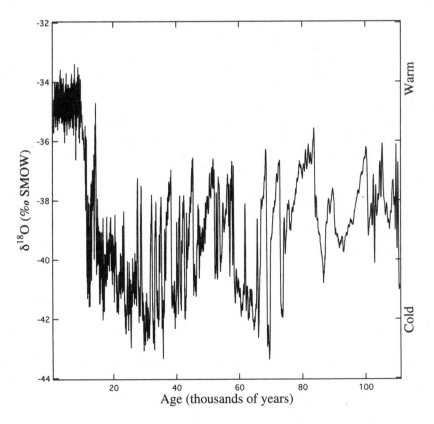

Figure 2.6: Variation in the oxygen 18/oxygen 16 isotope ratio in Greenland ice going back some 100,000 years. The oxygen isotope ratio is related to the local temperature. According to these data, the temperature over most of this period has been marked by very large and rapid changes. At times it's been as much as 15 degrees C colder than the present. By comparison, the climate over the last 12,000 years has been warm and stable.

We still have much to learn about climate change, of course, but one surprising discovery from this work is that atmospheric carbon dioxide could help prevent or delay glaciation in the future. Research by climatologists A. Berger and M. Loutre of the Institut d' Astronomie et de Geophysique in Belgium suggests that variations in the average amount of sunlight received recently by the northern hemisphere are quite exceptional.[69] They compared the near-term changes (due to the Milankovitch cycles) 5,000 years back to 60,000 years into the future, to those of the past three million years. They found that only five intervals in the past three million years were as moderate as our present.[70]

More recent climate simulations confirm Berger and Loutre's findings that the buildup of greenhouse gases from industry should postpone the next ice age cold spell. It's estimated to commence about 50,000 years from now without elevated carbon dioxide. According to the models, the level of carbon dioxide we have already achieved delays the onset to 120,000 years.[71] Continued rise in levels would delay it even further.

According to another study, Earth barely averted the start of the next ice age a few thousand years ago.[72] The rapid rise of farming about 7,000 years ago during the mid-Holocene may have produced just enough greenhouse gases—that is, carbon dioxide and methane—to prevent the early arrival of the ice age.[73] These days, we're used to hearing that a rise in greenhouses gases is all bad news. In reality, a modest rise of greenhouse gases may be good news for Earthly life.

You read that right: higher carbon dioxide levels could inoculate the planet, and us, against a "near-term" glaciation. Modern industry has helped us keep carbon dioxide levels well above the floor needed to set off an ice age.

We're now discovering that over the last several decades, the rise of carbon dioxide—which, you may recall, is plant food—has been greening our planet.[74] (See plate 6.)[75]

We have always affected our planet locally, but we've now reached the point where we can hope to understand the global climate system just as we're starting to affect it.[76] If we're smart, we should be able to decode Earth's countless hi-fi data recorders, sync our actions with natural patterns, and keep Earth habitable for the long haul.[77]

CHAPTER 3

PEERING DOWN

It may be that plate tectonics is the central requirement for life on a planet and that it is necessary for keeping a world supplied with water.
Peter D. Ward and Donald Brownlee[1]

We were writing this chapter for the first edition of this book when, on February 28, 2001, we experienced Seattle's strongest earthquake in half a century. We were both in tall buildings that rocked eerily, like trees in the wind. A friend in a ground-floor coffee shop saw a solid tile floor rippling like water, its solidity revealed as an illusion. Although it could be felt as far south as Salt Lake City and as far north as Vancouver, the earthquake's epicenter was thirty life-preserving miles underground. As a result, the earthquake—magnitude 6.8 on the Richter scale—caused only minor damage.

Earthquakes destroy property and kill many people every year; nevertheless, they make our planet more habitable and conducive to science. Without earthquakes, we likely wouldn't be here and, if we were, we would know far less about Earth's interior.[2]

Like a hammer hitting a bell, a strong earthquake generates waves in the solid Earth that radiate out and across its entire diameter. A seismograph anchored to the ground can detect waves from both strong, distant earthquakes and weak, local ones. Differences in density leave their marks on the wave paths, which then carry information about that part of Earth's interior. For instance, sharp discontinuities abruptly deflect the waves, as when light bends as it passes from air to a lens. Other wave features can reveal whether any of the traversed regions are liquid.[3]

But the real insight comes from collecting tracings from seismographs spread across Earth's surface. Using thousands of seismographs from thousands of earthquakes measured over the past several decades, geophysicists can produce a three-dimensional map of Earth's innards. The technique, called three-dimensional tomography, is a kind of geological CAT scan.

Earthquakes also help geophysicists probe small- and large-scale structures, such as an oceanic plate subducted beneath a continental plate. As the crustal plates slide past each other, sections from opposing plates often catch and build up stress. At some point, a sudden jolt relieves the stress and produces earthquakes. The deep earthquake near Seattle was produced within the Juan de Fuca oceanic plate, which is subducting beneath the North American plate. By measuring such deep earthquakes

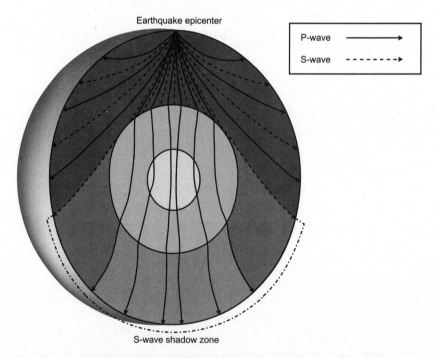

Figure 3.1: Cross section of the Earth showing its internal structure inferred from the study of P- and S-waves generated by earthquakes. This figure shows a hypothetical earthquake and the P- and S-waves it propagates through the mantle, outer core, and inner core. Seismographs, which are widely distributed on the continents around the Earth, are used to detect the waves. Note that only the P-waves can traverse the liquid outer core.

over time, geophysicists have developed three-dimensional pictures of sub-duction zones around the globe. Earthquakes also trace the mid-ocean ridges, where fresh crust forms like hot icing on a dry cake. Together, the zones and spreading ridges delineate the plate boundaries. If we had only half a dozen well-separated seismograph stations, we could still construct a map of Earth's major plate boundaries.

Today, seismographs are spread across Earth's surface. This would have been impossible earlier in Earth's history. About two hundred million years ago, only one supercontinent, called Pangaea, pierced the ocean sur-face of a watery, lopsided planet. Pangaea has since divided like a puzzle, with its pieces well distributed over Earth's face. To map our planet's liquid outer and solid inner cores, seismographs need to be nearly opposite an earthquake's epicenter. Like the continents, earthquakes are widely spread around the present Earth, occurring mainly along the plate boundaries. This arrangement is optimal for allowing us to probe Earth's structure.

Fortunately for science, earthquakes, unlike proverbial lightning, often strike the same place twice.[4] If geophysicists can measure the seismic waves of such repeating quakes at stations nearly antipodal to them—that is, on

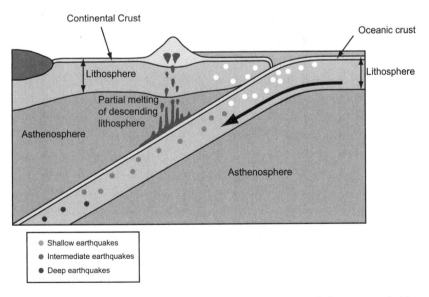

Figure 3.2: Subduction of an oceanic plate beneath a continental plate as revealed by earthquakes (in this case near the Japan trench). Earthquakes trace the three-dimensional shape of the sinking lithosphere—the rigid part of Earth's outermost layer, which includes the crust. The asthenosphere, which is easily deformed, is the part of the upper mantle. As viewed from the top, the earthquakes display a clear pattern of increasing depth as one moves away from the trench, towards the continental plate.

the opposite side of Earth—they can uncover secrets about the inner core. For example, for over three decades, researchers at a seismic station in Alaska measured the arrival times of waves generated by earthquakes in the South Sandwich Islands.[5] Since the waves arrived at different times, the researchers concluded that the solid core rotates a little faster than the rest of Earth. More recent research shows that the core rotation is changing.[6]

PLANETARY MAGNETISM

In the nineteenth century, physicists figured out that rotating a wire coil inside a set of magnets generates an electrical current in the wire. Today, we use this dynamo process to convert mechanical energy into electricity. The ingredients of a dynamo are a conducting medium, a magnetic field, and motion. From studying earthquakes and other events, geophysicists know that Earth's outer core is made of piping-hot liquid nickel-iron, which probably exceeds 3,000° C. Heat flowing through Earth's outer liquid core causes it to convect, like the heat transfer on a hot afternoon that forms cumulus clouds. Because it conducts electricity, the outer core sets up a dynamo generator. In an electrical dynamo, the magnets have permanent fields. But the Earth must regenerate its own field. Otherwise, it would decay after only a few hundred years. Such a self-sustained dynamo requires the planet to rotate fast enough to produce eddies in its outer core.

Geophysicists use Earth's fossil magnetic field to reconstruct its geologic past. Recall that Earth's field aligns ferrimagnetic minerals in grains as they sink to form undisturbed marine sediments. Additionally, whenever basaltic lava cools below the Curie point—the temperature above which a rock loses its magnetism—it freezes in the orientation of Earth's magnetic field at that moment. Such fossil magnetism, or paleomagnetism, remains preserved in a rock as long as it's not heated above the Curie point. We now know from studies of dated lava flows on land that our planet's magnetic field has changed polarity many times in the past—roughly once every million years.[7] The last one occurred 780,000 years ago. Geologists can use these global events to match widely separated strata of rock.

In the late 1950s, oceanographers began towing devices called magnetometers behind ships to map the ancient magnetic fields of the rocks forming the ocean floor. At first, they were searching for magnetic anomalies connected with large structures like volcanoes. But the resultant maps revealed a striking pattern of polarity reversals running parallel to and symmetric on both sides of the mid-ocean ridges.[8] We now know that these magnetic stripes result from reversals of Earth's magnetic field as fresh seafloor crust forms and spreads out on both sides of a ridge. The ocean floor acts like a giant magnetic tape recorder. To read it, all you need is a ship sporting a magnetometer. (See plate 7.)

Well, actually, you need a few other things. As geophysicist David Sandwell notes, "Indeed, the ability to observe magnetic reversals from a magnetometer towed behind a ship relies on some rather incredible coincidences related to reversal rate, spreading rate, ocean depth, and Earth temperatures."[9] Sandwell goes on to describe how these four scales conspire to produce measurable fields at the ocean surface:

> Most of this magnetic field is recorded in the upper mile or two of the oceanic crust. If the thickness of this layer were too great, then as the plate cooled as it moved off the spreading ridge axis, the positive and negative reversals would be juxtaposed in dipping layers; this superposition would smear the pattern observed by a ship. On Earth, the temperatures are just right for creating a thin magnetized layer.[10]

The remnant field variations on the seafloor would look attenuated and smooth if measured from too far away. For the signal to remain strong at the surface, the spacing of the magnetic stripes must have a typical spacing of about 6.3 times the ocean depth. The final two scales must be tuned as follows:

> Half-spreading rates on Earth vary from 6 to 50 miles (10 to 80 kilometers) per million years. This suggests that for the magnetic anomalies to be most visible on the ocean surface, the reversal rate should be between 2.5 and 0.3 million years. It is astonishing that this is the typical reversal rate observed in sequences of lava flows on land. . . . This lucky convergence of length and timescales makes it very unlikely that magnetic anomalies, due to crustal spreading, will ever be observed on another planet.[11]

This rare convergence creates a nearly ideal setting for accessing magnetic data from the ocean surface that's on the ocean floor. Geologists can then compare these data with data gathered from rock samples of ancient lava flows on land. The amounts of certain long-lived radioactive isotopes reveal a sample's age. And the orientation of a sample's acquired magnetic field reveals a detailed record of the Earth's field when it formed.

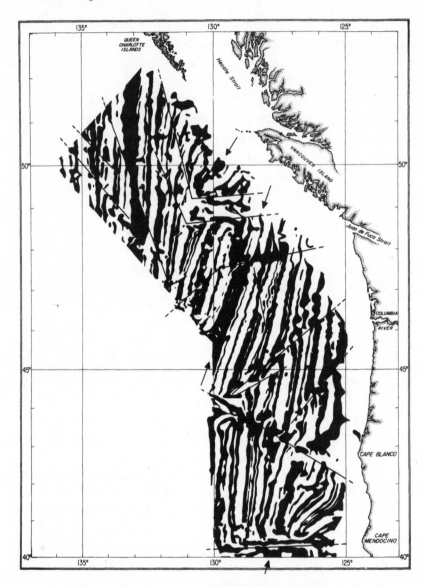

Figure 3.3: Geomagnetic reversals measured in the northeastern Pacific with a magnetometer towed by a research ship. The present magnetic field direction is shown as black, and the reversed direction as white. These magnetic patterns are used to reconstruct a detailed history of seafloor spreading. The locations of spreading ridges as well as faults are apparent in the figure. This figure (first published in 1961) played a significant role in the eventual adoption of plate tectonic theory.

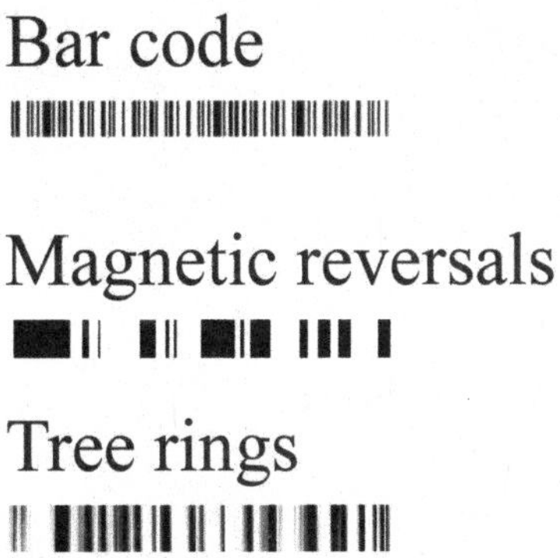

Bar code

‖ ‖‖‖‖ ‖ ‖‖ ‖ ‖‖‖ ‖‖ ‖ ‖‖‖‖‖‖ ‖‖ ‖ ‖‖ ‖‖‖ ‖ ‖‖

Magnetic reversals

■■‖ ‖ ‖‖ ■‖■ ‖‖‖ ‖

Tree rings

‖ ■‖■‖‖ ‖‖ ‖ ‖ ‖‖ ■ ■ ‖‖‖

Figure 3.4: Aperiodic patterns, as with the varying thickness of lines in a UPC barcode, convey much more information than regular or repeating patterns. This type of information is also present in nature, as with tree rings and geomagnetic polarity reversals.

By comparing seafloor magnetic tracings to these dated lava flows around the world, geologists can reconstruct a detailed history of seafloor spreading. The pattern is similar for all the spreading ridges around the globe. If the field were steady and unchanging, in contrast, it would be far less useful. Even a strictly periodic pattern would be inferior to the actual one since each magnetic reversal cycle would look like all the others. Scientists could not match a magnetic pattern on the seafloor uniquely to the same dated pattern on the land. Thankfully, these semi-periodic cycles generate unique patterns, much as tree rings do. This makes them far more useful for reconstructing Earth's past.

As it happens, Earth's magnetic field transmits information in all three spatial dimensions—a fact that German cleric Georg Hartmann discovered in 1544. As a result, Earth's field can move a compass's needle in three dimensions. If you place a compass on its side so that its needle can rotate up and down, it will assume an angle, called inclination. In the same way, rock samples from all the continents retain this 3D remnant imprint of the Earth's magnetic field. And more dimensions mean more information. These rock records work like a Global Positioning System (GPS); Earth's magnetic field serves as the network of satellites to define the absolute coordinate system. Unlike man-made GPS receivers, however, the natural version retains data about the deep past. That makes them a treasure trove for geologists.

Indeed, this evidence has been decisive in the history of geology. As it happens, these samples can reveal the locations of now-separated continents up to a few hundred million years in the past. It was this evidence that finally convinced skeptics of the theory of plate tectonics. The

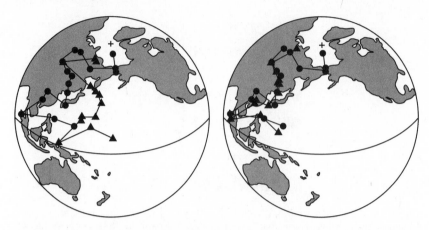

Figure 3.5: Apparent polar wander (APW) paths for the North American (circles) and European (triangles) continents superimposed on the modern globe. Geologists routinely measure declination and inclination values of the remnant magnetization of rock samples from all the continents. Since the inclination angle varies simply with magnetic latitude, it's fairly easy to determine the distance to the magnetic pole when a sample formed. Adding the sample's declination allows geologists to pinpoint where the magnetic pole was at that time. They can also prepare an APW path from the ancient magnetic pole positions derived from volcanic lava samples of various ages at a given site. By comparing polar wander paths from samples collected on different continents, geologists can reconstruct their motions over long periods. Such plots reveal the past locations of continents up to a few hundred million years in the past. The APW paths in this figure were determined from the fossil magnetism in rocks with accurately known ages. If the European APW path is rotated 38 degrees clockwise (as viewed from the North Pole), then the two paths overlap. This implies that the two continents were moving together over the time covered by the data, some 150 to 500 million years ago.

German meteorologist Alfred Wegener first proposed the theory's main idea, continental drift, in the 1920s. His theory nicely accounted for the jigsaw-puzzle fit of the outlines of some continents and the similar rocks found on now widely separated lands. But geophysicists at the time rejected Wegener's idea and clung to the so-called geosynclinal theory.

In the 1960s, however, magnetic data from land and the seafloor provided convincing evidence for continental drift and seafloor spreading—both supporting Wegener's theory.[12] The triumph of plate tectonics over the older geosynclinal theory ranks as one of the great paradigm shifts in science. And it only happened because Earth retained such precise evidence of its past.[13] Its magnetic field revealed its interior, and the magnetic reversals provided striking visual clues to the underlying mechanism.

We don't yet fully grasp what drives the magnetic reversals, but we do know the Sun displays them as well. (Earth's field even tells us something about the ever-changing state of the Sun's field.) The solar wind carries

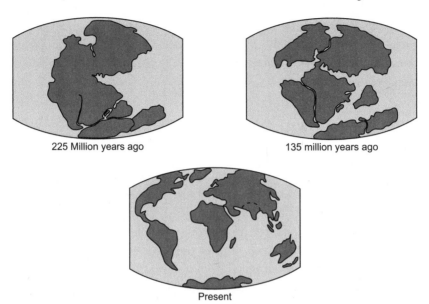

225 Million years ago   135 million years ago

Present

Figure 3.6: Continental drift from the Pangaea supercontinent to the present. These reconstructions of the relative positions of the continents are based primarily on measurements of the fossil magnetism in basaltic rocks. The Earth's magnetic field serves as a kind of Global Positioning System.

the Sun's field to the planets. For this and other reasons, this interplanetary magnetic field is constantly changing, interacting with Earth's field and causing small, rapid fluctuations. By monitoring these daily changes from ground stations over the last century, geophysicists found that the Sun's poles reverse every twenty-two years—twice the length of its main sunspot cycle.[14] Space physicists are now starting to grasp how the Sun's field affects the interplanetary magnetic field, which they can use to predict a sunspot cycle five to six years in advance.[15] This, in turn, helps NASA estimate atmospheric drag, which affects the lifetime of low-altitude satellites. During a sunspot maximum, the outermost region of Earth's atmosphere is more puffed up, increasing the drag.

Astronomers and geophysicists can now even measure the drift of the continents in real time. They do this by using widely separated radio telescopes aimed at the same quasar—an object so far away that it's in effect a fixed object. Astronomers use radio observations from different continents of several quasars over several years, to measure the relative motions of the continents. This experiment wouldn't work using nearby stars, since their motions in the Milky Way would overwhelm the weak signature of continental drift. As it is, however, astronomers can detect even tiny continental shifts of about one centimeter per year.[16] This is just one of the many ways our access to the distant universe helps us study our local environment.

THE HABITABILITY CONNECTION

When most of us think of earthquakes, we picture death and destruction. But ironically, they result from geological forces that make our planet life-friendly.[17] Heat flowing from inside Earth drives mantle convection, which, in turn, moves Earth's crust.

And again, this happy fate is far from inevitable. What does it take for cold, rigid, floating chunks of crust to keep getting pulled deep into a planet's gut for billions of years? A lot of things have to be just so for a planet to remain tectonically active for so long. On Earth, a crucial part of the recipe is liquid water on the surface.[18] Water reacts with the minerals in the crust, weakening and lubricating the crust so that it can bend without breaking.

A tectonically active crust makes Earth more habitable in a variety of ways. Let's explore them now, one by one.

THE CARBON CYCLE

A habitable planet needs a carbon cycle, and that cycle needs plate tectonics—the source of earthquakes. This cycle includes both organic and inorganic sub-cycles occurring on different timescales.[19] These cycles regulate the exchange of carbon-containing molecules among the atmosphere, ocean, and land. Photosynthesis, both by land plants and by phytoplankton near the ocean surface, is a leading character in this drama since it draws carbon dioxide from the atmosphere. Zooplankton, such as the forams mentioned in chapter 2, consume much of this matter. The carbonate and silicate skeletons of the marine organisms settle obligingly on the ocean floor, to be eventually squirreled away beneath the continents.

Another lead character in the carbon cycle drama is the chemical weathering of silicate rocks on the continents.[20] This occurs when rainwater, made acidic with dissolved carbon dioxide from the atmosphere, dissolves minerals in exposed rock. In time, the rivers carry dissolved silica, calcium, and bicarbonate ions (derived from carbon dioxide) to the oceans. Phytoplankton and zooplankton, and to a lesser extent corals and shellfish, then remove these dissolved chemicals from the ocean to build their skeletons. The carbon cycle is completed when the subducted carbonates are pressure-cooked deep in the crust. This releases carbon dioxide that eventually finds its way to the surface through volcanoes and springs.

Negative feedback loops keep the cycle in balance. Perhaps the most important such stabilizing feedback involves chemical weathering and temperature. Here's how it works. Suppose some large volcanic eruptions spike the amount of carbon dioxide in the atmosphere. Through the greenhouse effect, the upsurge in carbon dioxide raises the global temperature. The higher temperature and carbon dioxide level, in turn, speed up chemical weathering, thus removing carbon dioxide from the atmosphere. Slowly but surely, the carbon dioxide and temperature return to their pre-eruption levels.

Conversely, a drop in carbon dioxide would slow chemical weathering, allowing carbon dioxide to build up in the atmosphere. In either case, the loop comes full circle.

As we noted in chapter 2, a rise in carbon dioxide in the atmosphere spurs plant growth. Rocks fuzzy with plants weather about five times faster than bare rocks, allowing Earth to tuck carbon away under its surface more quickly. In other words, this accelerated weathering means Earth's climate system can adapt to perturbations far better than a lifeless world could. Thus, plate tectonics, together with plant life, make a planet much more nurturing for all life over the long haul.

CONTINENTS

A planet's carbon cycle needs both continents and oceans. Continents serve as a mixing bowl for minerals and water at the surface where energy-rich sunlight is available. The continents began to appear about a billion years after Earth formed.[21] As they grew, they and the crust that carried them extracted potassium, thorium, and uranium from the mantle. These elements have been the main sources of heat in Earth's interior for most of its history. Siphoning them from the convecting mantle weakens plate tectonics, however. If the continents and crust had grown more rapidly, they would have drawn more heat-producing elements from the mantle. This would have slowed down mantle convection and tectonics, making the climate less adaptable later.

Plate tectonics plays another life-critical role: it maintains dry land in the face of steady erosion. A large rocky planet like Earth wants to be perfectly round, with erosion slowly wearing down the mountains and even the continents, creating a true waterworld. Its interior must keep supplying energy to keep it from getting bowling-ball smooth. Without this form of recycling, such a place would probably become lifeless, since it would lack a way to mix the nutrients life needs in the sunbathed surface waters.[22]

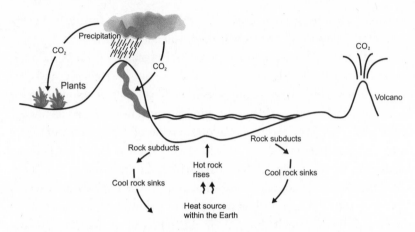

Figure 3.7: Plate tectonics works together with the carbon and water cycles to produce a planet hospitable to life. Heat from the Earth's interior migrates to the surface, setting the mantle in motion. The dynamic interior builds mountains and releases carbon dioxide into the atmosphere through volcanoes. Energy from the Sun evaporates water into clouds, which precipitates, returning water and CO_2 to plants, soil, and ocean. Transformed into other chemical forms, the carbon originally in the atmosphere makes its way to the seafloor. The carbon-enriched oceanic crust is then subducted into the mantle, and eventually reheated and returned to the surface as CO_2. The entire process keeps nutrients, water, and land available for life, and regulates the global temperature on timescales of millions of years.

MAGNETIC FIELDS

A terrestrial planet with plate tectonics is also more likely to have a strong magnetic field since both depend on the convective overturning of its mantle. And a strong magnetic field creates a magnetosphere, which shields a planet's atmosphere from the solar wind. If Earth's upper atmosphere had to directly face solar wind particles—consisting of protons and electrons—our atmosphere would lose water more quickly to space. That would be bad news for life.

Our magnetic field serves another life-essential role. Just as Star Trek's *Enterprise* uses a deflector shield to protect it from incoming photon torpedoes, Earth's magnetic field serves as the next line of defense against galactic cosmic ray particles. (The Sun's magnetic field and solar wind deflect the lower-energy cosmic rays.) These particles consist of high-energy protons and other nuclei, which, together with mesons, interact with nuclei in our atmosphere. The resulting particles can pass through our bodies, breaking up cell nuclei. Not good!

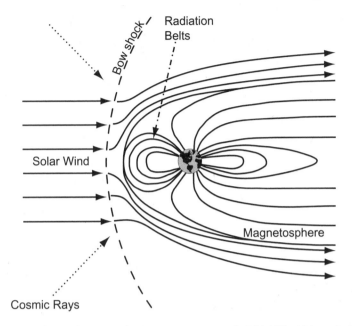

Figure 3.8: Earth's geology conspires to create a magnetic "shield," which protects the Earth's atmosphere from solar wind particles and low-energy cosmic rays.

PLANETARY SIZE AND MASS

Life also depends mightily on a planet's mass.[23] A less massive Earth twin would have weaker gravity on its surface and so would lose its atmosphere more quickly. What's more, its interior might cool too much to produce a strong magnetic field. And, as we'll show in chapter 5, smaller planets also tend to have erratic orbits.

In contrast, a more massive Earth-twin would likely start off with more water and other volatiles, such as methane and carbon dioxide.[24] It would also lose less of these over time. As a result, the planet might resemble the ice giant Neptune rather than Earth. In fact, Earth may be about as big as a terrestrial planet can get without ending up with a deep ocean and a thick atmosphere.[25] Life needs an atmosphere, but there can be too much of a good thing. For example, with a higher surface pressure, water would evaporate more slowly, drying out the interiors of any continents. The air on the surface would also be more viscous, making it harder for big-brained, mobile creatures like us to breathe.

In addition, given the same amount of water, more gravity at the surface would mean smaller mountains and shallower seas.[26] The planet would probably be covered by oceans and be too mineral-starved at the surface (and too salty throughout) to support life.[27] Even the gilled Kevin Costner character from the 1995 movie *Waterworld* would find such a world unappealing.

The surface gravity of a terrestrial planet grows with mass faster than one might at first guess.[28] Intense pressures compress the matter deep inside,[29] such that a planet just twice the size of Earth would have about fourteen times its mass and 3.5 times its surface gravity.[30] A larger fraction of gases like water vapor, methane, and carbon dioxide would likely end up in the atmosphere. Earth has kept continents poking above the oceans throughout its long history, in part, because some of its water has been sequestered in the mantle.

Larger terrestrial planets have another problem: They're more attractive targets, literally.[31] Asteroids and comets have a hard time avoiding larger planets, which exert a greater gravitational force and suffer far more high-speed collisions. Yes, their bigger surfaces distribute the greater energy over more area. This doesn't make the impacts less destructive, however, because surface area increases slowly with mass for terrestrial planets more massive than Earth—as we noted above, a bigger planet squeezes its contents together more densely.[32]

Not only would both smaller and larger terrestrial planets likely be less livable than Earth, but they would also offer poorer platforms for discovery. While smaller planets would have taller mountains for telescopes, they would have fewer earthquakes, delaying discoveries in geophysics. Smaller planets also provide a smaller, less effective platform for VLBI (Very Long Baseline Interferometry) radio observations—which require distant telescopes on different continents. This would mean a much lower and less useful bottom rung on the distance ladder for measuring the universe, which we'll discuss later. A planet larger than Earth would probably have more tectonic activity, but its smaller mountains and thicker atmosphere would hinder astronomy.

Of course, any atmosphere, including Earth's, will hinder astronomers. But large mountain observatories like the Hale 5-meter telescope in Southern California, the Keck telescopes on Mauna Kea or the Very Large Telescope in the Atacama Desert of northern Chile can still achieve a resolution rivaling that of the Hubble Space Telescope.[33] As a result, without leaving home, scientists can learn about both Earth's interior and the distant stars with far more ease than would observers on much smaller or larger planets. That's a good trade-off.

TECHNOLOGICAL LIFE

To have advanced technology, you need life-forms complex enough to develop it. Such life requires a much narrower set of conditions than simple life. We could say a species is technological if, at the very least, it can plan several years ahead, control and shape the basic elements with fire,[34] and alter its environment enough to enhance its survival and prosperity.

Controlling fire may be the linchpin for technology.[35] Early on, our ancestors used fire to cook and keep food warm, making it easier to digest. Later they learned to use fire to make pottery and smelt metal ores.[36] More recently, fire has allowed our species to sustain a population large and wealthy enough to allow many to focus on matters other than survival. Thus, a species that shapes metal to form tools to build plows, dams, bridges, and tall buildings would be technological, while one that spends all its time hunting wild animals, gathering wild fruits, and picking lice would not.

Some environments would prevent high technology altogether. A large aquatic animal lumbering around a waterworld, or Carl Sagan's giant floaters, drifting like gossamer through Jupiter's atmosphere, could not

Figure 3.9: Controlling fire is a prerequisite for high technology. Water-based animal life would not be able to develop it.

use fire or build electrical circuits.[37] But even the more benign terrestrial planets must have periods long and stable enough for its inhabitants to maintain widespread agriculture, which a large population depends on. As we showed in the last chapter, Earth's stable climate since civilization arose has been unusually long lasting.

THE FUELS OF DISCOVERY

When focused on our needs, we tend to think of plants and animals as sources of food and as beasts of burden. It's easy to forget how much of our technology and wealth depends on teeming life-forms that covered the Earth before we arrived. For instance, we likely owe most of the coal we burn today to the lush forests of the Carboniferous period about 300 million years ago. Petroleum was formed mostly from marine plants over roughly the same period. (Yes, dinosaurs also may have contributed, but only to a tiny fraction of the whole.) Without these fuels, the modern Industrial Revolution, and the wealth that followed, would not have happened.[38]

The wealth of industrial nations has allowed them to devote vast time and resources to science and technology. A planet with a feeble plant and microbe yield, in contrast, would take much longer to build up similar levels of coal and petroleum reserves.

It's curious that each new energy source has been abundant and long-lasting enough to hold us over while we developed the technology needed to reach the next level. Burning wood kept us going for thousands of years. Without wood, denizens of, say, a rocky, lichen-covered world would be hard pressed to find a rich source of long-lasting heat. Alternately, on an Earth with trees but no coal, we would have cut down all the forests long ago. In fact, this almost happened in England, until coal mining came to the rescue of their dwindling forests, providing a more abundant and cheaper form of fuel.

At some point, petroleum and then uranium entered the mix, and the polluted, dark lichen-covered trees of England returned to their natural, ashen hue. With luck, these fuels will sustain us until we can develop the next grade of energy—likely thorium fission, maybe even fusion.[39] Fortunately, each new grade of fuel is more densely packed with energy and demands less from the environment. And the carbon dioxide we've been producing as a result—contrary to the dire warnings of the Zero Carbon/Net Zero crowd—has given plant life a boost.

CONCENTRATED ORES

Technology requires not just concentrated energy, but also concentrated mineral ores in the crust—as Earth has. The more concentrated an ore, the less costly it is to mine. But this happy state is hardly inevitable for a planet. As Berkeley geologist George Brimhall explains:

> The creation of ores and their placement close to the Earth's surface are the result of much more than simple geologic chance. Only an exact series of physical and chemical events, occurring in the right environment and sequence and followed by certain climatic conditions, can give rise to a high concentration of these compounds so crucial to the development of civilization and technology.[40]

The same processes that make Earth habitable also concentrate mineral ores near the surface. Why? Because the crust is the interface between land, water, and air. Hydrothermal mineral-rich solutions flow through cracks in fractured crustal rock, precipitating minerals as they come into contact with cool rock surfaces. These dissolved minerals linger close to the surface because open fractures in rock are rare in the deep, highly compressed Earth. This prevents the liquids from flowing very far down. The rich cornucopia of Earth's mineral ores owes much to the dissolving power of water.

Consider the radioisotopes that help render a life-friendly planet. Despite some claims to the contrary, uranium, along with potassium and thorium, is most concentrated in the crust. Uranium is nearly one hundred times more concentrated in the continental crust than in the mantle.[41] Only the much greater mass of the mantle makes it Earth's main source of radioactive heat. The same forces that built the continents collected these elements in the crust; every overturn of the mantle deposits some radioisotopes there. If uranium were only as concentrated in the crust as it is in the mantle, we'd have no hope of mining it. And, as with other mineral ores, the same small bit of Earth's mass with the most easily mined uranium is also its most habitable.

Life has also done its part to concentrate ores in the crust. Take microbial mats—layered communities of bacteria that grew in salt evaporite lagoons in what is now Oklo in Gabon, Africa. Nearly two billion years ago, they left behind uranium oxide. At the time, there was far more of the relatively short-lived isotope uranium 235 in Earth's crust. In Oklo, it was so concentrated that nuclear reactions started, with water acting as a moderator.[42] Such reactions ceased long ago, however, since there's no longer enough of this isotope for natural sustained reactions. Today, however, Oklo is a major source of uranium for the French nuclear industry.

Those uranium ores were able to build up as they did only because the atmosphere had become oxygen-rich. This allowed uranium to exist in a water-soluble mineral form and be transported to bodies of water. Before oxygen built up in the atmosphere, there wasn't much uranium in the water for organisms in the microbial mats to concentrate. This is a good thing, because it would have been dangerous for life to concentrate uranium too early in Earth's history, when the uranium 235 isotope was more plentiful. Perhaps the rise of oxygen in the atmosphere had to await the decay of most of the uranium 235, lest it pose too great a threat to the life in the mats. Moreover, the productivity of the microbial mats determined how quickly oxygen built up in the atmosphere. In other words, this could turn out to be a process that set the timescale for the rise of oxygen in the atmosphere.

Brimhall also highlights the role of atmospheric oxygen for distilling certain metallic ores such as copper. So-called secondary enrichment of ores occurs because water has an easier time moving some metal ions in an oxygen-rich atmosphere. This process is equally vital for complex life and technology. A planet needs both a long history of plant life for its later residents to develop technology and an oxygen-rich atmosphere to burn their fuel.[43] "Another fascinating coincidence," Michael Denton observes, "is that only atmospheres with between ten and twenty percent oxygen can support oxidative metabolism in a higher organism, and it is only within this range that fire—and hence metallurgy and technology—is possible."[44]

Robert Hazen of the Carnegie Institution for Science has determined that there are over 5,000 mineral species on Earth; over half of them are mainly a gift of life itself.[45] In contrast, the solar system began with only a dozen minerals. Mars and Venus, Earth's nearest twins, are thought to have only about 500 and 1,000 mineral species, respectively. According to one writer interpreting Hazen's work, "our watery, living planet is probably much richer in mineral diversity than other rocky bodies in the Solar System."[46]

There's much more that could be said on this subject, but it should be clear from the above that life, technology, and a wide range of sciences depend on Earth's size, magnetic field, and plate tectonics, and on the resulting carbon cycle, continents, and the mixing of mineral nutrients.

CHAPTER 4

PEERING UP

The combined circumstance that we live on Earth and are able to
see stars—that the conditions necessary for life do not exclude those necessary
for vision, and vice versa—is a remarkably improbable one.
This is because the medium in which we live is, on the one hand, just thick
enough to enable us to breathe and to prevent us from being burned up by
cosmic rays, while, on the other hand, it is not so opaque as to absorb entirely
the light of the stars and block any view of the universe. What a fragile
balance between the indispensable and the sublime.

Hans Blumenberg[1]

In the science fiction classic "Nightfall," Isaac Asimov tells the story of Lagash, an inhabited planet that orbits Alpha, one of six stars in a multiple star system. Because of their crowded sky, the people of Lagash know neither darkness nor night.

The story opens just a few hours before one of Lagash's large moons is predicted to eclipse the one star still in the sky after the others had set. The event happens only once every 2,049 years—just often enough to give rise to myths, but not knowledge, of the stars. With only their hazy cultural memories, the people await the eclipse in fear of what darkness might bring. But despite their fear, they survive. As the eclipse comes to an end, the story wistfully concludes, "The long night had come again." The long night refers to the next two millennia, during which they will again be ignorant of the stars.[2]

"If the stars should appear one night in a thousand years," wrote Emerson, "how would men believe and adore, and preserve for many generations the remembrance of the city of God!" These bright denizens of the dark have inspired every great culture and religion. They also inspired the curiosity and obsessive observations that gave rise to astronomy. But how easily it could have been otherwise! If we lacked dark nights and an atmosphere transparent to visible light, we might know nothing of the universe beyond our tiny neighborhood. And by extension, we would know far less about our planet.

To grasp our good fortune, take the time at least once to stand on a mountaintop under the open sky, on a clear, moonless night. The air is so clear, and the stars so vivid, that only your lungs will remind your eyes that you're on a planet with an atmosphere. Our transparent atmosphere plays a part in one of the most eerie coincidences known to science. The range of wavelengths of light emitted by the Sun, transmitted by Earth's atmosphere, converted by plants into chemical energy, and detected by the human eye are independent facts. And yet they match.

Let us explain. Our eyes perceive light of certain wavelengths as colors, ranging from the shorter violet blue to green to the longer red. Looking at a diagram of the electromagnetic spectrum, we see the visible light emerging gracefully from the ultraviolet, segregating into the familiar colors of the rainbow, and disappearing seamlessly into the warm, invisible infrared.

The visible wavelengths rise and fall at the nanoscale. We measure the distances from their peaks in ten-billionths of a meter, called Ångstroms (Å). We see blue at about 4,800 Å, green at 5,200 Å, yellow at 5,800 Å, and red at 6,600 Å. Earth's atmosphere is transparent to radiation between 3,100 to 9,500 Å and to the much longer radio wavelengths. The radiation we see is near the middle of that range, between 4,000 and 7,000 Å, wherein the Sun emits 40 percent of its energy. Its spectrum peaks smack in the middle of this visible spectrum, at 5,500 Å.

The visible spectrum is but a tiny slice of the entire range. Now, add in the near-ultraviolet and near-infrared spectra—the forms of light that, together with visible light, are most useful to life and sight. Even this is a razor-thin sliver of the universe's natural emissions: about one part in 10^{25}. That is a much smaller portion than a single grain of sand from among all the grains of sand from all the beaches on earth: about 10^{21}. (See plate 8.)

As it happens, our atmosphere strikes a nearly perfect balance, transmitting most of the radiation useful for life while blocking most of the lethal. Water vapor in the atmosphere is likewise accommodating.[3] Even the fifteenth edition of the staid *Encyclopedia Britannica* enthuses about this. "Considering the importance of visible sunlight for all aspects of terrestrial life," it comments, "one cannot help being awed by the dramatically narrow window in the atmospheric absorption . . . and in the absorption spectrum of water."[4]

The oceans transmit an even narrower window of the spectrum, mainly the blues and greens, while halting the other wavelengths near the surface, where they nourish the marine life that figures prominently in Earth's biosphere.

One might suppose this is just due to our eyes evolving through natural selection to decipher the spectrum of light that happens to get through the atmosphere. But this explanation runs into problems.[5] As George Greenstein notes:

> One might think that a certain adaptation has been at work here: the adaptation of plant life to the properties of sunlight. After all, if the Sun were a different temperature could not some other molecule, tuned to absorb light of a different color, take the place of chlorophyll? Remarkably enough the answer is no, for within broad limits all molecules absorb light of similar colors. The absorption of light is accomplished by the excitation of electrons in molecules to higher energy states, and the general scale of energy required to do this is the same no matter what molecule you are discussing. Furthermore, light is composed of photons, packets of energy, and photons of the wrong energy simply cannot be absorbed.[6]

In other words, because of the properties of matter, the typical energy involved in chemical reactions corresponds to the typical energy of optical light photons. Otherwise, there would be no photosynthetic life. Photons with too much energy would tear molecules apart, while those with too little energy could not trigger chemical reactions.

Similar arguments hold for the wavelength range over which the atmosphere is transparent. Stars that don't emit the right sort of radiation in the right amounts won't qualify as useful energy sources for life.

The radiation a star emits, and its peak emission, depend on its temperature. Temperatures of normal stars range over a factor of about twenty. The Sun is a little below the midrange, while exotic stars emit most of their energy on the far ends of the electromagnetic spectrum, as gamma rays, X-rays, or radio waves.[7] A star like our Sun, which produces mostly photons that can power chemical reactions on Earth, will shine light that comes from just that region in its photosphere where atoms can form stable molecules. So we can't reduce one part of this coincidence to the others. Life can't use just any type of light from any type of star. Our Sun, it turns out, is near the optimum for any plausible kind of chemical life.[8]

To see the stars and the wider cosmos, though, a translucent atmosphere that merely allows light to reach the surface won't cut it. We need an atmosphere that is like a clean, clear pane of glass.[9] (See plate 9.) And even then, we can only enjoy the views because we have the good fortune of living on land. It would be a different story if we lived underwater.

But even dry land and a transparent atmosphere wouldn't be much use without dark nights. To get a dark sky, our planet must regularly rotate away from the intense direct light of the Sun.[10] If our day were the same length as our year, Earth would always keep the same face pointed toward the Sun, much as the Moon does Earth. The resulting large temperature swings between the day and night sides would be hostile to complex life (more on this in chapter 7). Any such life would stay on the day side, probably close to the twilight zone so as not to be boiling hot. Adventurers might attempt to travel to the dark side, but how far would they have to go before the Sun was far enough below the horizon for the travelers to see many stars? And by that point, how punishingly cold would it be? Unless the planet enjoyed total solar eclipses, like Lagash, the vast majority of inhabitants would never see dark skies.

We would suffer similar but less severe problems if we had several moons staring unblinkingly in the night sky like headlights on a busy highway. Our single Moon does interfere with astronomers who want to observe distant faint objects, but only when it's out. Moreover, night views are much better now than they were in the distant past, when the Moon was nearly three times closer to Earth, and thus nearly nine times brighter.

A translucent atmosphere might work for merely converting sunlight into chemical energy or hunting prey and avoiding predators. But because we are big-brained, mobile, surface-dwelling creatures, we need a certain type of atmosphere. It just so happens that that very atmosphere—mostly nitrogen and oxygen with some carbon dioxide and water vapor—is mostly transparent to optical radiation. Our biosphere needs all four of these chemical constituents. (Water does partially hinder our view of the distant universe—most evidently in the form of clouds—but the valuable trade-offs it provides more than compensate for this.)

Take the rainbow, for instance. Rainbows, like total solar eclipses, seem equal parts whimsy and mystery, summoning the artist's creativity and the naturalist's curiosity.[11] The early attempt to explain them was the first of our many tutorials on the nature of light. The scientists who followed those clues slowly learned how to unweave the white light of the Sun. Following René Descartes, Isaac Newton experimented with sunlight and prisms in 1666, groping toward the modern explanation for the rainbow and the science of spectroscopy. A rainbow is in effect a natural spectroscope as big as the sky. Once scientists learned how to use a prism to cast a rainbow, it was only a matter of time before someone scrutinized the spectrum of the Sun. But rainbows won't appear on just any planet. A good rainbow needs an atmosphere with some water vapor in it and, more specifically, a partially cloudy atmosphere, the golden mean between the uniformly cloudy and uniformly dry.

The amount of water vapor our atmosphere possesses gives us skies that, on average, are about 67 percent cloudy—more in some places, like Seattle, less in other places, like West Texas.[12] Clouds help balance the global energy by contributing to the global albedo, that fraction of sunlight reflected back into space. Earth currently reflects about 30 percent of the incoming sunlight. Four major surfaces play their own distinct roles in this process: land, ice, oceans, and clouds.

FEEDBACKS AND A CLEAR VIEW

A habitable planet must be flexible. If we were building one from scratch, we would want its albedo (reflectivity) types to adapt to changing conditions. If some part of Earth's climate system changes, its temperature will change unless another part of the system can compensate. For example, say the Sun gets a lot brighter over a few million years. If nothing counteracts this, Earth will heat up, possibly spelling disaster for life. The polar ice caps would melt, causing Earth to deflect less of the Sun's energy, leading to a vicious cycle of overheating that might evaporate the oceans. An adaptable environment, however, would compensate for the rising heat by producing more clouds, rain, and snow. These would reflect the excess solar energy back into space and encourage plant growth. Plants, in turn, sequester carbon dioxide, a greenhouse gas. We call such stabilizing forces negative feedbacks.

Earth's four albedo types operate on different timescales. Clouds are the rapid responders. Trees and plants react more slowly—from a single season to several decades. Major changes in sea level can take hundreds of even thousands of years. And geological changes are the slowest, taking a million years or more. With such variety, Earth's climate can respond to changes lasting anywhere from hours to billions of years. This is critical for life because the Sun does change its energy output over those timescales. A large sunspot group produces short-term changes against a backdrop of slower changes produced over the prominent eleven-year sunspot cycle and other, subtler, cycles that last hundreds of years. And over billions of years the Sun has slowly brightened as its core has heated up.

While we still don't fully grasp how clouds regulate the climate, we do know that a partly cloudy atmosphere is better than either a cloud-covered or a cloud-free one. Neither of these extremes could regulate climate by tweaking the cloud cover. The same is true for the contributions of land, ocean, and ice. Earth has the most diverse collection of reflective surfaces in the solar system, all providing climate controls.

The idea that Earth has feedback processes in which life interacts with nonlife to regulate the global climate is known as the Gaia hypothesis. James Lovelock and Lynn Margulis first proposed this in the 1970s, arguing that such interactions tend to make the environment more fit for life, mostly by taking the edge off climate change.[13] Lovelock has compared Earth's system, which he calls geophysiology, to the metabolism of a mammal or a redwood.[14]

Lovelock and Andrew Watson illustrated the Gaia hypothesis with a toy model called Daisyworld.[15] This world has an orbit just like Earth's and a host star just like the Sun. On Daisyworld, there are two species of daisies, one black and one white (representing different albedos), growing in large, distinct patches. Where there are no daisies there is brownish-gray soil. Both species grow best at the same temperature, and they both have the same lower and upper limits on their growth. Since the white daisies are more reflective than the black daisies, a patch of white daisies produces a cooler local environment than a patch of black ones.

Now, suppose the host star begins to brighten slowly. More light brings the white daisy patches closer to the temperature they prefer, so they multiply. The areas draped in black daisies also heat up, but since their local environment was already warmer, this temperature increase moves them beyond the temperature where they grow best—so their numbers diminish. Globally, then, it will soon be white daisies galore. The planet will reflect more light into space than it did before. This will dampen the heat

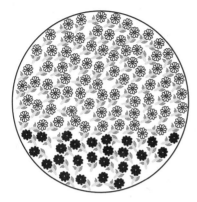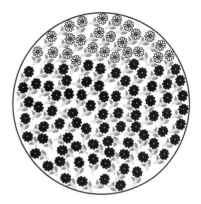

Figure 4.1: Two snapshots of a hypothetical Daisyworld. When the host star is faint, black daisies predominate; when bright, white daisies predominate. In this way, the changing ratio of white to black daisies maintains a stable planetary temperature, even as the host star slowly brightens.

swing that would otherwise occur, thereby keeping Daisyworld's average temperature more stable.[16]

Of course, Earth's biosphere is far more complex than Daisyworld. Recall that a key part of the carbon cycle is the rate of chemical weathering of surface rocks by carbon dioxide dissolved in rainwater, which forms a weak acid. This reaction can occur without life, but plants, trees, and tiny marine life really speed things up.[17] Since both rising temperature and the concentration of carbon dioxide increase chemical weathering, they speed up the removal of carbon dioxide from the atmosphere, which reduces greenhouse warming. Enhanced carbon dioxide also boosts growth of plants, which tend to be dark; plants also draw the curtains by evaporating more water and boosting cloud cover, both of which cool the surface.

Scientists now recognize other important links between life and the global climate. One such process involves cloud condensation nuclei (CCN), small particles in the atmosphere around which water can condense to form cloud droplets. Both natural and anthropogenic processes produce CCN.[18] Phytoplankton, such as *Emiliania huxleyi*, produce dimethyl sulfide, the first step in a chemical chain to build the CCN. Phytoplankton respond to a warming ocean surface by producing more dimethyl sulfide, which concentrates more CCN and makes marine stratus clouds more reflective. Sending more light back into space cools the ocean below. Higher carbon dioxide levels also stimulate production of dimethyl sulfide and CCN.[19]

All of this gives us an atmosphere that lets us see distant stars and galaxies from Earth's surface. Moreover, there's a cost in opting for a dryer, less cloudy planet to get rid of some of those pesky clouds that spoil an astronomer's stargazing efforts—and not just to habitability. We know from ice and marine cores that the glacial periods were much windier and dustier than the present.[20] So while the interiors of continents were less cloudy, the trade-off was a much dustier atmosphere. Astronomers would prefer a few cloudy nights interspersed with crystal-clear ones to nonstop dust and wind. Dust blocks starlight and makes observations far harder to calibrate. It's also murder on optical instruments. Observatories have strict rules how much dust and wind they will tolerate before they are forced to close.

While its precise makeup still eludes us, Earth's early atmosphere was clearly quite different from the present one; in particular, it contained more hydrogen. Oxygen was a trace gas, and the atmosphere probably had more carbon dioxide and methane. This early atmosphere may have resembled a warmer version of the present atmosphere of Titan, Saturn's largest moon. The Sun's ultraviolet radiation would tend to form more complex hydrocarbon molecules in such a reducing atmosphere, leaving behind a thick, oily, planetwide haze.[21] That haze probably cleared off with the rise of oxygen about two billion years ago.

Earth's complex climate system enhances both life and scientific discovery. Recall the natural recording processes described in the previous chapter that require the hydrological cycle. If our atmosphere lacked water, we would lose the hydrological cycle, get a much dustier atmosphere, and never have rainbows. As it is we get all these while enjoying an atmosphere that is still fairly transparent, and transparent in the same part of the spectrum where most stars emit most of their light. As a result, it combines otherwise competing factors to strike the best compromise for studying the universe.

Earth's present-day atmosphere, then, gives us great access not only to the past but to the wider universe while also nurturing life. And like the layering processes, standard Darwinian adaptation can't explain why we can decode this information—because this skill gave us no survival advantage until only very recently.

MAKING LONG-TERM PLANS

As human knowledge of the stars advanced, that knowledge began to provide some humans with survival advantages—namely, sailors navigating by the stars. But the advantage of a transparent atmosphere may help to protect the entire human race in the future.

Thanks to a combination of our transparent atmosphere and advanced observational instruments, astronomers can now catalog near-Earth objects or NEOs, which include both near-Earth asteroids and comets. NEOs have Earth-crossing orbits, which means they can hit us. As of October 2023, more than 33,000 NEOs have been discovered, with the total number spotted roughly doubling in just the past six years as technology has improved.[22] The threat of an NEO impact with Earth depends on its size. An NEO two kilometers in diameter is considered a "civilization ender," while one ten to fifteen kilometers in diameter is a "K/T event," like the one that probably killed off the dinosaurs sixty-five million years ago.[23] K/T events are thought to occur once every fifty to one hundred million years. As for two-kilometer NEOs, over a 10,000-year span there is about a 1 percent chance of one of these impacting (and a 50 percent chance for a 200-meter impact). In the latter case, huge tsunamis could flood the coasts.

The 1908 Tunguska event in Siberia is thought to have resulted from a roughly 50-meter asteroid or comet fragment exploding in the atmosphere. It's the only well-documented example of a destructive impact in recorded history. This 15-megaton explosion occurred over a mostly desolate part of Siberia. Were it not for this disturbing natural event—and the far-off comet Shoemaker-Levy 9 impact with Jupiter in 1994—we might not take impact threats seriously.

Comets seem to impact Earth less often than asteroids, but comets may do more damage because of their greater velocity. Hale-Bopp, the intrinsically brightest visible comet since the sixteenth century, inspired the world in 1997. (See plate 10.) Such a sight frightened many ancient peoples. Today, we fear comets, but for a different reason. If Hale-Bopp had hit the Earth, it could have been one hundred times greater than the K/T extinction event. That would wipe out everything but the hardiest microbes.

That's the bad news. The good news is that we can protect Earth from such threats if we can know far enough ahead of time about an NEO headed our way.[24] Astronomers can predict the precise position of an NEO a few decades into the future. Comets are a different matter, since astronomers usually discover them only a few months before they pass through the inner solar system. But that would still give us some time to prepare. And as our technology improves, so will our capacity to respond.

Just think if our atmosphere had been cloud-covered or translucent: We would have no advance warning of impact threats. Nor would we have developed a space program, which is a prerequisite for deflecting NEOs.[25]

If our atmosphere had been thicker, but still partly transparent, the extra distortions of telescopic images would have mostly thwarted our efforts to catalog these faint objects. So once again we see the eerie link between life and discovery.

If large comets and asteroids were now frequently bombarding Earth, our planet would be dreadful. But ironically, these rogue bodies were vital early on, for both the life and the science to come. As valuable as it is for us to view extraterrestrial objects from a distance, sometimes a little bit of the sky comes down, giving us more direct access.

DATA FROM HEAVEN

On January 18, 2000, a fiery ball streaked across the early morning sky. Its explosions broke the silence of the frigid Yukon and woke its sleepy residents. Defense satellites had tracked it from space. Its smoke trail lingered in the sky for a full day like the grin of the Cheshire cat. This had been the largest bright meteor, called a bolide, detected over land in ten years. A few days later, a man near Whitehorse, Yukon, searched the ice-covered Tagish Lake and found several fragments of the space visitor.[26] Today, scientists called meteoriticists will pay top dollar for these tiny rocks. They're not just rare and hard to find, but also brimming with unique information.

Astronomers are pretty sure they know how most meteorites make their way to Earth. Meteorites are shards from large, broken-up parent bodies in the asteroid belt, which rings the Sun between the orbits of Mars and Jupiter. Its largest member is Ceres—a hefty 1,000 kilometers in diameter. There are millions of smaller asteroids in the belt, creating enough traffic that collisions are still fairly common. A single collision can produce thousands of rock-sized fragments. Some wander into unstable zones, where Jupiter's gravity can perturb them. Once perturbed into less circular—that is, more eccentric—orbits, the fragments start to visit the inner solar system. Those that cross Earth's orbits can hit us.[27] When a fragment approaches Earth, it heats up from friction with the atoms in our atmosphere. We see it as a meteor. If it's big enough, it makes it to the ground as a meteorite.

The Tagish Lake fragments belong to a previously unknown type of meteorite. Astronomers can identify a class of meteorite with a specific asteroid or class of asteroids by comparing their spectroscopic signatures. If they have similar so-called "reflectance spectra," researchers conclude that they have a common composition, and hence, a common source. Based on such analysis, researchers think the Tagish Lake meteorites come from

D-type asteroids, which tend to lurk in the colder, outer reaches of the asteroid belt.[28]

Now, what does it take to see meteors? Apart from the asteroid and comet fragments themselves, we need an atmosphere. Even if we could live without an atmosphere, in its absence, we would have missed the discovery that meteor showers occur on the same dates every year and that they are associated with old comet orbits. In addition, we would lose out on using observations of meteor trails to calculate the trajectory of the incoming fragment.

These have two important uses. First, astronomers can backward extrapolate the path of the fragment and calculate its orbit. This research has shown that many meteors are caused by fragments from the asteroid belt. Second, they can forward extrapolate its path and estimate the landing spot of any surviving pieces. This method has been used to recover many freshly fallen meteorites—including the specimens from Tagish Lake.[29]

So, what's the big deal with meteorites? For starters, they contain the most pristine samples of the early solar system. Such material is not available from the homogenized and processed planetary bodies. The larger the body, the more heated and hence the more dynamic its interior. Some meteorites come from asteroidal parent bodies that have clearly differentiated; the metals—mostly iron and nickel—sink to the center, leaving the stony stuff behind in the crust. These don't preserve information about conditions before the asteroid formed. Carbonaceous chondrites, in contrast, which are the most primitive type of meteorites, presumably come from undifferentiated parent bodies. That makes them the most treasured.

Until the early 1970s, most meteor experts believed that the violent processes that formed the bodies in the solar system would melt and homogenize any pre-solar interstellar grains beyond recognition. So, they were surprised to discover that some micron-sized grains in primitive meteorites exhibit isotopic ratios inconsistent with an origin in a processed parent body. This means that some small bodies in our solar system contain matter that predates the birth of the planets and even the Sun. These grains hold unique clues about the chemical history of the Milky Way galaxy itself, and their discovery led to a new field of astrophysics. The Tagish Lake meteorites are a fine example since they contain more interstellar dust grains than any other specimens.[30]

Astronomers can not only connect meteorites to specific types of asteroid types, but they can also connect interstellar grains to specific sources in the Milky Way. For instance, most of the silicon carbide grains seem to come from Asymptotic Giant Branch (AGB) stars—very bright, cool stars near the end of their lives. Most of the other types come from supernovae, AGB stars, and maybe some novae. Astronomers can know this, in part, because some of the isotope ratios found in the interstellar grains can also be measured in the atmospheres of stars.[31]

ISOTOPES

Isotopes are like siblings in a family of elements. They have the same "last name"—they're the same element and have the same number of protons—but they have different "weights" due to varying numbers of neutrons. For instance, most of the hydrogen in the universe, called hydrogen 1, has one proton in its nucleus and one electron. Hydrogen 2, known as deuterium, has one neutron to keep the one proton company in the nucleus. Hydrogen 3, known as tritium, has two neutrons in its nucleus. These are all hydrogen atoms. The small differences, however, can give them unique properties and roles, much as siblings might look alike but have their own talents and quirks.

Interstellar grains in meteorites came together at the birth of our solar system—a mere pinpoint in the history of the Milky Way galaxy. Still, they're thought to sample all the previous history of the galaxy and a broad region of space. Somehow, they formed in the winds of Asymptotic Giant Branch (AGB) stars or in the debris of supernova explosions, wandered through space for millions or billions of years, merged in the dense environment of a giant molecular cloud, survived the early trauma of planet formation, and ended up on a researcher's shelf. As a result, we can hold a three-pound rock that contains thousands of original pre-solar grains, each possibly from a different star!

Interstellar grains, then, provide us with those invaluable isotopic ratios. But there is an associated investigative tool astronomers draw on. The study of stars can tell us the atomic abundances—that is, how much of each element is present. These two sources of information, isotope ratios and atomic abundances, complement each other. With the two together, we can determine many more details about the buildup of elements within galaxies and stars than we could with only one of these.

Figure 4.2: Astronomer Donald Brownlee (University of Washington) holding a two-hundred-gram Mars meteorite, with Kuni Niishiumi (UC-Berkeley) holding a three-hundred-gram Moon meteorite, in the Asteroid Café in Seattle. Both rocks were found in North Africa.

ASTEROIDS, COMETS, JUPITER, AND LIFE

Delivering primitive meteorites to Earth's surface requires a complex sequence of lucky breaks. At least some small bodies in the inner solar system must *not* be incorporated into larger ones. Otherwise, their unique identities would be erased by a planet's hot, dynamic interior. According to simulations of the early solar system, the asteroid belt has lost about 99.9 percent of its original mass. Probably several Moon- and Mars-sized bodies once lurked between Jupiter and Mars, helping to perturb most of the smaller bodies into the Jupiter and Saturn resonances.[32] Once there, the asteroids' orbits become more eccentric, allowing some to visit our neighborhood and others to visit Jupiter's orbit. Jupiter will usually then fling those that approach it out of the solar system.

But some of the scattered debris ended up on the Earth, blessing it with almost all its life-essential volatile elements—including carbon—and simple organics.[33] The details are complicated. Over time, Jupiter's gravity swept up the larger perturbing bodies from the region of the asteroid belt. The precise way that Jupiter formed is critical in all this. On the one hand, if it had formed a bit earlier, been a bit more massive, or had a more oblong orbit, it would probably have left too few asteroids to provide enough carbon for Earth. Same problem if Jupiter hadn't cleared the asteroid belt zone of planets: those planets would have hoovered up too many asteroids. On the other hand, if Jupiter had formed later or had a lot less mass, it could have left too many asteroids lumbering through the solar system, leaving the larger ones to pummel Earth too often for life to take hold.[34]

Earthly life's relationship with asteroids is . . . complicated. We needed them early on to deliver water and organics, but not too many, since they have an unfortunate tendency to extinguish life. Fortunately, that sequence is just what our planet got. It left our biosphere thriving and us a remnant of asteroid remains to figure out how the solar system and elements formed. Just enough meteorites are delivered to Earth to make it practical to study them.[35]

Meteorites, though the handiest source of encoded information about cosmic history, aren't our only source. Comets also deposit debris. Unlike asteroids, however, small comet fragments don't survive the journey to Earth's surface. Instead, they fall to Earth as comet dust. So, high-flying aircraft, like NASA's U-2 research planes, must nab the dust before it mixes with Earth's dust. Most planetary systems may have comets, but we suspect few have, as on Earth, just enough asteroids to keep meteor-collectors happy without wiping them out.

Who would have guessed that the asteroid belt, which first seemed like a failed experiment in planet-building or a menacing planetary junkyard, should play a starring role not only in Earthly life but in scientific discovery as well?

By now, we hope you've noticed a trend. The path to such a peaceable planetary state is marked by very specific twists and turns—many of which depend on each other. Earth, then, is likely a very rare place. This will become obvious when we compare it to the other planets in the solar system and then to those beyond it.

CHAPTER 5

THE PALE BLUE DOT IN RELIEF

A very great part of the surface of Venus is no doubt covered with swamps. . . .
The temperature on Venus is not so high as to prevent a luxuriant
vegetation. . . . The organisms [at the poles] should have developed into higher
forms than elsewhere, and progress and culture, if we may so express it, will
gradually spread from the poles toward the equator. Later . . . perhaps not
before life on the Earth has reverted to its simpler forms or has even become
extinct, a flora and a fauna will appear, similar in kind to those which now
delight the human eye, and Venus will then indeed be the "Heavenly Queen"
of Babylonian fame, not because of her radiant lustre alone, but as the
dwelling-place of the highest beings in the Solar System.
Nobel laureate Svante Arrhenius, 1918[1]

Man has speculated about life on the other planets for as long as we've been aware that there were other planets. But such questions really picked up steam in 1543, when Copernicus suggested that *Earth* was a planet that, like the others, revolved around the Sun. No longer were they just mysterious, wandering points of light. Planets were *places* like Earth. And such places might have people.

Only recently, however, could we compare those places to Earth. It started with the Mariner probes in the 1960s and continues to the present. With probes in orbit around Mars and Jupiter, our knowledge of the other planets and their moons has exploded. It has altered our perception of how Earth compares to the Sun's other children. It has also debunked speculations that, in retrospect, look fanciful. The other planets and moons in our solar system, along with newly discovered extrasolar planets, also give us a means to consider what is ordinary and what, if anything, is special about Earth. Despite all we've been told about this supposedly unremarkable pale blue dot, Earth is an extraordinary host for both life and scientific discovery.

THE PLANETARY MENAGERIE

Astronomers group bodies in the solar system into planets and dwarf planets, their satellites (or moons), asteroids (or minor planets), diverse denizens of the Kuiper Belt, and comets. We can also put the planets into two groups: terrestrial planets—Mercury, Venus, Earth, and Mars; and the gas giants—Jupiter, Saturn, Uranus, and Neptune. Some moons are about the size of terrestrial planets, and Titan, one of Saturn's moons, even has a thick atmosphere.

Let's begin with a tour of Earth's neighbors.

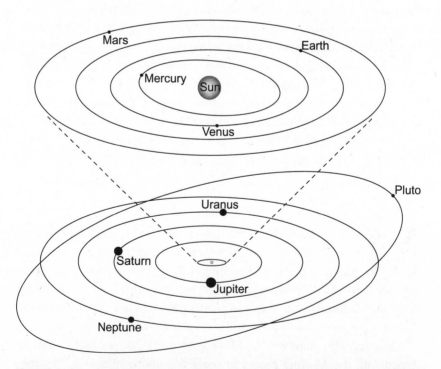

Figure 5.1: Our solar system contains eight major planets in fairly circular, stable orbits in roughly the same plane, and the dwarf planet, Pluto. Mercury, Venus, Earth, and Mars are rocky terrestrial planets. The asteroid belt lies between Mars and Jupiter. Jupiter and Saturn are gas giant planets. Uranus and Neptune are also mostly gas with some rock and ice. Comet-like Pluto is largely rock and ice. The orbits but not the planets are drawn to scale.

MARS: OUR MOST EARTHLIKE NEIGHBOR

Mars's climate is more like Earth's than any other body in the solar system, but that's not saying much. The pressure on its surface is less than 1 percent that of Earth's. And although some water condenses on its polar regions, Mars's surface is dry and dusty, so much so that Death Valley, California, is an oasis by comparison.

There are certain features of the Martian surface that at least resemble Earth's. Both polar caps wax and wane over a Martian year. And although its south polar ice cap is mostly frozen carbon dioxide—dry ice—the larger, northern one has a hearty portion of water ice, roughly half the volume of Greenland's ice cap. If melted, the water from the northern cap would form lakes and seas, though over a much smaller percentage of the planet's surface than is the case on Earth. These hypothetical Martian oceans would also be tinted red rather than blue due to all the iron oxide in the Martian dirt.

More notably, Mars lacks the features that make historical geology on Earth so richly informative. This absence also makes the red planet a nasty place to live. High-resolution orbiter images suggest that wind shreds Mars's surface, stripping away deposited layers and erasing important information in the process. (See plate 11.) The wind on Mars isn't typically very fast (less than 70 miles per hour), but the surface dirt lacks cohesion or plant cover, so it's quite vulnerable. Over thousands of years, persistent winds can turn a delicately layered plain into a chaotic dune field. And since Mars has precious little precipitation, one large dust storm could easily destroy or sublimate the ice laid down over several years on the northern polar cap.

Even if ice is still amassing in the Martian caps, it must do so very slowly, yielding thin annual layers. The dust deposited on the caps also reduces the ice's ability to reflect sunlight back into space, causing it to warm more than pure ice. While we can see some layering on high-res images of the Martian caps from orbiting probes, it almost surely lacks the fidelity of Earth's polar ice deposits, which accumulate steadily from one year to the next.[2]

Coincidentally, the present Martian and Earthly rotation periods and axial tilts are quite similar. But Mars lacks a large moon, and partly as a result, its axial tilt has ranged from fifteen to forty-five degrees over the past ten million years. The Earth with its large Moon, in contrast, has a modest and very stable tilt.[3] Even at the favorable angle it enjoys now, Mars's polar caps aren't great for recording data. At higher tilt angles, the polar ice would sublimate even more.[4] And we've already discussed the problems that such a wobbly tilt poses for complex life.

Comparing Mars's climate with Earth's reveals the peerless quality of Earth's ice deposits. Oceans surround Greenland and Antarctica. These provide a constant source of moisture for ice deposits far into the interior of these giant islands. The parched Martian climate, in contrast, doesn't guarantee net ice growth from year to year. Mars's vast oceans of sand also erode its delicate polar caps, reversing the familiar Earthly scenario of water waves washing over a sandcastle.

Carbon dioxide, which changes directly from ice to gas at Mars's surface, is the most abundant part of the Martian atmosphere, and is a poor substitute for water as a stable matrix for layered deposits. Mars's orbit is also quite eccentric, so it swings near and far in its annual course around the Sun. This, plus the wide wobbles in its axial tilt, make it less likely that annual ice layers will be deposited and preserved on its surface. In contrast, our planet's layered deposits are so sensitive that they record evidence of Earth's small changes in eccentricity and axial tilt.

Mars's geology is quite the contrast with Earth's. Starting in the late 1990s, the Mars Global Surveyor magnetometer detected a weak fossil field in the crustal rocks. Mars had a planetary field in the past, though it was probably short-lived. It now lacks a dynamo-generated field from the circulation of a fluid metallic core.[5] Some of the remnant field patterns are like Earth's striped crustal fields. This has led some to speculate that Mars may have briefly enjoyed plate tectonics.[6] In any case, its small mass led to a rapid shutdown of its planetary magnetic field and whatever geologic activity it had.[7]

The shutdown was probably hastened because heat-producing radiogenic elements, such as potassium, uranium, and thorium, were sequestered in its crust within about a half billion years after the planet formed.[8] Mars's crust is two to four times thicker than Earth's, and probably contains over 50 percent of the planet's heat-producing elements (compared to 30 to 40 percent in Earth's crust). Once removed from the mantle, the radiogenic elements can no longer contribute to its convection unless the crust is subducted back down into the mantle, which doesn't happen on Mars. This problem is worse on smaller planets, because their crust takes up a larger share of the planet's mass, causing them to cool more quickly.

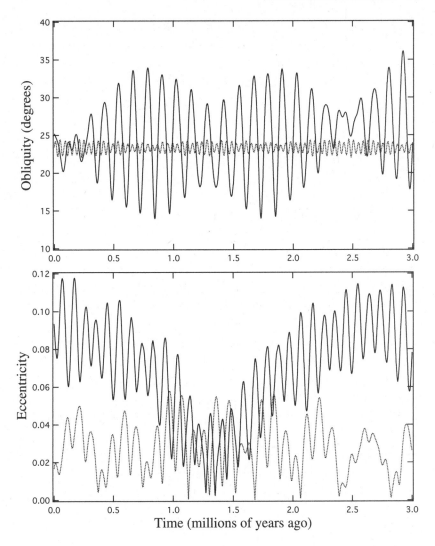

Figure 5.2: Variations of the obliquity (tilt) and eccentricity of Earth (dotted) and Mars (solid) over the past three million years. Even this long time span does not fully capture the range of Mars's variability. In the past ten million years, Mars's obliquity reached about 45 degrees, and over the past one billion years it probably reached 60 degrees. These large variations cause large climate swings.

We doubt Mars's ore deposits are as diverse and rich as those on Earth's surface. That said, some ores likely formed during Mars's early wet era and during its longer-lasting volcanic episodes.[9] Mars comes closer than any other planetary body in the solar system to matching the diversity of Earth's mineral ore deposits. And its crust may even be richer in iron than

Earth's for several reasons. First, less iron was likely sequestered into the core when it formed. Also, Mars has a thin atmosphere and is closer to the iron-rich asteroid belt than Earth is. As a result, more meteorites—some of which are nearly pure nickel-iron—have struck the planet's surface over the past few billion years. Other key metals, though, such as copper, are probably not so handy. In any case, Mars's atmosphere lacks enough oxygen to allow for fires on its surface, so Martians would enjoy neither Boy Scout campfires nor the metallurgy that led to high technology here on Earth.

VENUS: LOVER OF LIFE?

Unlike Mars, Venus has a thick atmosphere. It has often been called Earth's sister planet. But for life, the planet is one nasty hostess. Far from providing a loving home for a lush biosphere, as some astronomers believed a few decades ago, Venus's surface is a scorching 900°F. So, it has no water cycle. The name the ancients gave this planet is not quite right.[10]

Like Mars, Venus lacks a magnetic field, but for different reasons. It may still have a liquid metallic core. But with a day nearly the same length as its year, it probably rotates too slowly to provide much of a magnetic field.[11] As a result, the planet can't shield its atmosphere from the solar wind.[12]

Venus, while nearly the same size as Earth, has different internal dynamics. Earth appears to be the only planetary body in the solar system that still has plate tectonics.[13] Venus's thick, immobile crust prevents the tectonic movement we see on Earth; instead, Venus, like the other terrestrial planets in the solar system, undergoes so-called stagnant lid convection.[14] The steady convective overturning of Earth's mantle and crust slowly releases internal heat. An Earth-size planet with stagnant lid convection, in contrast, can build up heat and, like a pot with a partially sealed lid, release it episodically—and catastrophically. We know from counting the craters on its surface that Venus seems to have undergone such resurfacing about 500 to 800 million years ago. The event erased any information recorded on its surface and belched a profusion of greenhouse gas into its atmosphere.[15] Contrast this with the steady crustal recycling on Earth, which preserves information for long periods on its floating continents while regulating the amount of carbon dioxide in the atmosphere.

So why is Venus's history so different from Earth's? Perhaps stagnant lid convection is all there is to it. But it might also have something to do with the initial state of its rotation and/or its lack of a moon. Astronomers believe the last large bodies that hit the terrestrial planets would have come from random directions and set the direction and the starting rates of their rotation. As a result, a retrograde rotation, in which a planet's rotation is the opposite of its orbit, is just as likely as a prograde one, in which a planet orbits in the same direction it rotates. This really matters, since it affects how a planet's liquid core, rotation, and orbit interact. Gravitational torque from the Sun—and for Earth, from the Moon—moves the solid portion of a planet relative to its fluid core. Imagine a soft-boiled egg with a rigid shell, a firm white, and liquid yellow yolk that sloshes when the egg is spun.

The fluid motions of a planet can get intense when the liquid part resonates with the solid part.[16] This is like the resonance between a child on a swing and an adult's perfectly timed gentle pushes. In the case of a planet's spin that's in resonance with the fluid in its core, a given fluid mass element gets an extra push from the solid parts of the planet at the same spot in each circulation cycle. Friction converts this motion into heat at the boundary of the core and mantle, at the expense of the planet's spin energy. This extra heat may eventually cause extensive volcanism, and with it, the release of abundant greenhouse gases.

According to simulations, Venus's insides would heat up far more than Earth's during such a resonance passage.[17] This is because Venus's rotation is retrograde. It spins clockwise on its axis, but orbits the Sun counterclockwise, as viewed from the north. This retrograde rotation weakens the tidal forces on Venus. Our Moon does more to slow Earth's spin, even though Venus is closer to the Sun. As a result of this enhanced slowdown of Earth's spin, and the modest time it spends in resonance, it suffers less from heat pulses coming from its interior.

Even so, Earth apparently passed through a major resonance about 250 million years ago, when its day was a few hours shorter. Some astronomers and paleontologists speculate that this may be what stoked volcanism and led to mass extinctions. Others, however, blame a large impact. The resonance crossing could have been even worse had Earth, like Venus, lacked a large moon.

WHAT ABOUT THE OTHERS?

Of course, our solar system has five other planets. For hosting life, though, Venus and Mars are as good as it gets outside of Earth. Mercury is the only other terrestrial planet, but that's about all it has going for it. It's hot enough to melt lead and zinc on its day side. It also lacks an atmosphere and has an axial spin dragged almost to a halt by the Sun's gravity.

As we move out into the solar system, the prospects for life, and science, only get worse. The outer planets are quite spread out, making it hard to observe the other planets. And from, say, Jupiter or Neptune, the inner planets would be largely lost against the bright glare of the Sun. Observers on these outskirts would have a much harder time pulling off the first two tests of Einstein's general theory of relativity. We mentioned the bending of starlight in chapter 1. The second was the advance over time, or "precession," of Mercury's perihelion—the closest point in its orbit around the Sun. It was a problem for Newton's theory of gravity but made sense in Einstein's account.[18]

Then there's the time problem for scientists. The outer planets take decades to centuries to make one lap around the Sun. That would mean that valuable observations acquired over six months on Earth could take more than a human lifetime on the outlying planets. Mastering fire and metallurgy would also be a nonstarter, and good luck setting up a telescope or a particle accelerator in the gaseous atmosphere of one of the gas giants.

In any case, as previously discussed, these planets aren't promising candidates for life. Beyond the faint and flickering hope that we might find a few microbes in the Martian crust (and these potentially imported from Earth—see Appendix B), there's no hope of finding alien life on any of the planets in our solar system.

But that doesn't mean the alien hunters have given up on our solar system.

EUROPA: ANOTHER ABODE FOR LIFE?

Moving from the oven to the icebox, some astrobiologists are focusing on the second of the four Galilean Moons of Jupiter, Europa, which is about the size of our Moon.[19] The high-resolution images returned by the Galileo probe in the 1990s revealed a cracked, icy surface with evidence of recent melting and possible recent upwelling of fluids from beneath its surface. What kind of fluid? Very likely, water.

Many scientists have speculated about life in Europa's ocean. But much of it is driven by the assumption that where there's liquid water, there must surely be life.[20] The presence of water, while a necessary condition for organic life, is far from a sufficient condition.

First, if Europa's icy surface hides an ocean underneath, it's probably about one hundred kilometers deep. That's twenty times deeper than the typical basin of Earth's oceans. The pressure at the bottom would be about 2.5 times the typical pressure in Earth's basins. Simple life can't tolerate just any pressure. Some microorganisms can grow at pressures up to one thousand times the pressure at sea level on Earth, but Europa's ocean bottom may exceed this limit by about 30 percent.[21]

Even if a hardy bug could survive this pressure, Europa has other problems. For instance, its oceans have very little energy to power life. Sunlight can't penetrate the thick ice. What liquid water there is results from tidal energy percolating up from the moon's interior. We don't know how the heat makes its way to the ocean, but it's probably not from the vents as on Earth. Earth's vents result from plate tectonics. They're near mid-ocean ridges, where ocean water circulates through freshly minted crustal rock. When mixed with the right minerals and gases, the heated water can provide chemical energy. On Europa, the vents are more likely to be tethered to old channels lacking such chemical richness.[22] Plus, as we saw in chapter 2, the microbes around Earth's deep-sea vents benefit from Earth's robust land/air/water biosphere. Any hypothetical microbes around Europa's deep-sea vents would lack any such vital aid. Indeed, if a really robust microbe survived in Europa's ocean, we suspect it would be a lonely existence, since the moon lacks the interfaces of land, air, and water, where most Earthly creatures prosper. Even on Earth, which is far better situated for microbial life around its deep-sea vents, the vent systems are bit players in the game of life. The surface waters produce about 100,000 times more biomass than all the vents combined. So even if microbial life could persist around Europa's deep-sea vents, in the face of the challenges just noted, Europa could support, at most, one tenth of 1 percent of Earth's biomass, and probably far less than this.[23]

Third, Europa's oceans may be a planetwide Dead Sea, too salty to sustain life.[24] On Earth, shallow seas evaporate, leaving salt behind in the continental crust. If you've visited an underground salt mine, you have a sense of how much salt is stored just beneath Earth's surface. A terrestrial planet with dry land can lock away a lot of excess salts, keeping it out of the oceans. But Europa has no continents, so most of its original salt stays in its ocean.

What's more, since its tidal heating is episodic, Europa slips in and out of orbital resonances with other nearby moons. So, its ocean volume varies considerably over long periods.[25] This is significant because, during such episodes, its ocean grows much saltier—because the salt is not incorporated in the ice. It's left behind in the ocean where the ocean's salinity rises as its volume shrinks. This would kill off organisms that are teetering on the edge of what they can tolerate.[26] So, if you're hoping NASA will find the Europan lost city of Atlantis, or anything much more than a few lonely and super-hardy microbes, don't hold your breath.

What about Europa as a geological platform for discovery? Based on the number of craters visible in the images, planetary geologists estimate the surface of Europa is only about ten to twenty million years old. Thus, information on Europa's surface likely gets recycled on this timescale, which is about ten times faster than the recycling rate of Earth's seafloor. We can glean some information about the history of Europa's orbit from the pattern of cracks on its surface, but nothing like the data we can extract from Earth's ice and seafloor deposits.

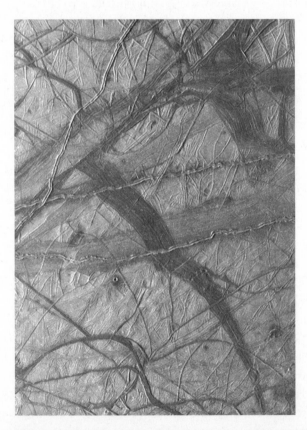

Figure 5.3: A 100-by-140 mile region of the icy surface of Jupiter's second Galilean Moon, Europa. Many fractures and ridges are present throughout its surface. The relatively small number of impact craters visible on this moon implies a surface that recycles on a timescale of millions of years. The Galileo orbiter took this image in November 1996.

EARTH IN CONTEXT

We can best appreciate our home planet by comparing it to all other known planets. The exercise also reveals how crucial trade-offs and thresholds are for doing science.

For instance, a planetary body with little or no atmosphere, such as the Moon or Mercury, offers a clearer view of the universe from its surface than does Earth. A Mercurian would have sharper, more constant views of the stars over a broader range of the electromagnetic spectrum. But that doesn't make Mercury a better platform for science overall. Here's why. First, our atmosphere is extremely stargazer friendly, allowing great advances in astrophysics. This is especially the case for mountaintop observatories. Higher up, the atmosphere is thinner, and some of these observatories, by correcting for image motion caused by the turbulent atmosphere, even surpass images from the Hubble Space Telescope. What's more, on clear, stable nights, astronomers can measure the absorption of light by the atmosphere well enough to calculate the brightness of a star as if there were no atmosphere. This trick requires little more than measuring the changing brightness of a star as it makes its nightly trek across the sky; it would not be practical in a dusty atmosphere such as Mars's. Space-based observatories have helped mostly with exotica, such as X-ray and gamma ray bursts that don't emit much radiation in the optical range.

Second, and more crucially, without an atmosphere, science would pay a heavy price elsewhere. We'd have no hydrological cycle at the surface, which is the basis for living growth layers and nonliving layered deposits. And the far heavier impact tilling from asteroids and comets would degrade any fine sediments. We'd lack concentrated ores and fossil fuels and oxygen in the atmosphere for fires. We'd also lack rainbows, those sublime, ephemeral clues to the astronomer's indispensable tool, the spectroscope.

In sum, the benefits to science of having a transparent, oxygen-rich atmosphere, in contrast to having none, greatly outweigh the costs.

Our atmosphere's advantages over a murky atmosphere are even more obvious. Atmospheres on gas giants like Jupiter or Saturn are opaque—all downside and no upside. These planets also lack a solid platform for scientists to live and work and to store information—as we described in Chapter 2. The traces from the comet Shoemaker-Levy 9 fragment impacts on Jupiter in July 1994 are long gone; only our photographs remind us of that grand event.[27]

Despite the Moon's dead geology, geologists have obtained enough seismic data to map its interior—and to detect quakes. By contrasting this data with Earth's, we get some idea of what's happening inside the other planets and moons in our solar system.[28]

The Moon no longer has active volcanoes, so why does the Moon still have quakes? They probably result from the tidal deformation caused by Earth's gravity, much like the Moon deforms Earth to give us the tides. In general, smaller bodies will have fewer quakes and fewer impacts than larger ones. Mars, for instance, should have fewer quakes not related to impacts than Earth does, but more than the Moon.[29] Europa's crust suffers severe stress under Jupiter's gravity. This should generate lots of quakes— though they would not be as useful as Earth's for doing geology. Seismic shear waves can't travel through its submerged ocean. The ice may also produce too much seismic noise, though we can't say for sure until we land a craft there.

Finally, there's Earth's orbit and axis tilt. (See plate 12). We discussed in chapter 2 the role these features play in making our planet life friendly. The axial tilt and eccentricity of Earth's orbit are about as stable as they can be. If Earth's orbit were more oblong or its axis more tilted, the temperature everywhere would vary more.[30] Life on Earth would be radically different—and likely far less diverse. All told, Earth might not support any animal life.

Earth's orbit and tilt are also great for (Earthling) science. Its orbit is so close to perfectly circular that the ancients could use the Sun's movement across the sky to mark the passage of time. A more oblong orbit would have made it much harder to model the motions of the planets. As we'll see in the next chapter, modeling the planet's movements was a crucial rung on the ladder to Einstein's general theory of relativity. Today, "paleoastronomy" benefits from the regular orbits of Earth, Moon, and giant planets, which gently impress their regular patterns on Earth's layered deposits. Among other things, these patterns allow astronomers to calculate the orbits of the Earth-Moon system and other planets tens of millions of years into the past. And Earth's stable axis tilt allows it to preserve delicate information about the past climate in its polar ice year after year.

Not all bodies in the solar system keep such a regular schedule. The orbits of Pluto and the main belt asteroids near Jupiter-resonances are highly chaotic, as are the tilt of Mars and the axial orientation of Saturn's irregular moon, Hyperion. In a chaotic system, future motions are very sensitive to initial conditions, like the proverbial flap of a butterfly's wings in China that leads to a thunderstorm in Chicago. A less fanciful example is a double pendulum, which has a joint in the middle of an otherwise stiff rod. You can try to start it the same way each time, but if you're off even slightly, it will quickly veer onto a different route.

Systems show some chaos if they're more complex than a simple two-body system. A single planet orbiting a star is like a simple pendulum, whereas several closely spaced planets are more like a double pendulum — they'll tend to experience large and even lethally rapid changes.[31] In general, the less massive planets have more chaotic orbits.[32] So, the orbits of the giant planets are the least chaotic, followed by Earth, Venus, Mars, Mercury, and finally Pluto. A planet needs to have a certain minimum mass (depending on its planetary neighbors) to maintain a stable orbit and climate over the long haul. Although Mercury is not expected to collide with Venus or get ejected from the solar system before the Sun becomes a red giant and swallows it, its orbit can be quite elongated.[33] Its more chaotic orbit and tilt would severely limit the information one could glean from any layering processes. What's more, even if it had an atmosphere, its more erratic climate would fail to preserve layered deposits very well.

Contrast that with Earth. Paleoclimatologists, who study our planet's ancient climate,[34] can find clues of subtle changes in the orbits of the other planets in deep marine cores and then extrapolate Earth's orbit back about fifty million years.[35] We can study the chaotic history of the orbits of the other planets only because we live on a stable platform. Earth's records may one day give us a precise look into the history of the solar system going back hundreds of millions of years.[36]

DISTANT WANDERERS

It was not until 1995 that we could compare our solar system with others. That was the year astronomers first detected a planet orbiting another sunlike star.[37] It caught many astronomers off guard since it wasn't what they'd expected — or predicted.[38] Hundreds of other discoveries in the last few decades have also defied predictions. (As of July 26, 2023, there were 5,483 confirmed exoplanets.[39])

These other systems differ profoundly from our solar system. First, some of the giant planets, called hot Jupiters, orbit very close to their stars. They make a full lap around their stars in three days or less. Second, those that take more than two to three weeks to make that journey tend to have highly elongated, or eccentric, orbits. So, they vary widely in their distance from their star. A perfectly circular orbit has an eccentricity of 0. The eccentricity of a closed orbit can approach 1.0—say, 0.9568—but not reach it. The exoplanet with the highest eccentricity known is HD 20782 b, clocking in at a whopping 0.97! Jupiter's eccentricity, in contrast, is a mere 0.05.

Our solar system, unlike almost all other known systems, has the regularity of a Swiss watch. Its eight planets are in nearly circular orbits, with an average eccentricity of only 0.06.

Moreover, we now know that our solar system lacks the two most common types of planets: super-Earths and mini-Neptunes. Super-Earths are up to ten times more massive than Earth. They can range from rocky

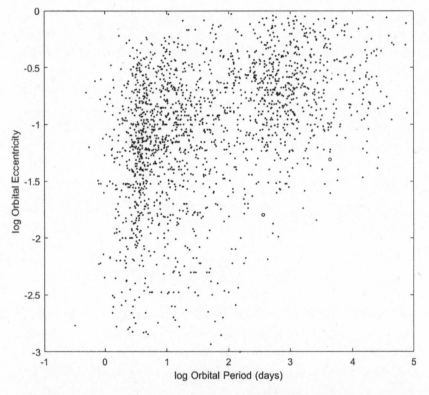

Figure 5.4: Orbital eccentricities and periods of all confirmed extrasolar planets having eccentricity measurements as of August 3, 2023; the plot includes 4,724 exoplanets. Jupiter and Earth are shown as open circles; they combine low eccentricities with relatively long orbital periods. (Negative infinity on the vertical axis would be a perfect circle.) The pileup of planets with very low eccentricity at small periods are the so-called "hot Jupiters." Planetary systems with such hot Jupiters are probably hostile to life, as are systems with giant planets in highly eccentric orbits. Plot based on data from the NASA Exoplanet Archive. (Orbital period and eccentricity are plotted on logarithmic scales.)

with a modest atmosphere like Earth to having a thick atmosphere like Neptune. Mini-Neptunes are between 1.7 and 3.9 times the size of the Earth and have thick hydrogen-helium atmospheres.

The most popular theory for the origin of hot-Jupiter systems is that these planets migrated inward from a more outlying orbit. Several mechanisms could account for this.[40] But they all hinder the formation of habitable terrestrial planets. A giant planet will scatter any terrestrial planets in the Goldilocks zone as it passes by.[41] And giant planets that perturb each other into more elongated orbits are better at perturbing smaller planets. In general, the massive planets are big bullies who push around the smaller kids in the neighborhood.[42]

If the trends continue as we map more extrasolar systems, it will mean systems like ours—with many planets in stable circular orbits—are rare.[43]

WHAT ABOUT MOONS?

The disappointment of finding all these menacing hot Jupiters and eccentric Jupiters has been tempered by the hope that these gas giants may have some large moons that are friendly to life.[44] But for all sorts of reasons, such environments are bound to be much less life friendly than our home world. Alas, *Pandora* remains firmly in the fantasy genre, not even coming close to sci-fi. First, since they're near giant planets, far more comets slam into them.[45] (See plate 13.) This is not just theoretical. We see the effects in the impact craters of Jupiter's Callisto and Ganymede—116 craters in eleven crater chains, laid out like gigantic pearl necklaces.[46]

Second, a giant planet's stronger gravity, and the greater orbital speed of its moon, would mean that comets would collide with such moons at much greater speed than they would on Earth.

Third, a moon near a gas giant with a strong magnetic field like Jupiter's would encounter more particle radiation. As a result, the moon would need a really robust magnetic field of its own. Based on our survey of the moons around Saturn and Jupiter, however, the rule seems to be magnetic fields that are much weaker than Earth's.[47] If this pattern holds for terrestrial moons around hot Jupiters in other solar systems, that's bad news for ET.[48]

Fourth, a moon's rotation would synchronize with its orbit in short order, like our Moon, causing one side to always face its host planet. For a moon with a thin atmosphere, this would mean greater extremes between day and night. For environments like the Galilean Moon system, orbits in

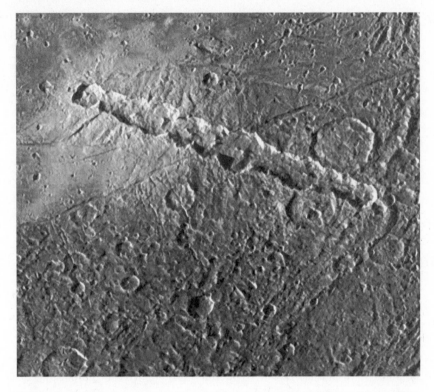

Figure 5.5: One of eleven known impact crater chains on the surfaces of the Galilean Moons, Ganymede and Callisto. A comet probably created this chain on Ganymede after it was torn into pieces by Jupiter's gravity as it passed too close to the planet. Such crater chains are a stark illustration of the dangers to life on moons around giant planets.

resonance can produce internal heat through tidal stressing.[49] However, changes in orbit over long periods—from 100 million to one billion years—prevent tides from being a sustained source of heat.[50] If Europa's ocean freezes over during such lulls, it would probably devastate any complex life there.

Okay, but what if a moon orbits far from its host planet? Wouldn't that make it less hostile to life? Yes, but at a cost. The amount of sunlight it receives from the parent star would fluctuate more, since the moon's orbit around its host star would be more elliptical.

A Jupiter-like orbit would create further problems for life on the giant planet's moon. Even if a giant planet can migrate inward and park itself in its star's circumstellar habitable zone, there's no guarantee its moons will come along for the ride. As a giant planet migrates inward, the gravity from its host star will compete for its moons. At some point, the planet may lose its moons to its greedy host star. According to simulations, the known hot Jupiters would have kept few if any of their moons as they migrated to their present spots.[51] How far a giant planet migrates before it loses a moon depends on its mass and type, as well as the mass of the moon, and the initial direction of the moon's orbit; massive moons in large prograde orbits tend to be lost more easily. Tidal interaction might even draw some moons into their host planets or fling them out of orbit, only to pose a collision risk later.

Finally, the makeup of such a moon, even if it survives, will likely not resemble Earth's Moon. Our solar system, and, presumably others, formed out of a flattened disk of gas and dust. After hydrogen and helium, water vapor was probably the most abundant gas. The hotter interior of the neb-ula prevented some gases, like water vapor, from condensing to become part of planetary embryos. But planets in cooler outer regions, outside the "water condensation boundary," could collect lots of water. (In our early solar system, that boundary was just inside the orbit of Jupiter.) That's why most of the outer moons in our solar system are so rich in ice.

Even if a system with a large moon orbiting an outer gas giant migrates into its star's habitable zone, its waterlogged moon will likely have too much water and carbon dioxide to have much hope for hosting life. If the giant planet can form in situ Sun-ward of the water condensation bound-ary, then its moons will have the opposite problem.

Capture of a moon by a giant planet is another pathway. The moon would need a source of volatiles to be habitable. But almost all the moons that we know were captured this way are just tiny rocks, and their orbits tend to be highly elliptical. The only exception is Triton, which orbits Neptune.[52]

Some giant planet-moon systems may avoid one of these problems, but we very much doubt they would avoid all the problems. We haven't even discussed perhaps the most serious problem. Giant planets, accord-ing to some models, can't form large moons.[53] In sum, an Earth-size moon around a giant planet is a far poorer candidate for life than a system like our Earth and Moon.

Moons around giant planets also aren't great for doing science. Observers on both hemispheres would have to put up with the reflected light from other moons in the system. And observers on the hemisphere facing the planet would experience a dark sky only briefly each month as it passed into the planet's shadow — unless its rotation wasn't tidally locked. Those on the other side would get more dark nights, but as we discuss in the next chapter, the motions of the stars and other planets in the moon's night sky would be bewildering. It was hard enough for us to figure out the true motions on our relatively simple platform. Our hypothetical moon dwellers circling a gas giant circling a star would have a much harder puzzle to solve.[54]

If they did figure out the true geometry of their home world, however, moon dwellers would have two advantages to Earthlings. Their orbit around the host planet would serve as a baseline to measure parallaxes to other bodies in their planetary system (we'll explain this later). Prior to the space age we had to satisfy ourselves with measuring parallaxes to nearby asteroids and terrestrial planets using Earth's surface as the baseline. Second, their host planet-moon system would serve as a mini-analogue of their star-planet system.

Nevertheless, the measurements from a moon wouldn't be straightforward, since the targets would be moving while their home world orbits about their host planet, and the host planet would be moving about its host star at the same time. To succeed, such observers would first have to figure out the far more complex geometry of their system. They would have to overcome far greater geometric hurdles to exploit their advantages, and then one of those would be undercut by further calculation hurdles. So, for scientific discovery, the costs of living on a moon would outweigh these meager benefits.

AT THE HEAD OF THE PACK

To recap, Earth's long-lasting water cycle, plate tectonics, oscillating magnetic field, continents, stable simple orbit, transparent atmosphere, and large moon at just the right distance, together provide the best overall lab bench in the solar system. Earth's surface strikes a balance between the permanence required to preserve patterns written on it and the dynamic, yet gentle, circulation that subtly sways these recorder pens without tearing up its paper-thin crust. Its crust records and stores information while maintaining the best known spot for life. Continents amid oceans of water, powered by plate tectonics, are the best overall habitat for observers. No other places yet discovered hold a candle to this blue dot, however pale it may appear to some.

CHAPTER 6

OUR HELPFUL NEIGHBORS

We can . . . be thankful that the solar system in which we live has been
unreasonably kind throughout the long history of human efforts to
understand its dynamics and to extend that knowledge to the rest of the
universe. At each step along the way, it has served as a perspicacious
teacher, posing questions just difficult enough to prompt new observations
and calculations that have led to fresh insights, but not so difficult that any
further study becomes mired in a morass of confusing detail.

Ivars Peterson[1]

For the ancients, the planets were a fickle lot. They seemed to drift slowly and reliably across the sky, only to reverse course as if trying to defy expectations. So capricious seemed their movements that they were named for this quality. The word "planet" derives from the Greek word for wanderer. And it's easy to forget that before they referred to planets, the names Mercury, Venus, Mars, and so on were the Latin names of the capricious Olympian gods of Greek and Roman mythology: Mercury, the messenger; Venus, the goddess of love and beauty; Mars, the god of war; and Jupiter, the thundering king of them all.

The planets have long played a role in astrology. In that ancient craft, they represent a person's "energies," such as the soul, will, and mind. On these wandering celestial bodies hang our individual stories, hopes, and destinies. While such flights of fancy are easy to dismiss, perhaps our race shares some real, if misguided, intuition that these strange objects in the heavens play a central role in our existence. Only recently, however, has that intuition found a scientific justification. Although the other wandering bodies in our solar system aren't encouraging sites for complex life, they *have* played a profound role in making Earth hospitable not only to life but to the emergence of science as well.

The mere presence of other planets in the solar system fostered the development of celestial mechanics and modern cosmology.[2] Danish astronomer Tycho Brahe's (1546–1601) observations of the paths of the planets against the background stars allowed Johannes Kepler (1571–1630) to formulate his three famous laws of planetary motion. Kepler built on that with his Third Law, which says the square of the orbital period of a planet is proportional to the cube of its mean distance from the Sun. To discover this law, Kepler needed the ability to see several planets. It also helped that the planets span a large range of distances from the Sun. And to discover his First Law—that the orbits are not circles, but ellipses—at least one such planet needed to have a discernibly eccentric orbit. For Kepler, Mars served this purpose.[3] If, like most of the extrasolar systems discovered to date, Earth had lacked such a plenitude of wandering neighbors, we might never have discovered these laws.[4]

Kepler's three empirical laws were the foundation for Isaac Newton's more general physical laws of motion and gravity. And Newton's laws, in turn, became the foundation for Einstein's General Theory of Relativity two centuries later.

The most habitable locale we know of happens to be near a star with several other planets whose orbital periods are much shorter than a human lifespan. A free-floating planet in interstellar space is not only a poor home for complex life but also a poor place to discover these universal laws. Even geniuses like Kepler and Newton needed a planetary playpen to discover the laws of motion and gravity and to realize that they apply throughout the cosmos. And once astronomers understood the motions of the planets, and had Einstein's General Theory in hand, they were on their way to understanding the rough structure and history of the universe.

Ivars Peterson, in the conclusion to *Newton's Clock: Chaos in the Solar System*, also notices this coincidence while discussing chaos:

> A deep-seated puzzle lies at the heart of this newly discovered uncertainty in our knowledge of the solar system. Was it an accident of celestial mechanics that the solar system happens to be simple enough to have permitted the formulation of Kepler's laws

and to ensure predictability on a human time scale? Or could we have evolved and pondered the skies only in a solar system afflicted with a mild case of chaos? Are we special, or were we especially fortunate?[5]

For Kepler to formulate his laws, it helped that neither the orbits of the planets nor the orientation of Earth's axis looks chaotic over a human life-time. For contrast, imagine yourself perched on the surface of the moon Hyperion as it tumbles along in its orbit around Saturn. It would be as disorienting as sitting on an erratically swirling chair on a rotating Ferris wheel on the edge of a spinning amusement park, while believing that everything is revolving around you.

Now, you might think that since the other planets enhance Earthly science, they would do the same for the other planetary bodies as well. What's good for Earth should be good for Mercury, right? Well, not quite. Mercury completes three rotations every two orbits.[6] Venus has a slight mismatch between its year and its day—to say nothing of its nearly opaque clouds. Even if a planet's dynamics aren't strongly chaotic, its motions might still be too complex for any hypothetical residents to discover the planetary laws.[7] Unlike the other options, the length of Earth's year is quite different from the length of its day, making it easier to separate the effects of revolution and rotation.[8]

Moons would offer even more bewildering views. Earth, astronomers say, is only one motion removed from the Sun. The Moon is two motions removed from the Sun, since it both revolves around Earth and around the Sun. Adding another nested layer of motion to an observer's platform would make the task of figuring out the true geometry of the orbits tortur-ous. Despite the simplicity of our Earthly platform, it still took well over a thousand years and several geniuses—including Copernicus, Brahe, Galileo, and Kepler—to get from the geocentric cosmology of Aristotle and Ptolemy to a grasp of the true geometry of the solar system.

EARTH'S ATTIC

The planets inspired Kepler, but it was the Moon that inspired Newton to apply his Earthly laws to the broader universe. Without the Earth-centered motion of the Moon, it would have been much harder to make the con-ceptual leap from falling bodies on Earth's surface to the motions of the Sun-centered planets. By linking the motions of the Moon and planets to experiments on Earth's surface, Newton gave a physical basis to Kepler's Third Law. Otherwise, the Third Law would have remained a mathematical curiosity, more a sign of a clever mathematician with too much time on his hands than a deep truth about the universe.[9] As it is, astronomy gave birth to physics.

Our Moon, like the other planets, also played a crucial role in helping us unravel the secrets of celestial mechanics. Apart from the Sun, the Moon is the only other body in the solar system that appears as more than a nondescript point of light, one that an observer on Earth's surface can resolve with the naked eye. The Moon provides a conceptual bridge, allowing us to imagine Earth as a similar wandering planetary sphere. Since Earth's rotation is not yet locked with the Moon's orbit—as it is for the Pluto-Charon system—everyone gets to see the Moon revolving around Earth. Otherwise, the Moon would be visible from only one hemisphere, where it would appear motionless in the sky.

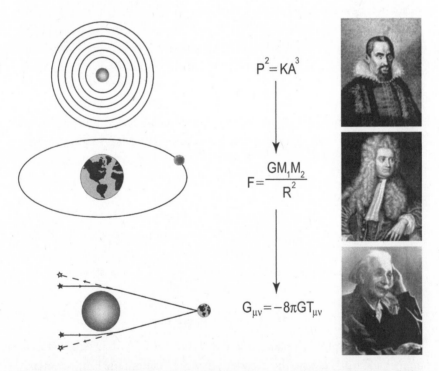

$$P^2 = KA^3$$

$$F = \frac{GM_1 M_2}{R^2}$$

$$G_{\mu\nu} = -8\pi G T_{\mu\nu}$$

Figure 6.1: The configuration of the solar system was crucial in each step of the development of gravitation theories. Kepler based his Third Law of planetary motion on Brahe's naked-eye observations. Newton based his law of gravitation, in part, on the orbit of the Moon about the Earth and on Kepler's third law. Einstein built his general theory of relativity on Newton's Law, and Einstein's theory was tested first with total solar eclipses.

Consider observers on a planet without a moon. To see their home as a planet, like the other planets in their system, they would have to make a huge mental leap. The surface of a planet, up close, looks nothing like one tens of millions of miles away. Distant planets look like stars to the naked eye.[10] Certainly, telescopes helped wean astronomers from Aristotle and Ptolemy. But to see details on the other planets, we also needed an atmosphere that allows resolution much better than one minute of arc. Only Venus reaches nearly this angular size on its closest approach to Earth.

As we noted in chapter 1, the Moon also makes Earth friendlier to life. It stabilizes the Earth's tilt on its axis so that it now varies a mere 2.5 degrees. Without the Moon, our temperature swings would be far greater. Climate swings would be larger and, thus, easier to measure. But they would be more likely to destroy past layered deposits. One pole would trade a large volume of ice with the other each half-year, without ever accumulating more than a year's worth of information. And the large changes in the tilt over a few thousand years would eliminate any long-lived ice deposits that might have built up near the equator. For science, climate variations must be large enough to be measurable in the climate record but not so large as to destroy it.

So, the Moon's stabilizing influence preserves the polar ice, that great cache of historical information, while the extra astronomical cycles it provides help climatologists calibrate their ice and marine sediment cores.

Finally, the higher tides resulting from the Moon's gravity subject more land area to periodic flooding. Without this effect, we would have fewer fossil tidalites today to reconstruct the history of Earth's rotation. If the Moon were much larger and induced stronger tides, in contrast, it would have slowed Earth's rotation more quickly. Earth's day might now be the same length as its month, leading to large disparities in temperature from day to night and perhaps endangering the preservation of ice deposits. The Moon would loom larger than the Sun, and the overall tides would be weaker, since only the Sun would induce changing tides on Earth. Greater stability, then, doesn't make a place better for life or science.

And as we saw in chapter 1, if the Moon were much larger or smaller at its current distance, we wouldn't enjoy perfect solar eclipses and all the scientific benefits such eclipses have provided.

The relationship between Earth and its Moon is so intimate that it's probably best to think of our planet as the habitable member of the Earth-Moon system. This partnership not only makes our existence possible but also provides us with scientific knowledge we might otherwise lack.

THE HISTORY OF THE SOLAR SYSTEM

Earth's Moon is about the same size as Europa, but it's as bone dry as Europa is wet.[11] Its crater-scarred surface preserves traumas it has suffered over its history. Its face appears ancient, yet ageless. The Moon's crater patterns allow us to reconstruct its cratering history—a key source of information about the early history of the inner solar system.[12, 13]

Far more craters should have formed on Earth over the same period, since it has a much larger surface and much stronger gravity than the Moon. But its recycling has erased almost all the older craters.[14] Does this count against the correlation of life and discovery? Not really, both because the positive trade-offs for both life and discovery more than compensate for the loss, and because much of the Moon's cratering record is plainly visible from Earth's surface.

As our nearest planetary neighbor, the Moon was the first to be mapped. We don't need to leave home to produce detailed maps of the Moon's near side, as anyone with a small telescope knows. We even have some lunar meteorites, though we might not have been able to identify their source without the lunar samples brought back by the Apollo astronauts. Since they come from all over the lunar surface, however, these meteorites better sample the Moon than the Apollo missions did.

What's more, we can translate the rich cratering record on the Moon's surface to reconstruct Earth's cratering history; so, Earth's crustal recycling doesn't penalize us as much on this score as you might guess.

Most of the Moon's visible maria, or large dark areas, formed between 3.8 and 3.9 billion years ago, during the so-called late heavy bombardment.[15] The impact rate was high enough to erase most of the lunar surface before this period. Nevertheless, there's a unique treasure buried just beneath the Moon's surface. A large impact on one of the inner planets can blast a lot of stuff into space, and its neighboring planets and the Moon sweep up most of it. Therefore, fragments of Mercury, Venus, Earth, and Mars should be preserved on the Moon, most dating from 3.9 billion years ago or earlier. That means Earthly meteorites buried in the lunar regolith may contain remains of early life from about 3.8 billion years ago. The Moon's central spot in the inner solar system and its lack of an atmosphere make it an ideal collector of planetary detritus.[16] The Moon is Earth's attic, where relics from the early history of the inner solar system are stored and preserved, waiting patiently for someone to climb up there and collect them.

MIRROR, MIRROR . . .

The Moon's stable surface has also taught astronomers about real-time changes in our global climate. As the Moon orbits Earth, it goes through its familiar phases. Careful observers will notice that the unlit part of the crescent Moon's face is not completely dark. This is called earthshine, and, as the name implies, is caused by sunlight reflecting off Earth and illuminating the Moon. The Moon acts like a huge mirror in space that allows us to see our reflection.

We can't discern any details about Earth's surface with this method, but we can derive a surprisingly accurate measure of its albedo—that is, the fraction of light that its surface reflects.[17] It's comparable to the best estimates derived from much more costly artificial satellite observations. Long-term measurements of Earth's reflectivity from space also require that instruments remain stable for several decades—no easy task. The Moon's surface is a very stable mirror, which allows reliable measurements over century timescales. Indeed, it allows us to reliably measure Earth's albedo from a single observatory on the ground. Knowing the average albedo and its change over time is critical to understanding our climate.

Figure 6.2: Three-day-old Moon. The sunlit regions have been overexposed to reveal the detail in the regions illuminated by "Earthshine." Astronomers observe Earthshine to monitor changes in the Earth's albedo and to learn what the Earth's spectrum would look like to a distant observer as well as what a distant Earth would look like to us.

The lunar mirror also gives us a way to obtain a spectrum of Earth as it would look to a distant observer. This is helpful in the search for other Earths.[18]

A LUNAR TELESCOPE

The Moon also helps astronomers study distant stars. As the Moon moves across the sky, astronomers can measure the angular sizes of stars and discover new binary stars by timing how long it takes for the Moon to cover, or occult, them. Because the Moon looms large in our sky, it occults many stars along its path. In this way, the Earth-Moon system acts like a giant telescope, allowing astronomers to resolve objects normally too small or close together to measure from the ground. A slow-moving moon like our own allows us to capture more details. This method works best with a large moon that lacks an atmosphere—which produces a crisp, knife-sharp edge on its limb. A moon with an atmosphere would also be far less useful for measuring our albedo and studying the Sun's chromosphere during a solar eclipse.

True, our Moon would be even better optimized for capturing details of any given star if it were substantially farther from Earth, but there's a trade-off with distance: the smaller a moon appears, the fewer stars it occults over a month.

Earth and its Moon function as one system, a double planet, as it were. They're near perfect opposites but make a great pair. The Moon lacks water, active geology, and an atmosphere, while Earth has all these in great abundance. Together, they allow for far deeper and more comprehensive discoveries than either would if it had remained single.

THE LOCAL DISTANCE LADDER

Of course, figuring out the laws of motion and gravity alone will not unlock the mysteries of the universe. For that, we need to see far beyond our neighborhood. The Moon served as the first stepping-stone in establishing the size scale of the solar system.[19] The story begins with Aristotle and the Athenians, who already knew that Earth was a sphere from noticing its curved shadow on the Moon during lunar eclipses.[20] (See plate 14.) To reach the first rung on the so-called cosmological distance ladder our ancestors first had to figure out the shape and size of Earth.[21]

Around 200 B.C., Eratosthenes of Cyrene calculated Earth's size by measuring the Sun's zenith distance—that is, the angle from directly

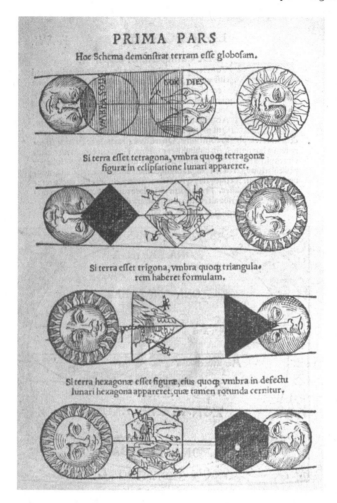

Figure 6.3: In this diagram from a popular sixteenth-century astronomy textbook (1540), Peter Apian illustrates an argument known at least since Aristotle: the shape of Earth's shadow on the Moon during a lunar eclipse shows that the Earth is a sphere. Apian compares the shadow produced by a spherical Earth with the shadows that would be cast if the Earth were a square, a triangle, or a hexagon.

overhead—at two cities along the Nile, which is helpfully aligned north-south. He noted that on a certain day of the year, the Sun shone into the bottom of a deep well in Syene, in southern Egypt. In contrast, the Sun was seven degrees south of the zenith in Alexandria, in northern Egypt. He knew the distance between the two cities and that the Earth was a sphere. He also assumed that the light rays from the Sun were parallel to each other. With these premises, he estimated Earth's circumference to be about 29,000 miles. That's only about 16 percent over the correct value of (about) 25,000 miles.

The next step on the distance ladder is the distance to the Moon. Hipparchus of Nicaea was the first to try to measure this, using observations of the total solar eclipse of March 14, 189 B.C.[22] He noted that the Sun was totally eclipsed in Hellespont, but only 80 percent obscured in Alexandria. This difference is a parallax effect; the much closer Moon appears to shift much more than the Sun. You can see the same effect by holding your thumb six inches in front of your face and alternately closing one eye at a time. Your thumb will seem to jump against the background. This is because the angle of view changes from one eye to another. The closer your thumb, the greater your thumb seems to move. To determine the distance to the Moon, one needs to know how much the Hellespont/ Moon/Sun angle differs from the Alexandria/Moon/Sun angle, and the linear distance between the two observers. Hipparchus figured the Moon was about seventy-five Earth radii from us—300,000 miles—not far from the modern value of sixty Earth radii—240,000 miles.[23]

Even before Hipparchus, in the early third century B.C., Aristarchus of Samos had devised a way to measure the distances to the Moon and the Sun.[24] He argued that the Earth, Moon, and Sun formed a right triangle, with the Moon at the right-angle position, when the phase of the Moon is in its first or third quarter. In principle, you can estimate the Sun's distance from Earth by measuring the angle between the Sun and the Moon during one of these two lunar phases. It's best to try this in the daytime, when both the Sun and Moon are high in the sky.[25] But it's quite low-res: Aristarchus reckoned the Sun to be about twenty times as far as the Moon; today, we know it's some 390 times farther than the Moon. His method could provide only a lower limit. To get a useful distance, he would have needed to measure the angle between the Sun and Moon to within a quarter the width of the Moon.[26]

Still, Aristarchus's result was good enough to convince him that the Sun was much larger than the Moon, and to support his heliocentric view. Combining the work of Eratosthenes and Aristarchus, the ancient Greeks could estimate the absolute sizes and separations of the Earth, Sun, and Moon. Some of these were far off the mark by today's standards. Still, using only their naked eyes, they put Earth in the proper perspective compared to the two major lights in the sky. Aristarchus's method of measuring the relative distances to the Moon and Sun would be impractical from any other moon-bearing planet in the solar system; the angle formed between

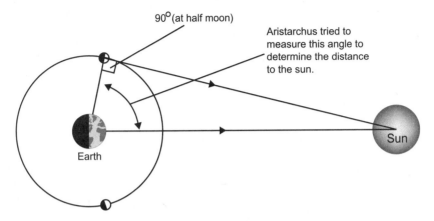

Figure 6.4: Aristarchus's method for estimating the distance to the Sun requires measuring the angle between the Sun and the first quarter (or third quarter) Moon. Using this method, the Greeks determined that the Sun is much larger than the Earth.

a moon during first or third quarter and the Sun would differ from 90 degrees even less than they do for an earthbound observer.

Earth again provided a useful baseline in the nineteenth century, when astronomers tried to establish the absolute size scale of the solar system. Kepler's Third Law had given them the relative distances of the planets, expressed in terms of the Earth-Sun distance—the Astronomical Unit, or AU. So, all they needed was a precise distance from Earth to one other body in orbit around the Sun. Aristarchus had given us only a crude distance to the Sun. The great hope among astronomers of the seventeenth to twentieth centuries was to measure accurately the parallax of a body in the solar system other than the Moon.

As mentioned above, parallax is the effect of seeing an object appear to move when you look at it from two different places. To measure the parallax of an object, you need to compare it to distant objects. The Sun is too bright and big to measure its parallax easily. A smaller and dimmer object is better, because you can see how it shifts against the more distant stars in the background.

Fortunately, the solar system offers lots of targets. Astronomers looking to measure the scale of the solar system first trained their telescopes on Mars, since it has a close approach to Earth every couple of years. The earliest serious attempts, with widely separated observers, began in the seventeenth century, with the best premodern measure in 1877. Observers took advantage of those times when Mars passed close to bright stars. But Mars's atmosphere and angular size limit the precision of its parallax.

Even better is a large asteroid that has a highly elongated orbit and so periodically comes close to Earth but has a quite different mean distance from the Sun. Eros has been the best of the lot. In 1931, an international team of astronomers observed the asteroid during a close approach to Earth. This resulted in the best measure of the distance from Earth to the Sun until the advent of radar astronomy decades later.

All this was hard work, of course, but much easier to pull off from Earth's surface, which, among the options, is especially well suited to measure the size scale of the solar system. This is due to Earth's size, the distribution of its continents, and the clarity of light sources observed through its atmosphere. And our Moon and some asteroids served as handy intermediate rulers in this well-appointed measuring kit. As a result, Earthlings could determine our place in the solar system.

PLANETARY PROTECTORS

Once again, those features of our environment that are congenial to scientific discovery also promote Earth's habitability. We've briefly discussed the role of the Moon and giant planets in all this. Jupiter and Saturn are likely the lead actors in the great drama of life's history on our planet. They have delivered water, winnowed the total mass of the inner planets, and cleaned up the asteroid belt.[27]

We suspect the other terrestrial planets have played a role in Earthly life as well—but that's a matter for future research. Still, we can surmise that other planets in the inner solar system reduce the number of asteroids and comets hitting Earth, for a simple reason: Anything that hits one of these others is no longer around to slam Earth. How much protection the other planets add depends on their combined surface areas and their proximity to Earth. Thus, Venus, the closest planet, and nearly the same size as Earth, offers the greatest protection in the inner solar system. Mars, though a little farther away, is closer to the main asteroid belt; it has surely taken a few hits from asteroids and comets on our behalf. The Moon has only about 7 percent of Earth's surface area, but since it's quite close, it's played some role as well.[28] To get a sense of what Earth might have endured without this defensive line, just look at the Moon's scarred face through a small telescope.[29] Yikes!

There's a second, and admittedly speculative, way the other planets may have helped Earthly life. Early in the solar system's history, Earth probably experienced several bad encounters with giant asteroids. Some may have vaporized its oceans and sterilized the entire planet,[30] but other planets could have served as temporary refuges for life.[31]

Too many planets, however, tend to make a system less stable. So, the best system for life and science will be one that best balances these trade-offs.

Our solar system, it turns out, manages these many trade-offs beautifully. Indeed, we're just starting to grasp how much Earthly life and science depend on the precise and finely tuned structure of our solar system.

SECTION 2

THE BROADER UNIVERSE

CHAPTER 7

STAR LIGHT, STAR BRIGHT

*Within this unraveled starlight exists a strange cryptography. Some of
the rays may be blotted out, others may be enhanced in brilliancy. Their
differences, countless in variety, form a code of signals, in which is conveyed
to us, when once we have made out the cipher in which it is written,
information of the chemical nature of the celestial gases. . . . It was the
discovery of this code of signals, and of its interpretation, which made
possible the rise of the new astronomy.*

William Huggins[1]

When ancient peoples looked at the starry night sky, they saw
vivid pictures, and wove those pictures into mythological
tales. Many of these pictures, or constellations, recur in otherwise diverse cultures. These include Leo, the lion; Taurus, the bull; and
Scorpius, the scorpion. Many people have also believed that the movement of these constellations across the sky shaped events on Earth. The
Greeks and Romans even went so far as to identify stars (and planets) with
various gods.

In stark contrast, the austere biblical account of the creation of the stars
describes them as mere created lights:

And God said, "Let there be lights in the firmament to separate the day
from the night; and let them be for signs and for seasons and for days and
years; and let them be lights in the firmament to give light upon the earth."
And it was so. (Gen. 1: 14–15)

While human beings have speculated about stars for millennia, we only recently figured out what they are. Two centuries ago, spectroscopy was a young science, and the mathematical description of radiation emitted by heated bodies was still decades away. At the time, thermometers were the only way scientists could measure temperature, and to figure out what something was made of, you had to test it in a lab. That's a problem when it comes to stars since you can't stick a thermometer in them or pour them into a test tube. This led Auguste Comte (1798–1857), the father of positivism, to conclude that the temperature and makeup of stars would lie forever beyond our ken.[2]

But Comte was mistaken. Indeed, stars not only have yielded these and other facts about themselves; they also have served as space probes, conveying information about their local environments over vast distances. Enough information to keep thousands of astronomers busy for decades.

Somewhat counterintuitively, this is true in part because, compared to living things and even planets, stars are simple. Most are nearly spherical, allowing astronomers to describe their structure with a few equations. Two of their most basic properties are luminosity—absolute brightness—and surface temperature. Around World War I, American astronomer Henry Norris Russell and Danish astronomer Ejnar Hertzsprung each realized that more luminous stars tend to be hotter. The diagram showing this, with luminosity on the vertical axis and temperature on the horizontal, has come to be called the Hertzsprung-Russell (or H-R) diagram. H-R diagrams have used several different kinds of temperature indices. The simplest is the photometric color index, determined by observing a star's brightness through two filters of quite different colors. Another index is the spectral type, determined from optical spectra.

When astronomers started classifying stars, they put the stars with the strongest hydrogen (H) absorption lines at the beginning of the alphabet, thinking the strength of the lines would translate into temperature. But they later learned that the strength of the hydrogen line doesn't correlate neatly with a star's temperature. With some patient detective work, astronomers figured out that the spectral-type sequence for stars is, in order of decreasing temperature, O, B, A, F, G, K, M, for which Russell mercifully provided a mnemonic: Oh be a fine girl; kiss me.[3]

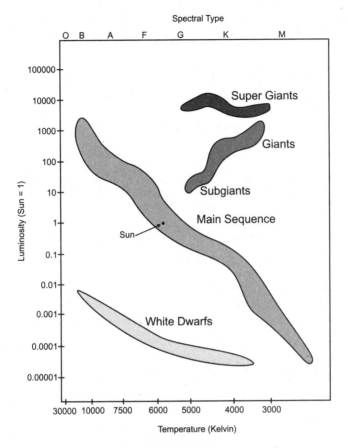

Figure 7.1: The Hertzsprung-Russell diagram for nearby stars. When stars are plotted according to their luminosity and temperature, they conform to a clear pattern. For stars on the main sequence—about 90 percent of nearby stars—the brighter ones are hotter. Such orderly information allows astronomers to learn a great deal about stars, even very distant ones.

PROBING THE HEART OF A STAR

With a star's basic observed properties in hand, astronomers can infer its internal structure with the help of equations based on simple laws of physics. These laws provide built-in tests that allow astronomers to verify their calculations. First, they can measure small amplitude oscillations on the surface of the Sun, oscillations caused by sound waves traversing its interior. Waves of different lengths sample different regions, with the longer waves sampling the deeper layers. As a result, they give astronomers enough clues to test their models of the Sun's interior. Like the information we gain about Earth's interior from waves generated by earthquakes, sound waves bring information about the Sun's invisible interior to the visible surface. We can't send probes to the Sun and plant seismographs on it. But because the Sun is a relatively simple ball of gas, it's easier to model than Earth's complex innards. And because its visible surface is a hot, glowing gas, we can detect its oscillations from a distance.

Astronomers can detect the waves with spectroscopes because they affect the radial motions of the gas—that is, motions along our line of sight—at the Sun's surface. The waves also alter the compression of the gas as it undulates, slightly changing its temperature and, hence, the brightness of a given patch of the Sun's surface. Given this, astronomers also monitor the Sun's oscillations by measuring its overall short-term changes in brightness. Applying what they've learned, astronomers have extended their knowledge to distant suns mainly by measuring their rapid changes in brightness.

These effects are often subtle, but some types of stars exhibit larger changes due to bulk surface motions called pulsations. One type, the Classical Cepheid, is especially useful to cosmologists (we'll cover these in chapter 9). Small, highly dense white dwarfs, which are Earth-sized dying stars, pulsate and rotate with periods measured in minutes. Some

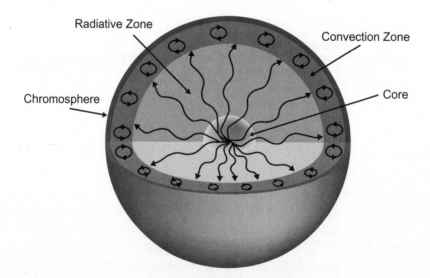

Figure 7.2: Cross section of a Sun-like star. Over most of its volume, the energy generated in the core through hydrogen fusion migrates toward the surface via radiative transport. Neutrinos are generated in the core. Over the outermost 20 percent of its radius, the star transports the energy by convection.

display a rich pattern of pulsation periods, revealing all their important structural properties.

NEUTRINOS

Neutrinos, nearly massless neutral particles that hardly interact with matter, provide another test. According to the solar interior models, they should be produced at a certain rate in the core of the Sun to account for the energy leaving its surface. While the optical photons we observe from the Sun betray its surface temperature, the neutrinos, which we can also detect from Earth, tell us something about its core temperature, thus serving as a kind of long-range thermometer.

As we'll discuss later, solar neutrinos also reveal something about the nature of the universe.

STAR SPECTRA

A star's spectrum abounds with information about its makeup, temperature, gravity, and magnetic fields at the surface. As noted in chapter 1, the optical spectrum of a sunlike star is a smooth continuum interrupted by thousands of sharp absorption lines, where less light is emitted. (See plate 15.) Two forms of matter produce these lines: neutral and ionized atoms, and simple molecules in the chromosphere—the thin, colorful layer of the Sun's atmosphere visible during total solar eclipses. The continuum is produced mostly by hotter gas in the underlying denser photosphere. An absorption line results when electrons absorb photons in a particular energy level in atoms of a particular element. Its strength depends mostly on the number of atoms of the element in the gas—its "abundance"— and the surface temperature of the star. The more atoms of an element present in a star's atmosphere, the darker its absorption line will be in the spectrum.

Astronomers can derive such high-quality data from the Sun's spectrum not just because the Sun is so close but also because its spectrum turns out to be nearly optimal for extracting information. Their success depends on how precisely they can measure the absorption lines. This, in turn, requires that there be some small windows of the spectrum's continuum among the forest of absorption lines. Our Sun nicely accommodates this need.

Other types of stars are less accommodating. Stellar spectra change their appearance drastically from O to M spectral types. Hotter stars have

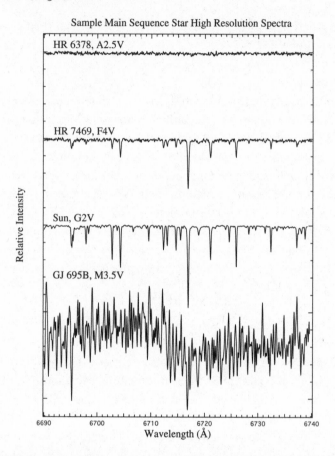

Sample Main Sequence Star High Resolution Spectra

Figure 7.3: Comparison of high-resolution spectra of four nearby main sequence (dwarf) stars. The 50 Ångstrom slice corresponds to about 2 percent of the spectral range shown in Plate 15. The dips in the intensity tracings correspond to absorption lines (visible as the dark vertical lines in Plate 15). Hot stars like HR 6378 have far fewer absorption lines than cooler stars like the Sun. But the strong molecular lines in very cool stars overwhelm most other spectral features and obliterate the smooth continuum visible in hotter stars. The Sun's spectrum is a golden mean. It displays features of both hot and cool stars, which allow astronomers to extract more information from the Sun than from either type alone.

broader and fewer absorption lines and well-defined continua, but the dearth of lines limits how much information astronomers can extract. Absorption lines become stronger and more plentiful in the spectra of cooler stars, in part because molecular lines start to stand out. In the coolest stellar atmospheres molecules produce thick forests of absorption lines.[4] Since molecules dominate the spectra of stars cooler than the Sun, they swamp the underlying atomic lines and obscure the continuum. In other words, the extra information the absorption lines of a given molecule provide fails to compensate for the wealth of information they hide.[5]

The Sun then, in contrast to hotter and colder stars, is nestled in a spectroscopic Goldilocks zone, as if made to order for astronomers. Its spectrum, combined with the information derived from meteorites, gives us the complete recipe of the solar system's birth cloud.[6]

The many sharp absorption lines in a star's spectrum also function as velocity markers.[7] By monitoring the Doppler shifts in the wavelengths of the thousands of absorption lines in a star's spectrum, astronomers can detect tiny changes in the star's velocity along their line of sight.[8] We've all heard a Doppler shift from an ambulance siren or a passing train's whistle. The sound's pitch drops as it passes the observer.[9] The state of the art in Doppler measurement of starlight can detect bulk motions as small as one to two meters per second—a walking pace! It works best for stars at least as cool as the Sun; the absorption lines in the spectra of hotter stars tend to be too sparse and too broad. With this method, astronomers are finding many planets orbiting other stars.

POINTS IN SPACE

If the value of stars to science ended here, it would already be impressive. But stars also tell us things about the broader universe. It really helps that their sizes are many orders of magnitude smaller than their separations. (The distance from the Sun to the nearest star, for example, is about one hundred million times the Sun's diameter.) This means a star's light in the night sky is narrowly concentrated, allowing its position to be measured precisely and thereby allowing mariners and astronomers to use stars as reference points. Even today, the high-resolution James Webb Space Telescope shows nearly all stars as unresolved points of light.[10]

Of course, there's one major downside to seeing a star as a point—we can't study details on its surface. But there are several upsides. First, we can study the Sun's surface as a representative midrange star. Second, most stars have uniform surfaces, so knowing their surface details would not add much important information. Third, we don't need to resolve stars to determine their basic properties because, as noted above, they're spherical, symmetric, relatively simple, and, thanks to their stellar spectra, can keep few of their secrets to themselves.

Today, astronomers, armed with fast computers, can simulate how thousands of stars interact, representing stars as mathematical points.

Because of the great distances between them compared to their sizes, stars are very close to this mathematical ideal. The calculations would probably be intractable if stars were much larger, since they would distort each other on every close encounter, making it necessary to have detailed knowledge of the interior structure of all stars.[11] Like the virtual tag each virtual star carries with it, a real star contains and transmits information about its makeup, age, mass, position, and velocity. The Milky Way galaxy is so weirdly accommodating to our efforts to measure its chemical and dynamical properties that it's like a gigantic simulation, sampling the local gravity field and transmitting encoded information to us, its interpreters.

TESTING PHYSICS

Stars are also remarkably useful testing grounds for the laws of physics. In chapter 1, we saw how the Sun helped test some weak-gravity limits of general relativity via the bending of starlight. Astronomers also use a more exotic type of star, the pulsar. A pulsar has sunlike mass compacted into a sphere about ten kilometers across. A pulsar is a leftover from the explosion of a massive star, which blows its outer layers apart, leaving behind an extremely dense, furiously spinning neutron core.

We know about pulsars because they emit highly directional radiation. Charged particles circulating around a pulsar's strong magnetic fields emit photons along narrow cones. This strong beaming allows us to measure a pulsar's rotation period precisely, using radio telescopes. In fact, some pulsars, especially those in binary systems, may be the most precise clocks in the universe — probably even more precise than atomic clocks. Thanks to these stable pulses, radio astronomers can study many types of phenomena that affect the pulse arrival times.[12] Specifically, astronomers can determine the characteristics of bodies orbiting pulsars,[13] test various aspects of general relativity,[14] and learn about the properties of matter at nuclear densities. Even the way the radio waves from a pulsar interact with the free-floating atoms along our line of sight reveals much about the intervening interstellar matter.

Perhaps the most spectacular discovery made using pulsar timings is the ever-elusive gravity wave. Gravity waves were one of the predictions of general relativity in 1915. Remember our discussion of the deflection of starlight near the Sun in chapter 1 due to the Sun's gravity distorting the space around it? Well, gravitational waves are produced when massive bodies orbit around each other and distort the space. It's much like the waves on the surface of a pond made by twirling a stick in the water. Gravitational waves, like light, travel vast distances across the universe. Their existence was first inferred indirectly in 1989 via timing observations of a pulsar in a binary system.[15]

Recently, astronomers used pulsar timing to detect what's called the "gravitational wave background." This is the combined gravity waves from sources throughout the universe. It would be like all the waves in a pond were made by tossing hundreds of pebbles in it all at once. Four groups of radio astronomers announced their historic results on June 29, 2023.[16] While the findings are preliminary, the fact that all four groups found the expected signature of gravitational waves is strong evidence that they exist.

These discoveries are not the same as the waves first found in 2015 by the Laser Interferometer Gravitational-Wave Observatory (LIGO), which detects individual in-spiral events of compact binaries consisting of various combinations of black holes and neutron stars.[17] The gravitational waves found using pulsars have much longer wavelengths and are the sum from many thousands of sources.

The idea is that gravitational waves passing between us and a pulsar cause the space to contract and expand periodically. This, in turn, alters the travel time of the pulsar signal to Earth. Each research group monitored dozens of pulsars and compared arrival time variations between pairs of pulsars as a function of their angular separation on the sky.

If physics had required stars to be much larger or to produce far fewer sharp absorption lines in their spectra, or prevented pulsars and white dwarfs from forming, the universe would have been a far less measurable place.

The properties of stars are, as we have seen, also delicately balanced for life. And of course, stars are key ingredients of habitable zones in the universe.

COZY LITTLE CIRCLES

In the late 1950s, astronomers introduced the concept of the circumstellar habitable zone (CHZ), mentioned in chapter 1.[18] It's the region around a star where liquid water can exist continually on the surface of a terrestrial planet for at least a few billion years. Modelers have tended to focus quite narrowly, considering only a planet's distance from its host star, the composition of its atmosphere, and how these relate to heating its surface. The zone's inner boundary is usually marked as the point where a planet loses its oceans to space through a runaway greenhouse effect.[19] The outer boundary is defined as the point where oceans begin to freeze or carbon dioxide clouds begin to form. Both outcomes increase a planet's albedo and trigger a vicious spiral of coldness until the oceans freeze over.[20]

To understand this Goldilocks zone, we must know about all the processes that help maintain liquid water on a planet's surface. These include the greenhouse effect, myriad biological processes, ocean circulation, clouds, ice sheets, plate tectonics, and various phenomena that depend on it, such as the carbonate-silicate weathering cycle. To make the problem more tractable, researchers base their models on slightly perturbed versions of the present Earth. Thus, they assume an Earth-size and earthlike planet in a circular orbit in a planetary setting like ours. Astronomers have slowly enriched their models. For example, by the 1970s astronomers recognized that since the Sun is gradually brightening, this would cause the zone to move outward over time. And starting in the late 1990s modelers began treating plate tectonics as a changing rather than steady process in equilibrium.[21] While these changes have made the models more realistic, other crucial factors that hinge on the distance of the host star are still missing.

THE DEVIL'S IN THE DETAILS

Consider the asteroid belt. You might think that the farther from our asteroid belt, the better, assuming you remain within the zone. But the impact threat to a planet varies with distance from its host star in more than one way.[22] Planets closer to the main asteroid belt will collide with asteroids more often. (The peak of the asteroid distribution in our solar system is at two astronomical units, AUs, from the Sun, about three AUs inside Jupiter's orbit.)[23] It also helps if a planet's orbit is nearly circular, lest it wander too close to the main asteroid belt at its farthest point from the Sun. Mars, in some sense, defines the inner edge of the main asteroid belt. It sweeps up lots of asteroids, making the inner edge sharper than it otherwise would be.[24]

Since arriving at Mars in 2006, the Mars Reconnaissance Orbiter has detected over a thousand freshly formed craters. Most are between 1 and 60 meters in diameter, though the largest is 150 meters across. There are about 80 impacts per Martian year that form craters bigger than 11 meters.[25] The current flux—impact rate per unit area—of impactors on Mars is about twice that on the Earth.[26]

Without Mars, Earth would be the closest planet to the main asteroid belt and bear the brunt of the impacts. You might assume that Mercury

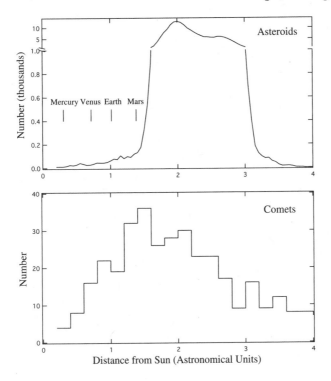

Figure 7.4: Distribution of the perihelia (closest approach to Sun) of nearly 200,000 asteroids and four hundred comets. The aphelia (farthest distance from Sun) of the four terrestrial planets are also shown. Asteroids are about 2.5 times more common near Mars compared to Earth. The comet perihelia distribution peaks near Mars. (The vertical axis of the top diagram is split for clarity.)

and Venus are safer from such catastrophes since they're farther from the main asteroid belt than Earth is. But as Kepler showed, the inner planets orbit the Sun faster than the outer planets. This applies to asteroids and comets, too, which orbit the Sun at greater speeds when they're close to it. And since Mercury and Venus are closer to the Sun, asteroids and comets in their vicinity move much faster. So, when one of them does strike, it's likely to be a real whopper.[27] Among the terrestrial planets, then, there's an optimum distance for minimizing the threat from impactors, a distance probably quite close to that of Earth's.[28]

Next is a planet's atmosphere. All else being equal, the farther a terrestrial planet is from its host star, the more carbon dioxide its atmosphere needs to keep it warm. A planet near the outer boundary of the habitable zone needs far more carbon dioxide in its atmosphere to boost the greenhouse effect and keep liquid water on its surface.

Cranking the carbon dioxide level way up, however, leads to other troubles. A thick carbon dioxide atmosphere isn't a problem for some forms of life, but large mobile creatures require an oxygen-rich atmosphere with a relatively low fraction of carbon dioxide.[29] Moreover, life on a planet near the inner edge of the habitable zone, like Earth,[30] will be more fecund and diverse than one farther out.[31] The sunlight's energy maintains a large number of photosynthetic organisms, which, in turn, support a lush and diverse biosphere. While some autotrophs produce energy without sunlight, they're minor contributors to the parts of the biosphere relevant to complex life. A planet that produces biomass more slowly than Earth may not be able to ready itself for complex life before its host star leaves the main sequence. Such a planet also would be more vulnerable to sudden external shocks to its ecosystem. So, it may be no accident that Earth resides very close to the inner boundary of the Sun's habitable zone. In fact, the zone may be much narrower than many assume.

Moreover, just putting an Earth-twin in the right place in a model doesn't mean it can form there or remain in a stable near-circular orbit. Other planets in the system will have a say. But a planet's initial distance from its host star has an enormous influence on its final properties. While a protoplanetary disk is present when a planetary system starts to form, gas and dust close to the star heat up more than does material farther out in the disk. The more volatile—that is, easily evaporating— elements remain gases in the hot regions, while the "refractory" elements, which tend to have higher melting points, can condense to form solid grains.[32] We probably have asteroids and some comets to thank for the few volatile compounds, such as water and carbon dioxide, that abound in Earth's crust and outer mantle. If Earth had formed closer to the asteroid belt, its greater initial carbon and water endowment would probably have left a deep ocean, a thick carbon dioxide atmosphere, and a dead world.[33]

Besides having just the right amount of carbon dioxide in its atmosphere, a planet also needs just the right chemicals in its core. Only certain cores can generate a life-protecting magnetic field. The makeup of a terrestrial planet's core probably depends on how it forms.[34] The inner terrestrial planets are more likely to have relatively larger iron-nickel cores and less sulfur mixed in, since sulfur is volatile. Adding sulfur reduces the core's melting point, much as salt reduces water's freezing point.[35] Too little sulfur, and a liquid core will not only need to be hotter to remain liquid, but it will also likely freeze solid when it cools. Too much sulfur, however, and a pure iron solid core may not form at all. To have a strong magnetic field, a decent fraction of a planet's core must circulate.

Potassium is another key volatile element. Its long-lived radioactive isotope, potassium 40, is a key heat source that helps keep the Earth's mantle convecting and the crustal plates moving. As noted in chapter 3, plate tectonics keeps the continents above water. There may also be some potassium in the core, where it could help power the geodynamo. Its abundance in planetary cores probably goes up with distance from the host star. Just how much is in Earth's core, and how it's incorporated, are still debated, but they both depend on several factors. And it's not just a simple one-to-one relation between distance and abundance in the core. Any sulfur in the core greatly increases how much potassium the core can be sequestered. The same goes for oxygen at higher temperatures.[36]

Each of these factors varies differently with distance from the host star. Some improve habitability at greater distances, while others do the opposite. The devil is also in the details when it comes to the rate of asteroid impacts. How other factors hinge on distance from the host star is quite complex. When taken together, these factors greatly narrow the best estimates for the Goldilocks zone for complex life and probably also for simple life. In short, a planet's true Goldilocks zone depends on far more than the intensity of light from its host star.

THE HOST STAR

And then there's the question of the star itself. Stars play two life-support roles: They supply most of the chemical elements and a steady supply of energy. Under the pressure of gravity, stars fuse the nuclei of atoms in their hot interiors to build the chemical elements. Stars spend most of their lives fusing the nuclei of hydrogen atoms (protons) in a phase of a star's life called the main sequence. Because hydrogen is so abundant, the main sequence is the longest lasting phase of a star's life. While on it, a star's luminosity doesn't change much. The Sun, which has been a main sequence star for about 4.5 billion years, has brightened by about 30 percent since its hydrogen first ignited.[37] How long a star spends on the main sequence depends largely on its mass, and the relationship is counterintuitive. Stars with twice the Sun's mass have twice the fuel, but they only spend about a billion years on the main sequence since they burn through their fuel much faster. Stars with half the Sun's mass last about 100 billion years. Once it leaves the main sequence in about six billion years, our Sun will grow several thousand times brighter in fairly short order. (And no, your home's AC system will not be able to handle that.)

There's been liquid water somewhere on Earth's surface for most of its history. Thus, Earth must have stayed within the Goldilocks zone even as the Sun brightened and the zone moved outward. The region over which all the time-stamped habitable zones overlap over some extended period is called the circumstellar continuously habitable zone (CCHZ). Let's call it the super Goldilocks zone. It's narrower than the regular Goldilocks zone, especially if the time interval in question is a large fraction of the main sequence lifetime of the host star.[38] Of course, the position of this zone varies from star to star. Low-mass main sequence stars are less luminous than the Sun, so they have small, close-in super Goldilocks zones; the opposite is true for more massive stars.

But, again, there's more to this game than just the light energy a planet receives at a given distance from its host star. Stars more than 1.5 times the Sun's mass are probably not viable hosts for planets with complex life because they spend too little time on the main sequence before they become red giants, and while on the main sequence, their radiant energy changes faster than the Sun. Such rapid changes lead to a less stable climate. Also, because of complex orbital dynamics, a star that brightens quickly triggers another threat—asteroids.[39] The scene around a star approaching the end of its main sequence lifetime will be like the early, violent stages of planet formation. So, more massive stars will endanger their planets even before they leave the main sequence.

WHAT ABOUT LOW-MASS STARS?

Near the opposite end of the scale, low-mass main sequence stars—M dwarfs—would be very low-rent districts. Any planet in the super Goldilocks zone would be close to its host star, and so suffer strong tides. This would quickly grind down its rotation—like Earth's Moon.[40] This is bad for life. If the planet's atmosphere is thin, it will freeze out on its dark side; a perpetually shadowed and cold region acts as a sort of cold trap. Think of cold traps used in vacuum pump systems to extract water from the air. High levels of atmospheric carbon dioxide could prevent this, but at the expense of animal-like life, which needs air high in oxygen and low in carbon dioxide.[41] Even with a thick carbon dioxide atmosphere, however, life could only tolerate temperatures in a narrow band where the light and dark sides meet.[42] And, since starlight is weak and the temperature low at the terminator, life processes would be weak and slow as well.

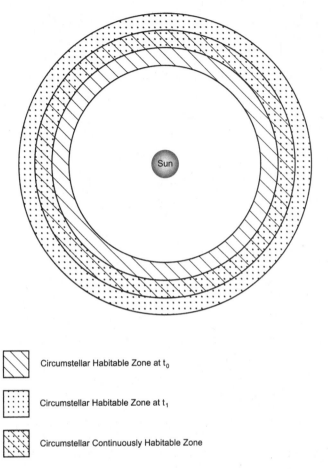

Circumstellar Habitable Zone at t_0

Circumstellar Habitable Zone at t_1

Circumstellar Continuously Habitable Zone

Figure 7.5: The circumstellar habitable zone (CHZ) is that temperate region around a star where liquid water can exist on the surface of a terrestrial planet for extended periods. Since the luminosity of even a stable star like the Sun changes over billions of years, however, a star's CHZ will move outward. The circumstellar continuously habitable zone (CCHZ) is the overlapping region of various instantaneous CHZs.

Even worse for life on an M dwarf's planet: all the water on its surface would likely end up sequestered on its dark side, frozen, leaving the side facing its sun hot and dry.[43] These problems could be mitigated, if like Mercury, the orbit of a tidally locked planet were quite eccentric; this would prevent the same side from always facing the host star. Of course, a highly elliptical orbit will lead to large temperature swings on the planet, regardless of its rotation. So, a planet in the super Goldilocks zone of an M dwarf star will either suffer from unevenly distributed heat or extreme temperature swings over the course of its year.

What if we place a planet-size moon around a gas giant planet, which, in turn, is in the super Goldilocks zone of an M dwarf? This would avoid the problem of having only one side locked in and facing its sun, since its full surface would receive light over its month. But this still won't work, for the reasons we gave in chapter 5. In particular, it's not clear that the gas giant can even retain a large moon as it migrates inward into the habitable zone of an M dwarf, where the star's gravity struggles mightily for possession of the moon. And even if the moon does survive, the size of the moon's orbit around its giant host would be a significant fraction of the distance to the host star, creating large variations in temperature on the moon's surface. Finally, while terrestrial planets are quite common around M dwarfs, giant planets are rare.[44]

M dwarf stars pose other problems for life. Like the Sun, they exhibit flares. Some are stronger than solar flares, and because M dwarf stars are far less luminous, a flare's intensity compared to the star is that much greater. A strong flare on an M dwarf star can spike the relative X-ray radiation by a factor of one hundred to one thousand compared to strong flares on the Sun; this would also increase the harmful ultraviolet radiation reaching the planet's surface.[45] Not only would such flares threaten surface life, but they probably also would quickly strip away a planet's atmosphere. The large starspots associated with flares would cause the star's brightness to vary by about 10 to 40 percent over longer timescales, mimicking an eccentric planetary orbit. Starspots and flares decline steadily as a star ages. So, while the passage of time would reduce these problems, at any age an M dwarf host star will be a less constant source of energy than a star like the Sun.[46]

Such bursts also could damage a planet's existing ozone layer. That's because, in its quiescent state, an M dwarf star produces less ultraviolet radiation compared to optical radiation than the Sun does. The steady flow of the Sun's UV radiation maintains the ozone shield in Earth's atmosphere, which offers some protection from modest increases in the UV radiation from other sources. So, life on a planet in the habitable zone of an M dwarf star will be more susceptible to harms from stellar flares and nearby supernovae.[47]

Ultraviolet radiation is also crucial for building up oxygen in a planet's atmosphere, forming prebiotic molecules, and removing excess hydrogen.[48] It was the steady dissociation of hydrogen-rich light molecules such as methane and water in Earth's atmosphere, and subsequent loss of the hydrogen, that allowed oxygen to become so abundant in our atmosphere.

The process still took about two billion years on Earth, about half its present age. This was probably a good thing for us, since an early oxygen-rich atmosphere would gum up the production of prebiotic molecules. The early energetic pre–main sequence phase of an M dwarf star can result in the complete loss of a planet's atmosphere, its desiccation, or the rapid buildup of a thick oxygen atmosphere, depending on the properties of the planet.

The red spectra of M dwarf stars means that very little blue light will reach the surface of its orbiting planets. Although photosynthesis doesn't require blue light, it generally becomes less effective without abundant light blueward of 6800 Å. Some bacteria can still use infrared light, but not to produce oxygen. Any marine photosynthetic organisms would have a hard time using red light for energy, since ocean water transmits blue-green light far better than blue or red light.

One of the most interesting exoplanetary systems is TRAPPIST-1.[49] which consists of a cool M dwarf with seven orbiting planets, averaging close to an Earth mass. This planetary system would fit inside the orbit of Mercury with lots of room to spare. But since the host star is far less luminous than the Sun, two of the planets (d and e) may actually reside in its habitable zone. All the planets are in orbital resonances, like the three inner Galilean moons of Jupiter—Io, Europa, and Ganymede. The planets' close spacing causes a lot of tidal forcing between them, resulting in slightly noncircular orbits. This might prevent them from getting tidally locked in orbit around their host star, but their rotation would still be quite slow. What's more, the tidal force would heat up the planets' interiors, as we see on Io with its active volcanoes.

Observations with the James Webb Space Telescope (JWST) rule out both an atmosphere for planet b and a thick CO_2 atmosphere for planet c.[50] And so far, it looks like the planets in this system are poor in volatile elements.[51]

RARE SUN

Our Sun is often described as a hopelessly mediocre star, but that claim is superficial at best. The average star mass and luminosity are much smaller than the Sun's. The Sun is among the 9 percent most massive stars in the Milky Way galaxy.[52] (See plate 16.) It's also highly stable. Its light output varies by only 0.1 percent over a full eleven-year sunspot cycle, perhaps a bit more on century timescales. This is a more stable average than sunlike stars of similar age and sunspot activity,[53] preventing wild climate swings on Earth.[54] Taken together, then, these anomalies suggest that the Sun is atypical in ways that make Earth more fit for technological life.

A LATENT RULER

As we argued earlier, the Sun's surface temperature and absolute brightness, or luminosity, are nearly optimum for extracting information from its spectrum. Other virtues of the Sun, and our orbit around it, are important for scientific discovery as well. The method of stellar trigonometric parallax is a vital tool in the astronomer's tool chest. Earlier we showed how the Earth and Moon have been used as rulers to measure distances in the solar system. Earth's orbit also serves as the baseline ruler for measuring distances to nearby stars. Because Earth moves many tens of millions of miles around the Sun during a year, nearby stars appear to wobble in reflex motion relative to distant background stars; this apparent motion is called annual stellar parallax.

Hypothetical Mercurians or Venusians would have a smaller baseline—since these planets have smaller orbits around the Sun—while Martians would have a larger one and Jovians—residents of Jupiter—larger still. But a larger baseline comes at a cost—time. Jupiter takes about twelve years to orbit the Sun. It would take several decades to measure two or three back-and-forth wobbles of a star. For measuring the distances to other stars, then, the ideal orbit of a planet should be large enough to provide enough parallax but not so large that it takes too long to complete a useful series of measurements in an observer's working lifetime.

Let's assume, however, that all the problems noted above could be overcome. Planets around the common M dwarfs still would offer their inhabitants a much poorer measuring rod. Since an M dwarf star is such a feeble light source, a planet would need a very close orbit to maintain liquid water on its surface. Astronomers on a planet around a star with 20 percent of the Sun's mass would have at their disposal a baseline only about 10 percent of ours. This means they could measure only 0.1 percent as many stars as we can from our solar system (assuming the same level of precision), and almost all of them will be other low-mass stars. Put another way, for measuring parallax, it would take observers from a thousand M dwarf systems to observe the volume of space we can survey from our single planet.

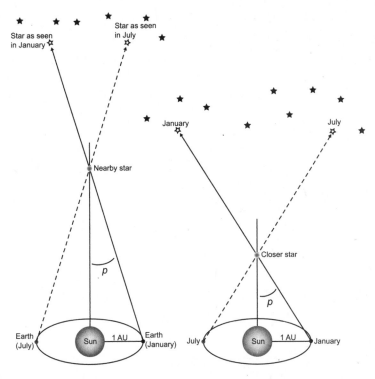

Figure 7.6: The Earth's orbit serves as a ruler for measuring distances to nearby stars. Because the Earth's position around the Sun changes during a year, nearby stars appear to move relative to distant background stars. The closer the star, the more it appears to shift against the distant background. The effectiveness of the method of trigonometric parallax depends on the luminosity of the host star. Hypothetical inhabitants of the much more common red dwarfs could only survey a much smaller volume of space with this method. (Even the nearest stars are much farther away than illustrated here, and the angle *p* is much smaller than shown.)

Because we orbit a star that is among the 9 percent most massive, we can measure the parallaxes of the even rarer, and therefore more distant, very massive O and B stars. This is vital for understanding the physics of stars, since the parallax combined with the apparent brightness of a star gives us its luminosity—its absolute brightness. Luminosity tells us the total power output (energy emitted over some time interval) of a star. This, in turn, tells us about nuclear fusion and how it depends on temperature and pressure at a star's core.

Once astronomers figured out the absolute brightness of several nearby O and B stars from parallax measurements, they learned they could use such stars as astrophysical "standard candles." That is, astronomers can calculate their distances without having to know their parallaxes. The technique involves measuring the apparent colors of O and B stars. From there, astronomers can easily calculate the star's true colors, how much is obscured by dust, and distance. And because we can see them at great distances, O and B stars have been very useful probes of the Milky Way galaxy.

Astronomers first attained reliable stellar parallax measurements in 1838 and 1839.[55] The work remained arduous until astronomers began to use photography regularly in the latter half of the nineteenth century. Even by 1900, they had only determined the parallaxes of about one hundred stars; by 1950, though, they had measured several thousand.[56]

An environment needs at least three assets to make this possible and/ or practical:

(1) The observer's home planet must be far enough from the host star.
(2) The planet's atmosphere must not be too thick or murky.
(3) There must be enough stars in the host star's vicinity.

Earth qualifies on all counts.

If Earth were a lot closer to the Sun, it probably would be like Venus—an unlivable hothouse with a thick, opaque atmosphere. If it were much farther from the Sun, it would need a much thicker, more obstructive atmosphere to keep water flowing on its surface—indeed, so thick with CO_2 that it would be hostile to animal life. In both cases, views from the ground would be poorer than those we enjoy. Once again, the most habitable location provides the best overall setting for scientific discovery.

A HANDY ASTROPHYSICS LAB

While Earth's orbit around the Sun offers a very good baseline for measuring the apparent movement of nearby stars, the Sun itself serves as a kind of astrophysical lab. Yes, the Sun is vastly closer than other stars, so it's easier to study. But quite apart from that, as noted above, we can derive more high-quality information from the Sun's spectrum than from most other types of stars.

Although the Sun is in several ways anomalous, it helps that some of its properties are near midrange. In constructing theoretical models, astronomers must start with the Sun and then extrapolate to lower and higher temperatures and luminosities to describe very different kinds of stars. If the Sun were at either end of the temperature scale—either very hot or very cool—such extrapolations would be much more prone to systematic errors. As it is, the Sun's atmosphere has qualities of both hot stars—a well-defined continuum in its optical spectrum and absorption lines arising from transitions in ionized atoms—and cool stars—molecular absorption lines in its spectrum and a convection zone near its surface. If astronomers could only exist near a *very* rare class of star, such as a blue supergiant, then they would have a hard time extending the knowledge of their home star to the more common main sequence stars.

Astronomers have been able to decipher clues about the Sun's internal structure for some time; but only in the last couple of decades have they detected oscillations on the surfaces of other sunlike stars.[57] These oscillations are excited by the convective currents of its outer layers. What if Earth circled a much less luminous main sequence star? If theory holds, such a star would have smaller wave heights in its velocity and brightness oscillations,[58] so much smaller that they well might foil astronomers pursuing such measurements.

Low-mass stars also offer a poor prospect for the other major test of stellar models—neutrinos. In the early 1930s, physicists such as Wolfgang Pauli and Enrico Fermi first suggested that the universe might contain elusive particles now called neutrinos, produced prodigiously in the cores of stars. The particles would be so tiny, subtle, and noninteracting that most could pass through the entire Earth without hitting a single atom. Detecting them, even from an ideal star, would require huge, costly, underground detectors that provide many chances for neutrinos to collide with other atoms. But starting in 1965, scientists began to detect neutrinos from the Sun with just such detectors.[59]

This work would have been much harder had our Sun been only slightly less massive or less bright,[60] or if we had lived much earlier in the Sun's history, when it had a cooler core. Our perch near the inner boundary of the super Goldilocks zone helps, too. Even at Mars, the neutrino flux would be nearly 2.5 times less.[61] So, three aspects of our solar system conspire to give us the best chance for detecting the neutrinos from our host star: the type of star we orbit, its age, and our location within the super Goldilocks zone.

Neutrino detectors that have come online since the late 1980s have resolved a long-standing mystery. Though some had theorized that neutrinos might have no mass, these more recent experiments suggest that they do.[62] This is vital for cosmology, since neutrinos have long been a candidate for solving at least part of the so-called dark matter problem.[63]

By producing neutrinos we can detect on Earth, the Sun's core serves as an indispensable workshop to learn about the very small and the very large, and to directly test our models of the Sun.[64] The Sun is just barely bright enough for neutrino astronomy and helioseismology to be practical. Observers with our level of technology living on one of the far more common K or M dwarf stars would lack these valuable tests.

So, the Sun's local setting offers a curiously well-tuned habitat for complex life. At the same time, it discloses more vital information than would more common types of stars, while still giving us an excellent example of stars in general. Just when they needed to test their theories, astronomers have found that the experiment was already halfway prepared for them.

CHAPTER 8

OUR GALACTIC HABITAT

Ironically, our relatively peripheral position . . . is indeed rather
fortunate. If we had been stationed in a more central position — say, near
the galactic hub — it is likely that our knowledge of the universe of other
galaxies, for example, might not have been as extensive. Perhaps in such a
position the light from surrounding stars could well have blocked our view of
intergalactic space. Perhaps astronomy and cosmology as we know
these subjects would never have developed.
Michael J. Denton[1]

Look at the sky on a clear night far from city lights and you'll see a fuzzy band of white light that looks like a wispy, luminous stretch of clouds. The ancient Chinese saw it as a river in the sky. The Greeks and Romans explained it with the myth of Heracles, the son of Zeus, who bit the breast of Zeus's wife Hera, spilling her white milk across the black firmament. The Romans called this part of the sky the Via Lactea, the "Milky Way," and, at least in the West, the name stuck. (The word galaxy derives from the Greek words for Milky Way.)

Not all the explanations for the Milky Way were this fantastical. There were actually some good guesses explaining this marvel in the night sky. But no one really knew what it was until Galileo pointed a telescope at the band in 1609 and resolved "a congeries of innumerable stars."[2] We now know that it's composed of stars in our home galaxy, densely concentrated because when we observe that milky band in the night sky, we're looking edge-on into our galactic disk and so taking in a stretch of it many times thicker than when we look elsewhere in the night sky. Once believed to be the entire universe, our galaxy is just one of billions of galaxies, which astronomers previously identified as mere fuzzy patches within the Milky Way galaxy, patches they called nebulae.

Galaxies, like noses, come in many shapes and sizes. Despite such diversity, astronomers reduce them to three basic types: spirals, ellipticals, and irregulars—the latter a bit of a catchall. Ours is a spiral galaxy.[3] Most of its stars are located in its flattened disk, only about 1 percent as thick as its diameter. We live in the disk, very close to its midplane, about halfway between the galactic nucleus and its visible edge. Decorating the sky like heavenly pinwheels, spiral galaxies are named for the beautiful pattern formed by their young stars and bright nebulae. The spiral pattern is thought to be a density wave phenomenon, somewhat like cars concentrating on crowded highways. The concentration itself progresses at a different speed from the individual cars that make it up, so snapshots at different times reveal different cars. But the overall pattern remains the same. We reside between the Sagittarius and Perseus spiral arms, a bit closer to the latter.

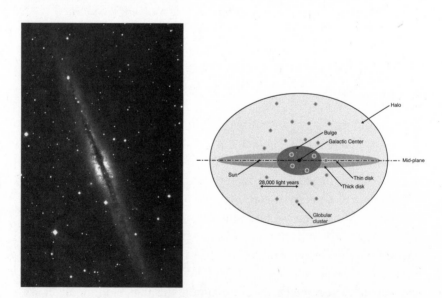

Figure 8.1: (left) Edge-on spiral galaxy NGC 891. Our galaxy would probably look much like this galaxy from the same vantage. Note the dust lane running across the full length of the disk. (right) The major components of the Milky Way galaxy. The Sun is located at the mid-plane in the thin disk, about half way from the galactic center to the edge of the disk. The globular clusters are spherically distributed around the galactic center and are the most visible members of the halo.

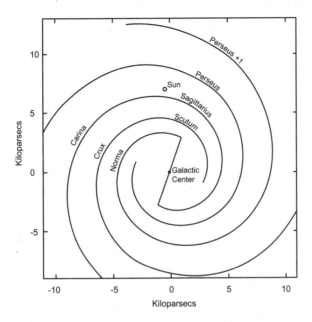

Figure 8.2: Face-on view of the Milky Way galaxy, a large, barred spiral galaxy. Shown are the outlines of its major spiral arms. We are located between two major spiral arms, Perseus and Sagittarius. The location is unusually good for learning about stars, galactic structure, *and* cosmology, rather than just one or two of these. One kiloparsec is about 3,200 light years.

Of the many denizens of our galactic neighborhood, the most obvious are the seemingly countless stars. Astronomers can see several star types in our patch of galaxy, from the feeble brown dwarfs to the brilliant blue-white O stars. They see stars in all their life stages, from the not-yet-born pre–main sequence stars to main sequence stars, long-deceased white dwarfs, neutron stars, and black holes. They see stars in isolation, pairs, triplets, and galactic or open clusters.

Astronomers can also see a motley assortment of matter between the stars. These include ghostly giant molecular clouds (GMCs) which can be millions of times more massive than the Sun; diffuse interstellar clouds; supernova remnants; and the winds from dying red giant stars as well as their descendants, the planetary nebulae. They sometimes can see a glowing interstellar cloud when bright stars heat its inside. Such a fluorescing cloud is called an H II region because its hydrogen gas is ionized.

There are also the reflection nebulae. These are illuminated from the outside when the light from a nearby radiant star reflects off its dust. When astronomers view a star behind an interstellar cloud, they see the spectrum of the cloud's atoms and molecules superimposed against the spectrum of the star as sharp absorption lines. Hot O and B stars are the best type of star to study interstellar clouds in this way, because they have much broader absorption lines than these clouds. This makes it easier to distinguish between the spectral lines formed in the star's atmosphere and those formed by interstellar clouds.

Astronomers usually subdivide the Milky Way into four populations: the halo, bulge, thick disk, and thin disk. Each is characterized by the ages, compositions, and movements of the objects within it. The halo contains only old metal-poor stars in highly elliptical orbits. (Recall that astronomers call every element heavier than hydrogen and helium metals.) The bulge, as fat as the disk is thin, contains stars spanning a wide range of metal contents, from about one-tenth to three times the Sun's. Even today, new stars are forming in the bulge. The orbits of its stars are also elliptical, but less so than those in the halo.

The flattened thick and thin disks overlap. The thin disk contains the greater diversity of objects, including most of the stars in the Milky Way, while the thick disk is more puffed up with older, more metal-poor stars. In the solar neighborhood, only a few percent of the stars are members of the thick disk; Arcturus, the brightest star in the constellation Boötes, is probably the most prominent nearby thick disk star. Even within the thin disk, older objects are more spread out on both sides of the midplane. The most concentrated are very young "zero-age" objects such as Giant Molecular Clouds, H II regions, and O and B stars, distributed in the thin disk with typical "scale heights" of about 150 to 300 light-years from the midplane. F and G dwarf stars reach heights of 600 and 1,100 light-years, respectively,[4] while the Sun reaches a maximum height of only about 250 light-years from the midplane.[5] If you think this seems like a dizzying array of galactic objects, you're right. And we study them all from our highly accommodating perch in the midplane of the Milky Way.

OUR GALACTIC PERSPECTIVE

Observers could not sample such diverse delights from every place in the Milky Way galaxy. The gas and dust in our neighborhood are diffuse compared to other stretches of the local midplane.[6] This gives us a fairly clear view of objects in the nearby disk and halo, as well as of distant galaxies. Since interstellar dust is concentrated in the midplane, we don't get a clear view of objects behind the Milky Way band, looking edge-on into the disk. (For a time, it was called the "zone of avoidance," because optical sky surveys did not show galaxies in this region.) Fortunately, this represents only about 20 percent of the area of the optical sky.[7]

At optical wavelengths we can only see a few thousand light-years through the dust in the disk; at or near the midplane, where we are, the nearby dust is better at hiding everything behind it. This means that a good bit of the galaxy is hidden from an observer near the midplane. Now, imagine moving the Sun perpendicularly a little farther from the midplane. A bit more of the previously hidden distant parts of the galaxy would now come into view, but it would increase the area of the Milky Way band in the sky. It would increase the zone of avoidance. Dust is not the only problem; more foreground stars would also interfere with our view of distant background objects. That's not a good trade. As we'll see later, this would profoundly hinder what we could discover about the cosmos as a whole.

Of course, since we're very near the midplane, dust absorbs some light, and foreground stars interfere with our view of background objects. We can learn more about how dust is distributed in our galaxy by observing other similar galaxies. Telescopic images of nearby spiral galaxies in the optical part of the spectrum clearly show bright spiral arms separated by darker regions. You might suppose that there are fewer stars between the arms, but, in fact, stars are only about 5 percent more concentrated in the arms. The arms appear so much brighter because they contain star nurseries, where the brightest stars are born and die. Their intense radiation illuminates the surrounding gas and dust. Therefore, while spiral arm dwellers would get a closer view of the rarer, very massive O stars, overall, this bright, thick dust would obscure their view of the local and distant universe.

The few examples of nearby superimposed galaxies give astronomers a direct measure of the transparency of spiral galaxies. A pair of galaxies are superimposed when a foreground galaxy partially blocks our view of the background one. The dust in the foreground galaxy absorbs and reddens some of the light coming from the background galaxy. Studies of such pairs show that the interarm regions are much more transparent than the dusty arms; they also show that the interarm regions become less transparent toward the center of a spiral galaxy.[8] Dust maps for our galaxy confirm this.[9] Observers near the galactic center would have to deal with more foreground contamination from stars.[10] (See plate 17.)

So, observers closer to the densely packed galactic center would have a much brighter night sky. This might be pleasing to the eye, but

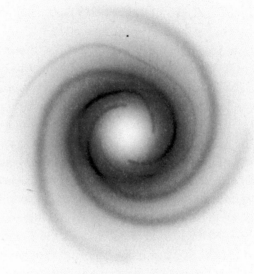

Figure 8.3: Face-on view of the Milky Way galaxy showing large-scale distribution of dust density. The bar and nucleus have been left out of the diagram. The Sun's location is shown as a black dot. Although a dust "hole" is shown in the center where the bulge dominates, there is some patchy dust there.

it would block far more than it would reveal. We're quite far from the bulge. Moreover, dust absorbs and scatters light across our line-of-sight to this galactic hub, greatly reducing its brightness. As a result, the otherwise bright galactic nucleus is invisible to the naked eye. Even so, on really dark nights we can see the bulge in the constellation Sagittarius as a brighter patch of the Milky Way band, since its light seeps through less dusty regions. Had we been at a quarter of our present distance from the bulge, the nucleus would appear several *million* times brighter.[11] The bulge wouldn't be as bright as the full Moon, but it would cover nearly half of the sky, and brighten the rest.

Astronomers in a globular cluster would face other problems. They would get closeup views of G, K, and M dwarfs and red giants, but they wouldn't see O, B, A, or F dwarfs, or interstellar clouds.[12] All the stars would have the same age and metal content. And, of course, the many bright foreground stars sprinkled all over their sky would ruin the views of the universe beyond the cluster.[13]

A view from the halo far from a globular cluster would offer a nice slice of the sky largely free of contaminating dust, gas, and stars. The view of the

Figure 8.4: The Great Cluster of Hercules, M13. This is one of the better known globular clusters to observers in the northern hemisphere. It contains several hundred thousand metal-poor stars about 25,000 light-years from Earth. Observers living in M13 would get a poor view of the universe outside their cluster home. In 1974, its multitude of stars inspired astronomer Frank Drake to transmit a message toward it, hoping it would be intercepted by an extraterrestrial civilization (see chapter 12).

Milky Way's spiral disk and bulge would be spectacular. But the trade-offs would be severe. Astronomers would lose close-up views of a great variety of stars, nebulae, and open clusters. And light pollution from the bright bulge and disk would hinder the view of faint objects. That's a heavy cost because most of what we know about stellar astrophysics depends on our ability to examine individual nearby stars over a wide range of masses, ages, and compositions.

Above all, from the solar system, we can measure the parallaxes of many nearby stars and discern their true brightness. Since stars are much more spread out in the halo, only a few M dwarf and perhaps some white dwarf stars would be close enough for our hypothetical halo dwellers to measure parallaxes without technology on par with the GAIA space observatory.[14] Important laws of stellar astrophysics, like the mass-luminosity relation, would be much harder to discover. Since young stars stay close to the midplane, observers in the halo, including those in globular clusters, would grasp little about how stars form. Near the midplane offers the best vantage point to understand stars, and it's only because we have a good grasp of the physics of stars that we can properly interpret the distant galaxies. After all, they're made of the same stuff we get to see up close in our galactic neighborhood. In short, halo dwellers would have a hard time fathoming distant galaxies, and the cosmos as a whole.

We don't have a bird's-eye view of the Milky Way, but our perch is quite nice. First, we live among a small collection of galaxies called the Local Group. The Milky Way and Andromeda (Messier 31) galaxies are its two largest members, followed by the galaxy in the constellation Triangulum (Messier 33). Both M31 and M33 are spiral galaxies, with the latter presenting us with a nearly face-on view. These teach us much about the overall structure of our home galaxy.

Second, we know we're living in a spiral galaxy because we can map its structure from the inside. Astronomers have learned to exploit the peculiar properties of the flattened rotating disk of gas and stars. So-called differential rotation allows radio astronomers to translate the measured radial velocities of interstellar gas clouds into distances from the Sun.[15]

For instance, imagine you're an ant riding on a spinning record and watching another ant on the record some distance from you. Although the record is spinning, the distance between you and the other ant does not change. This is because the spinning record is an example of solid body rotation; every part of the record is rigid relative to every other part, no matter what the record is doing in space. The Milky Way disk does *not* rotate like a solid body. Instead, it rotates differentially, so that stars closer to the center complete an orbit more quickly than those farther out. Hence, observers living around a star in the disk would see stars or interstellar clouds elsewhere in the disk moving toward or away from them—yielding blueshifts or redshifts, respectively. The regularity of this rotation allows astronomers to convert the observed position on the sky and radial velocity of a given cloud into a distance.[16] Observers in our spinning-record galaxy would not have this extra information.

Comparisons with other types of galaxies also help us understand our own home in the Milky Way disk. Less flattened galaxies, like irregulars and ellipticals, lack the simple rotational dynamics of disk galaxies. Elliptical galaxies, which have less gas and dust, contain stars with a wide range of orbits—most of them highly inclined and eccentric. For observers inside this kind of system, no law relates the distance of an object to its observed radial velocity. Irregular galaxies, as the name implies, have very irregular dynamics and patchy distributions of stars and nebulae.

Observers in such galaxies would have an even tougher time making sense of their neighborhood.[17]

Many galaxies are members of rich clusters containing thousands of gravitationally bound members. Life in a rich cluster of galaxies would be different from our home in the sparser Local Group. We'd get a close-up view of more nearby galaxies but at the expense of our views of the distant universe.

In short, settings in the halo, a globular cluster, the bulge, a spiral arm, an isolated galaxy, a dense cluster of galaxies, an irregular galaxy or an elliptical galaxy would be less revealing than ours.

We occupy the best overall place in the Milky Way for making a range of observations, and the Milky Way is itself the best type of galaxy to learn about stars, galactic structure, and the distant universe, the three major branches of astrophysics. Does our location also offer the best overall type of habitat in the galaxy? Indeed, it does.

HOME SWEET ZONE

Just as many other spots in the Milky Way are less amenable to science, so too are they less hospitable to life. In sci-fi stories, interstellar travelers visit exotic places in the Milky Way and meet interesting aliens. You name the place, and some intrepid writer has already put a civilization there: the galactic center, a globular cluster, a star-forming region, a binary star system, a red dwarf star, even the vicinity of a neutron star. It wouldn't surprise us to learn that there's a cheeky novelist out there with a motley assortment of alien toughs smoking cigars and playing five-card stud inside a black hole. Of course, most of these storytellers are just letting their imaginations run wild. But there was a time, not long ago, when prominent astronomers seriously speculated about intelligent beings on the Moon, Mars, Venus, Jupiter, and even the Sun.

The Apollo, Mariner, Venera, Viking, and Voyager missions put an end to those conjectures. Nowadays, we know a lot more about these environments and the stringent requirements for life. Canal-building Martians and Sun-dwellers now sound quaint rather than futuristic.

Understand, we like a good far-flung space western as much as the next fellow. But when a self-declared expert starts talking about civilizations all over the galaxy, not as whimsical fiction but as cosmic certainty, we get skeptical. For just as most of the solar system fails the strict requirements for complex or technological life, so too does most of our galaxy.

Like the circumstellar habitable zone in our solar system, there is also a *galactic* habitable zone (GHZ). (See plate 18.) It's all about forming earth-like planets and the long-term survival of animal-like aerobic life.[18] The boundaries of the galactic habitable zone are set by the needed planetary building blocks and by threats to complex life in the galaxy. Let's discuss each in turn.

ASHES TO ASHES, DUST TO DUST

The story of how the elements came to be assembled on Earth is told by modern cosmology, galactic chemical evolution, stellar astrophysics, and planetary science. The big bang produced hydrogen and helium and little else. Over the next 13 billion years, this mix was cooked within many generations of stars and recycled. Beginning with the fusion of hydrogen atoms, massive stars make ever-heavier nuclei deep in their hot interiors, building on the ashes of the previous stage and forming an onion-like structure. Exploding as supernovae, the massive stars then return atoms to the galaxy. But they return them with interest, by producing heavy elements that didn't exist before. As a result, our galaxy's metal content—that is, its metallicity, the abundance of heavy elements, or metals, relative to hydrogen—has gradually increased to its present value, which is close to the Sun's. Today, metals make up about 2 percent of the mass of the Milky Way's gas and dust in the disk. Stardust courses through our veins.

Cloud collapse is one of the least understood steps of star and planet formation. An interstellar cloud must contract by many orders of magnitude to form stars with planets. When contracting, a cloud will heat up as part of its gravitational energy is converted into the thermal energy of the motions of its atoms and molecules. The rising temperature increases the cloud's pressure support. It will stop contracting if it can't cool by radiating heat in the far-infrared parts of the spectrum, since infrared photons can escape the dense cloud. How well a contracting cloud can cool depends on its makeup. Hydrogen and helium lack very strong emissions in the infrared, so a cloud without metals must be fairly massive for its self-gravity to overcome the internal pressure. When metals are present, carbon and oxygen are the main coolants (via ionized carbon, carbon monoxide, and neutral oxygen). Thus, a metal-rich cloud is more likely to fragment into smaller cloudlets to form more low-mass stars.[19] Since massive stars make inhospitable hosts for life, metal richness may be an essential ingredient for a system's future habitability.

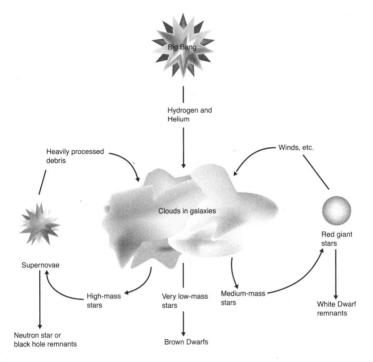

Figure 8.5: The big bang produced hydrogen and helium and little in the way of other elements. Early stars processed this primordial mix into heavier elements, which they returned into the interstellar medium. Low- and high-mass stars return matter in different ways and leave different kinds of remnants.

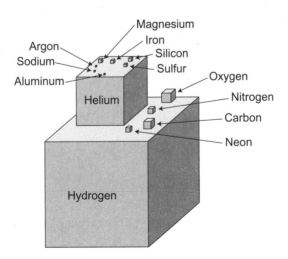

Figure 8.6: The relative proportion (by number) of the most abundant elements in the Sun. Hydrogen, helium, and the elements resting directly on top of the hydrogen cube dominate the composition of stars and Jovian planets. The terrestrial planets are composed mostly of oxygen and the other elements on top of the helium cube.

Gas giant planets like Jupiter, for all their visual interest, are mostly hydrogen and helium. At present, the most popular model for the formation of a gas giant is called core instability accretion.[20] According to this model, the future gas giant must first form a rocky core of at least ten to fifteen Earth masses. Then the growing planet's gravity can attract and retain the plentiful hydrogen and helium in the protoplanetary disk. This can lead to a runaway growth of the planet. Thus, a minimum amount of metals is required so that the rocky core forms quickly before most of the gas is lost from the system (on a timescale near ten million years). Though we still have much to learn, the discoveries of planets outside our solar system are helping us to get a handle on the value of this threshold metallicity.[21] Astronomers aren't finding giant planets around stars with less than about 20 percent the metal content of the Sun.[22]

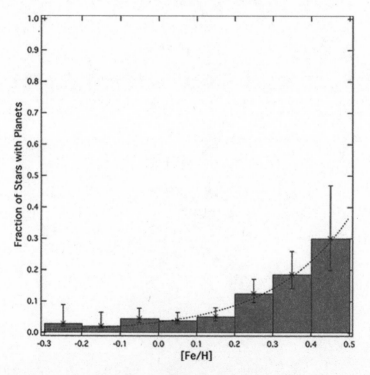

Figure 8.7: Relative incidence of Doppler-detected planets as a function of metallicity (quantified here as [Fe/H]). This notation is a logarithmic scale relative to the Sun. Zero is the same Fe-to-hydrogen ratio as the Sun, 0.3 is twice, and –0.3 is half. The dashed curve is the best-fitting power law to the binned data (shown with error bars). The relative incidence of stars with giant planets increases sharply for metallicity values greater than the Sun's.

Conversely, too much metal in the mix may pose other problems. On average, the higher the allotment of metals in its birth cloud, the more planetesimals and planets will form in a system. This greatly complicates their orbits. All the gravitational elbowing in this crowded situation may cause giant planets to migrate inward, where they would perturb each other and dump terrestrial planets into the host star or fling them out of the system.[23]

Metal-rich stars pose other burdens for life on any earthlike planets that orbit them. If the Sun were more metal-rich, more dangerous ultraviolet radiation would reach the Earth's surface.[24] The Sun's energy output would also vary more and produce more flares and sunspots.[25]

A metal-poor system poses a more basic problem. Earth-size terrestrial planets are less likely to form in a system condensing out of a more metal-poor cloud, such as a globular cluster.[26] Indeed, in the extreme case of a cloud without metals, not even fist-sized rocks will form, let alone Earth-size planets.

There's another reason that the amount of metals must be just right in a star system for it to be habitable: The metals seem to determine how many of the planets are gas giants versus terrestrial planets.[27] And it now looks as if you need both gas giants and terrestrial planets to form a habitable system. As a result, the ideal metal content for building a habitable planetary system may be quite narrow.[28]

Recall that relatively few stars much more metal poor than the Sun have giant planets. Suppose that the amount of heavy elements needed to form a minimum-mass habitable planet is only a little less than the upper threshold for forming giant planets in sufficiently modest numbers that they don't wreak havoc in the system.[29] If so, then few systems will have enough metals to form an Earth-size terrestrial planet and at least one gas giant planet of the right mass and in the right orbit to get a habitable terrestrial planet, but not so much metal as to lead to a deadly game of planetary bumper cars.

How many asteroids and comets form in a system likely also hinges on the portion of metals the system starts with. But this sweet spot—producing neither too many nor too few asteroids and comets—probably does not precisely overlap the metallicity sweet spot for planet formation. This further narrows the range that yields a habitable system.

THE METAL GRADIENT

The metal content of stars in the solar neighborhood is diverse, varying from about one-third to three times the Sun's.[30] But beneath this diversity exists an important trend on the galactic scale. Several lines of evidence show that the metal content of the disk goes down as we move from the center to the edge of the galaxy.[31]

As we move from the center to the edge of the galaxy, the gas gets less dense, and with it, the rate of star formation. Since stars are the source of most heavy elements in the galaxy, this explains why metallicity drops as we move outward. Early in the history of the Milky Way, the metals built up quickly in the inner galaxy as star formation ramped up and peaked after only about two to three billion years. Since then, the star formation rate has been declining, interrupted by sporadic episodes of activity. Today, metallicity in the solar neighborhood is still going up,[32] but it's expected to grow more slowly in the future as the supply of fresh gas dwindles. All this means that long ago, any region in our galaxy was more metal-poor than that area is now. To find enough heavy elements for a habitable system billions of years before our system began to form, we would need to look to our galaxy's inner regions. But our inner galaxy is one tough, crowded neighborhood with little patience for some planet trying to kick-start even primitive life.

How chemical elements are distributed has a bearing on life, too. The Earth is mostly iron by mass, with oxygen, silicon, and magnesium next in abundance. Surprisingly, the cosmically abundant and life-essential volatile elements hydrogen, carbon, and nitrogen are mere trace elements in the bulk Earth. Of course, what really counts for habitability is their abundance in the crust, where they're much more common. Astronomers believe asteroids and comets delivered most of these volatile elements to the Earth late in its formation, where they were incorporated into its crust. A terrestrial planet that forms like the Earth needs these volatile elements for habitability, but the quantity is critical, and the right elements are not always formed in the same proportion in all places and times in the Milky Way.[33]

The most abundant elements in the Earth were produced primarily in supernovae. There are two basic types, Type Ia and Type II supernovae, which together provide the various elements needed to build a habitable planet. A Type II supernova results when the core of a massive star collapses suddenly, releasing enormous quantities of energy that blow apart its outer layers. Many elements are synthesized in the ensuing maelstrom, and others produced earlier (in the so-called "onion layers") are also ejected.[34]

The origin of Type Ia supernovae is less clear. The consensus is that they result from the detonation of a white dwarf star in a binary star system. Unlike the progenitor of a massive star supernova, a white dwarf star does not have a prominent onion layer structure; it consists mostly of carbon and oxygen. The Type Ia supernova progenitor accretes matter from its companion until it reaches the so-called Chandrasekhar limit near 1.4 solar masses. At that point, runaway thermonuclear reactions in the interior convert much of the carbon and oxygen into heavier elements.[35]

Today, supernovae pop off in our galaxy about once every fifty years. But judging from the present abundance of metals, the supernova rate must have been much higher early on. Their average rate over the entire history of the galaxy must be about one every three years.[36] Both the supernova rate and the relative numbers of Type II to Type Ia supernovae have been declining since shortly after the Milky Way galaxy began forming.[37]

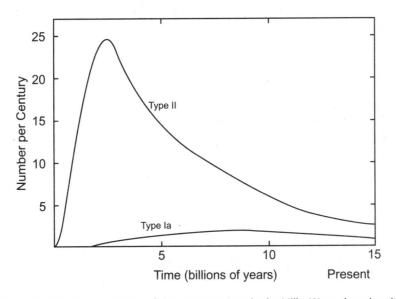

Figure 8.8: Calculated evolution of the supernova rate in the Milky Way galaxy since its formation. Type II (massive star) supernovae have always dominated over the Type Ia (white dwarf binary) supernovae. These reconstructions are based on galactic chemical evolution models, calibrated using the observed chemical abundance patterns of stars of varying ages. In reality, the evolution of the supernova rate was probably not as smooth as shown in the figure, but any other variations were probably secondary in magnitude. The results of these calculations show that our galaxy was a much more dangerous place early on.

Since supernovae return processed matter to the galaxy, stars forming today should differ from those formed in other times. In particular, there will be less oxygen, silicon, and magnesium relative to iron in interstellar medium as the galaxy ages. All else being equal, this implies that a terrestrial planet forming today will have a larger iron core relative to its mantle compared to one formed in the past. This may narrow the time frame for forming earthlike planets if this ratio turns out to be critical for maintaining plate tectonics.[38]

Another important trend is the concentration of the long-lived radioisotopes in the galactic disk. Compared with iron, radioactive isotopes of potassium, thorium, and uranium are becoming less abundant because the ratio of Type II to Type Ia supernovae is declining (only Type II events return these radioisotopes to the interstellar environment). So again, all else being equal, Earth-mass terrestrial planets forming today will, in 4.5 billion years, have only about 60 percent the internal heating from radioactive decay that Earth has today. "It is tempting," wrote planetary scientist Francis Nimmo and collaborators, "to speculate that the Earth is habitable in part because it possesses a 'Goldilocks' concentration of radiogenic elements: high enough to permit long-lived dynamo activity and plate tectonics, but not so much that extreme volcanism and dynamo shutoff occur."[39]

In sum, at first, the Milky Way had enough elements to build Earth-mass planets only in its inner regions. Additionally, the formation of giant planets and comets, too, depends on the supply of ashes from supernovae, and the relative abundance of magnesium, silicon, and iron leads to differences in the geology of a planet. The typical Earth-mass planet forming in the future will probably have a smaller concentration of geophysically important radioisotopes and, as a result, different volcanic and tectonic activity. Each of these factors limits the time and place in our galaxy where habitable planetary systems can form.

SURVIVAL OF COMPLEX LIFE

Even if you manage to get all the right atoms in the right place at the right time to build a terrestrial planet of the Earth's mass and orbit, it still might not earn the "habitable" label. A planet also must be largely free of threats to complex life over the long haul. As we've briefly noted, there are two major long-range extraterrestrial dangers: impacts by large asteroids or comets, and transient radiation events.

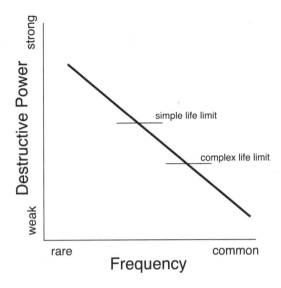

Figure 8.9: Most threats to life on Earth follow a simple rule: the more energetic, the rarer. Solar flares, supernova radiation reaching Earth, asteroid and comet impacts, and volcanic eruptions pose major threats to life over long timescales. Since the threshold for extinction is higher for simple life than it is for complex life, a planet might be able to support simple life but not complex life. This figure also implies that a planet that experiences more frequent low-intensity events of a given type, such as distant supernovae, will also experience more frequent high-intensity events of the same type (such as nearby supernovae).

Several large extinctions are evident in the geological record. After many years of debate, today most paleontologists are convinced that one of them, the Cretaceous-Paleogene (K-P) extinction 65 million years ago, was by a large asteroid or comet slamming into what is today the Gulf of Mexico.

Comets are quirky; they visit the inner solar system like a transient uncle who drops by unannounced from who knows where. Astronomers think comets spend most of their time in two reservoirs, the Kuiper Belt and the Oort cloud. Until the early 1990s, the Kuiper Belt was only theoretical, a vast swarm of icy bodies in fairly low inclination orbits just beyond Pluto's orbit. Today, we know the orbits of several thousand Kuiper Belt objects.

Infrared observations of young nearby stars indicate that most are surrounded by excess dust, suggesting that they too possess Kuiper Belt objects.[40] Astronomers have interpreted changes in the shapes of certain spectral lines in Beta Pictoris, a young star with a dust disk, as infalling comets. Some have interpreted the discovery of water vapor around the aged star IRC+10216 as evidence of a swarm of comets around it.[41] We infer the existence of the Sun's Oort cloud, its other major comet reservoir, from the long-period comets that visit our night skies. These comets have highly elliptical orbits—in fact, they're very nearly parabolic, so they spend most of their time far from the Sun, beyond the Kuiper Belt, typically at about twenty thousand AUs. Since comets in the Oort cloud are only weakly bound to the Sun, it doesn't take much to perturb them.

The creation of far-flung comet reservoirs around the Sun depends on gravitational deflections by the giant planets. Once a comet is in one of these huge, nearly parabolic orbits, it's very sensitive to galactic-scale perturbations. These include the galactic radial and vertical tides, passing stars, and close encounters with giant molecular clouds (GMCs).[42]

The galactic tides vary as the Sun oscillates up and down relative to the disk midplane on its trek around the center of the galaxy. Some astronomers have argued that the period of this vertical oscillation matches the Earth's cratering record, but this idea remains controversial.[43] Remember that a typical GMC contains about half a million solar masses of gas and dust. Extrapolating the Sun's trajectory shows that we haven't passed near a GMC any time during the last few million years, nor are we in danger of passing near one anytime soon. But stars can pass close to or even through the Sun's Oort cloud at any time. The expected combined effect of all these perturbers is an occasional spike in the comet influx into the inner solar system superposed on more semi-regular variations.

Comets' orbits around a star are sensitive to the star's position in the Milky Way galaxy. Stars in the disk are more densely packed the nearer they are to the center of the galaxy. Moreover, a planetary system forming out of a more metal-rich molecular cloud than the Sun did will likely form more comets. Therefore, planetary systems in the inner disk of our galaxy (a more metal-rich neighborhood than ours) probably start with more populated Kuiper Belts and Oort clouds. They should experience more frequent "comet showers" for this reason and because of stronger galactic tides and more frequent star encounters.[44] While the outer Oort cloud of an inner disk star will become depleted more quickly, it will also be replenished more quickly from its more tightly bound inner Oort cloud comets.

The threat posed by asteroids in our solar system depends on the details of Jupiter's formation and orbit (as discussed in chapter 4). Galactic-scale disturbances probably have little effect on asteroids. The initial number of asteroids should be proportional to the initial endowment of metals, but their final number will depend on local factors. There is, however, one possible external factor relevant to asteroids—short-lived radioisotopes produced by massive stars in the solar system's birth cloud. Chemical evidence from meteorites suggests that their parent bodies were heated early on. This extra heating hydrated the minerals in the asteroids, securing the water inside them. Otherwise, the Sun's heat would have sublimated the water. This is important because many asteroids carried their water to Earth, helping fill its oceans. Without enough short-lived radioisotopes in the asteroids, Earth might never have received enough water to become habitable.

High-energy radiation may work its destruction on unprotected life less dramatically than large impacts, but it does the job.[45] Earth's magnetic field and atmosphere shield surface life from most particle radiation and dangerous electromagnetic radiation. But certain extraterrestrial radiation bursts can damage the ozone layer in our upper atmosphere, resulting in more deadly radiation on the Earth's surface.[46] Such "energetic transient radiation events," in order of decreasing duration, include active galactic nucleus (AGN) outbursts, supernovae, and gamma-ray bursts. These and other radiation sources are more threatening in the inner regions of the Milky Way galaxy simply because stars are more concentrated there.[47]

Another grave danger lies at the heart of the Milky Way. Like a giant dragon sleeping in the heart of a mountain, a massive black hole dwells in the center of our galaxy, though at the moment it's fairly dormant. Like our Milky Way, most large galaxies in the nearby universe do not have active nuclei, but astronomers have discovered that almost all probably have a giant black hole lurking there. Presumably, a dormant black hole wakes up when a star or cluster wanders too close and becomes disrupted. The radiation emanates not from the black hole itself but from the hot accretion disk around it. Fortunately for us, massive early-type galaxies, that is, giant ellipticals, seem to be the favored host of the most massive black holes.

Black holes are fearsome beasts, distorting space, time, and common sense, understood to be so densely packed that not even light can escape their horizons. Massive black holes are thought to be the engines that drive distant quasars and the intensely luminous nuclei of nearby active galaxies, some with visible jets like M87, a giant elliptical cD galaxy in the Virgo cluster. An AGN produces both high-energy electromagnetic and particle radiation. Most of it is emitted along the rotation axis—that is, above and below the galactic plane, though many of the charged particles will spiral along a galaxy's magnetic field lines and fill its volume. So, not surprisingly, an AGN can pose a severe threat to life.[48]

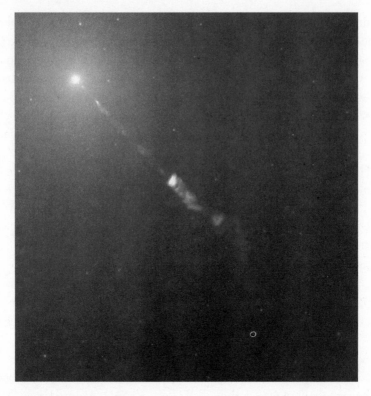

Figure 8.10: M87, a giant elliptical galaxy at the core of the Virgo cluster of galaxies, probably has a giant black hole in its nucleus. A "jet" is faintly visible coming out of the nucleus. Giant ellipticals like M87 offer more dangerous environments and poorer platforms for scientific discovery than flattened galaxies like the Milky Way.

The safest place to be during an AGN outburst is far from the nucleus, and probably close to the midplane, where the Earth happens to be. The Sun's nearly circular orbit and proximity to the corotation circle (the imaginary circle around the Milky Way's center, marking where the stars orbit with the same period as the spiral arms) make it less likely that it has recently or will soon cross a spiral arm.[49] Still, just because we're nestled between the major spiral arms doesn't guarantee we will avoid all supernova threats. Spiral arms are ragged, with spurs that protrude into the areas between the arms. For example, the nearest cluster of massive stars is the Scorpius-Centaurus OB association (Sco-Cen), at about four hundred light-years; it's not inside a main arm.[50] Fortunately, this is too far away to pose a serious threat to us. (Interestingly, it was closer to the Sun's present position five to seven million years ago, but the Sun was then somewhere else.[51])

The worst place to be is in the bulge, with scorching radiation and stars with highly inclined and elliptic orbits, which can come close to the energetic nucleus or pass through its jet.

Type II supernovae spew deadly radiation as well. Most are concentrated in the thin disk, inside the Sun's orbit, and especially along the spiral arms. The more energetic but less frequent Type Ia supernovae are more uniformly sprinkled in the disk and probably peak near the nucleus. Observations of supernova remnants indicate that the supernova rate peaks at about 60 percent of the Sun's distance from the galactic center, where they are 1.6 times more frequent than where we are.[52] Supernovae of both types were also more frequent in the Milky Way galaxy during its first few billion years. Estimates of the rate of life-threatening supernovae in the Sun's neighborhood vary; they average one every few hundred million years.[53]

Figure 8.11: M81, an elegant spiral galaxy in Ursa Major, is one of the nearest large galaxies. The arrow points to the location of supernova 1993J, which occurred inside the galaxy. The image was obtained in May 1993, about two months after the new star first appeared. All the other bright stars in the image are foreground stars in our galaxy.

Finally, there are gamma-ray bursts. In the late 1990s, astronomers discovered that these enigmas occur at great distances, making them among the most energetic transient radiation events since the big bang.[54] Wherever they come from, there's only one form of protection against them: location. Perhaps we're just lucky to have avoided a nearby direct gamma-ray burst jet, but it helps that we are not in a region that's low in metals and where stars are forming, such as the outer disk.[55]

We should also be glad we're not in a globular cluster, and not just for the reasons mentioned earlier. Because their orbits intersect the galactic disk nearly perpendicularly, globulars pass through the disk at blistering speeds—about 250 kilometers per second or more. At such speeds, hydrogen atoms in the disk hitting the atmosphere of the planet would produce deadly X-rays; impacting dust would deposit heaps of energy as well—even worse than a Sahara sandstorm.[56] Bulge and old-disk stars would suffer from similar if somewhat less extreme threats. In contrast, the solar system, with its very "cold" orbit in the thin disk (see below), is unlikely to suffer from such threats. Therefore, globular clusters are surely hostile to life, and bulge and old-disk star systems, while not quite as bad, aren't exactly the Ritz Carlton of the Milky Way either.

We've now reviewed all the galactic-scale factors—at least all the ones we can think of—that set the boundaries of the galactic habitable zone.

AN EXCLUSIVE COUNTRY CLUB

In light of all these factors, we see that this habitable zone is a rather exclusive country club for observers. Compared to our present location, the inner ghetto of the Milky Way suffers from greater radiation threats and comet collisions, and an Earth-size planet is less likely to form there in a stable circular orbit. The outer regions are safer, but stars there will be accompanied by only fairly small terrestrial planets, planets too small to retain an atmosphere or sustain volcanism or plate tectonics, and they will probably lack Jupiters. While we can't yet say how wide it is, our best guess is that the GHZ is a fuzzy annulus (or ring) in the thin disk at roughly the Sun's location, a ring whose habitability is compromised where the spiral arms cross it. If a system needs to be close to the corotation circle to be habitable, then this thin and often broken ring could be narrower still. (See plate 18.)

At the same time, the galactic habitable zone—and, more specifically, our location in it—offers one of the best overall places to be a successful astronomer and cosmologist. Even though we're near the midplane, there's little gas and dust obstructing the views in our neighborhood. The disk is highly flattened and so less obstructive, and we're far enough from the galactic center to keep it from obscuring too much of our view of the distant universe. Yet, despite our location inside the Milky Way, our old friends, carbon and oxygen, as carbon monoxide, have served as probes of its spiral-arm structure.[57]

Our model of this habitable zone is incomplete, as is our understanding of the threats to life and of its basic needs. But we're hopeful that advances in astronomy will answer many of the questions we've posed.[58] If the trend continues as it has for the last few decades, the estimated size of the zone will continue to shrink.

OTHER GALAXIES

Our Boardwalk suburb makes much of the rest of our galaxy look like Baltic Avenue by comparison, but let's not be too hard on the Milky Way. The best way to appreciate it is to compare it to other galaxies. About 98 percent of galaxies in the local universe are less luminous—and thus in general, more metal-poor—than the Milky Way.[59] So, whole galaxies could lack Earth-size terrestrial planets.[60] In addition, stars in elliptical galaxies have less ordered orbits like bees flying around a hive minus a bee's aptitude for reacting to impending collisions. Thus, they're more likely to experience close flybys lethal to their planets.[61]

But that's not the half of it. The stars in elliptical galaxies are more likely to visit their galaxy's dangerous central regions, and to pass through interstellar clouds at disastrously high speeds. In many ways, ours is the optimal galaxy for life: a late-type, metal rich, spiral galaxy with orderly orbits and comparatively little danger between spiral arms.

The way galaxies interact also matters for any life they contain. For example, the Andromeda Galaxy is predicted to have a close encounter with our galaxy in about three billion years. Such an event will dislodge most stars in the disks from their regular orbits. It may also feed fresh fuel into our galaxy's central black hole and bring it back to life, further eroding the already depressed housing market in the inner galaxy. So, our galaxy's habitable zone may only last another three billion years. On the bright side, lots of stars will be flung into intergalactic space. Any surviving inhabitants in such systems could remain safe, unless they get too close to the active nucleus of a large galaxy.

The local density of galaxies also affects habitability. As noted above, the Milky Way galaxy is part of the Local Group, a collection of more than eighty galaxies. The largest three are spirals, and the remainder are small dwarf spheroidal and irregular galaxies. Compared to other concentrations of galaxies in the local universe, ours is sparse. Rich clusters, like Virgo and Coma, each contain thousands of galaxies. Close encounters, called "galaxy harassment," are common among these galaxies, and they aren't restrained by good manners. The smaller galaxies—like the two Magellanic Clouds, both irregular galaxies orbiting our galaxy—can lose much of their gas during these close encounters. The motions of galaxies in clusters also heat up the gas between them. This, in turn, strips gas from the outer disks of the galaxies.

Massive cD elliptical galaxies are found in the centers of many rich clusters, presumably having grown at the expense of many hapless smaller galaxies. With their intense nuclear activity, such supergalaxies would be anything but super places to live. Overall, then, rich clusters are probably less habitable than sparse groups.

Precious few plots of galactic real estate are as amenable to complex life as ours, to say nothing of its value for observation. At our present state of knowledge, we can still say that there's no place like home. And, as we'll see, we can add to that another adage: There's no time like the present.

CHAPTER 9

OUR PLACE IN COSMIC TIME

The progress made in our understanding of the universe during
the twentieth century is nothing short of stunning.

Michael S. Turner[1]

In the 1920s, American astronomer Edwin Hubble began a careful study that led to a rediscovery of cosmic time. It began as a fairly mundane research project. Using the 100-inch Hooker telescope at Mount Wilson Observatory in California—then the world's largest—he was studying spiral nebulae, including Andromeda. Since at least the time of Immanuel Kant, scientists had wondered whether these football- and cigar-shaped objects were nearby and smallish, or distant and enormous. Kant had guessed that they might be "island universes" in their own right.[2] With cutting-edge telescopic power, Hubble resolved individual Cepheid variable stars in the spiral nebulae M31 (Andromeda), M33, and NGC 6822. This was a big deal because these stars were a "standard candle" that astronomers could use to measure distances. By measuring the light of single stars in these spiral nebulae, then, he could calculate how far they were from Earth.

With further study, Hubble noticed something odd. The spectra of many nebulae were shifted toward the red end of the electromagnetic spectrum compared to our Sun and nearby stars. That is, their wavelengths tended to be longer. He then compared these redshift data to his measures of distances and found that the more distant a nebula, the

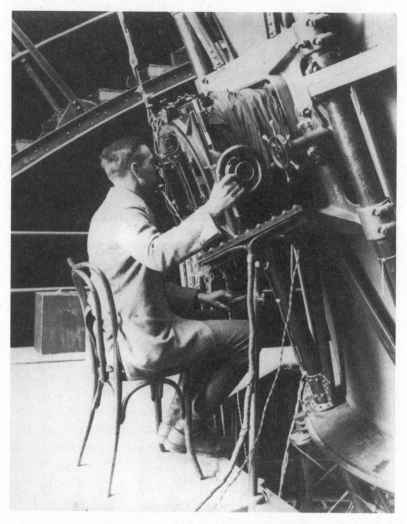

Figure 9.1: Edwin Hubble, seated in a bentwood chair, looking through the Newtonian focus of the Mount Wilson one-hundred-inch telescope (c. 1922).

greater its redshift.[3] This pattern suggested revolutionary answers to three contested questions. First, the Milky Way, contrary to the common view at the time, was not the universe. Second, the spiral nebulae are distant galaxies, perhaps as our Milky Way galaxy would appear from a distance. (See plate 19.)

Finally—and even more shocking—the universe was expanding. In a historical instance of the left hand not knowing what the right hand is doing, Albert Einstein, in his general theory of relativity, had predicted that the universe was either expanding or contracting. Unfortunately, Einstein found the notion so distasteful that he had introduced a "fudge factor," a variable called a cosmological constant, retrofitted to keep the universe in steady, eternal, equilibrium. As a result of Hubble's discoveries, and the works of Georges Édouard Lemaître and Alexander Friedmann— whose solutions to Einstein's theory implied an expanding universe—he repented of his cosmological constant. He even called it the "greatest blunder" of his career. Although Einstein didn't need to see Hubble's data for himself, he made a trip to California to visit Hubble. This provided the public with photo record of one of the century's greatest matches between scientific theory and observation.

Like Einstein, most astronomers at the time, including the young Hubble, believed in a static and eternal universe. Even after Einstein conceded his error in the late 1920s, many scientists weren't pleased with the prospect of an expanding universe, since it suggested that our world had a finite past—even a beginning. One critic of the idea, Fred Hoyle, dubbed such an event the "big bang," and the name stuck.

Consider the account C. F. von Weizsäcker gives of a debate he had with the physical chemist Walther Nernst in 1938:

Figure 9.2: Albert Einstein (1879–1955).

He said, the view that there might be an age of the universe was not science. At first I did not understand him. He explained that the infinite duration of time was a basic element of all scientific thought, and to deny this would mean to betray the very foundations of science. I was quite surprised by this idea and I ventured the objection that it was scientific to form hypotheses according to the hints given by experience, and that the idea of an age of the universe was such a hypothesis. He retorted that we could not form a scientific hypothesis which contradicted the very foundations of science. He was just angry, and thus the discussion, which was continued in his private library, could not lead to any result.[4]

Most scientists trusted their eyes more than Nernst's definition of science and came to accept the expanding universe model. It helped that Hubble's discoveries came right when theoretical cosmologists started to ponder the universe as a whole.

It also marked a striking convergence of theory and discovery. Earlier in the 1920s, the Belgian Georges Lemaître and the Russian Alexander Friedmann proposed expanding models of the universe derived from Einstein's equations. Friedmann saw that general relativity implied, as he put it, that "at some time in the past (between ten and twenty thousand million years ago) the distance between neighboring galaxies must have been zero."[5] Lemaître, a Catholic priest and physicist who studied under Arthur Eddington, was the first to describe an early version of what we now call the hot big bang model. "The evolution of the world," he explained, "can be compared to a display of fireworks that has just ended: some few red wisps, ashes and smoke. Standing on a well-chilled cinder, we see the slow fading of the suns, and we try to recall the vanished brilliance of the origin of the worlds."[6]

STELLAR STANDARD CANDLES

There's no evidence that Hubble knew of Friedmann's and Lemaître's proposals, but one scientist's work did influence him. In 1908, Henrietta Leavitt at Harvard College Observatory was diligently measuring the periodic changes of light from Cepheids in photographic plates of the Magellanic Clouds. (The Large and Small Magellanic Clouds are the Milky Way's largest satellite galaxies and are only visible from the southern hemisphere.) In due course, she found that those Cepheids with longer periods were brighter. Because all the Cepheids in a cloud are nearly the same distance from the Earth, the correlation between period and apparent brightness translates into a relation between period and luminosity—that is, *absolute* brightness. Like cosmic lighthouses, Cepheids communicate through the simple ebb and flow of light: *Slower is brighter.*

Satellite galaxies where stars were still being born provided the critical clues for discovering this link between period and luminosity (the P-L relation). Classical Cepheids are massive stars that last only a few million years. As a result, we find them not in clusters whose stars are billions of years old but instead where stars are forming. Leavitt and other astronomers building on her work calibrated the P-L relation by observing a few nearby cepheids.[7] This allowed Hubble and other astronomers to work out the distances to the spiral nebulae. Those Cepheids with the longest period, nearly fifty days, were among the brightest known stars. For these reasons, Cepheids were the first useful standard candles to reach beyond the galaxy. And like much of the evidence on which modern cosmology is based, if we were in a different setting, we might never have seen these extragalactic measuring sticks.

We mentioned in chapter 7 that astronomers have used massive main sequence stars as standard candles. And, early in the twentieth century, astronomers such as Harlow Shapley had some success in estimating distances with the low-mass variable stars known as the RR Lyraes.[8] But none could compete with the Cepheids. Some are over four hundred times brighter than the RR Lyrae variables.[9] As a result, Hubble could measure the brightest Classical Cepheids in the Andromeda Galaxy. Since then, astronomers armed with better telescopes and detectors have found them in ever more distant galaxies. The greater reach has allowed them to refine the relation between distance and redshift.

Over the past few decades, astronomers have concluded that Type Ia supernovae are also great standard candles. Recall that one of the end states of a moderate mass star in a binary system is a Type Ia supernova, which can pop off a billion years or so after the birth of its progenitor. This means that we can see Type Ia supernovae in galaxies where stars are no longer forming. They're so brilliant that they outshine all the other stars in their host galaxies for a few weeks. And since they occur in both nearby and distant galaxies, and in spirals and ellipticals, these standard candles cover a major swath of the visible universe.

Once astronomers find a Classical Cepheid or Type Ia supernova in a galaxy, they can establish its distance. Just how well they can do this depends largely on how standard the candle is and how well they have calibrated it. To pull this off, they must first establish a distance ladder. The Earth's surface is the lowest rung. Next comes the Moon's and then Earth's orbit, which serves as the baseline for measuring stellar parallax. Once astronomers nail down the distances to a few bright secondary candles from their parallaxes and other, less direct means, they can measure much greater distances[10] and calibrate other types of standard candles.[11]

DISCOVERING THE ECHOES OF CREATION

While most astronomers were convinced by Hubble's findings that the universe was expanding and so had a finite age, a few still held out for an eternal past. Eddington, speaking for many, said that "philosophically, the notion of a beginning to the present order of Nature is repugnant."[12] In the early 1930s he still resisted Lemaître's hypothesis of a dense, hot beginning. Instead, he opted for a universe with a cosmological constant, one that had fallen out of balance from an eternal static state in favor of expansion. Eddington's model suffered from obvious logical problems, so most astronomers had grudgingly accepted Lemaître's model.

But the idea of an eternal universe was not yet ready to go gently into that good night. A bid to rescue the idea came in 1948, when Hermann Bondi, Thomas Gold, and Fred Hoyle each proposed steady-state models. They postulated that matter spontaneously appears in the space between the galaxies as the universe expands. This would maintain a constant density of matter on large scales and account for an expanding universe without the troublesome idea of a beginning. Steady state, as the name implies, describes a universe without major change when averaged over long timescales.

For several decades, no direct evidence allowed astronomers to decide between big bang and steady-state models. In 1965, however, two engineers at Bell Telephone Lab, Arno Penzias and Robert Wilson, noticed excess noise in a radio antenna at seven centimeters wavelength. They found it coming with equal strength from all directions of the sky and couldn't credit it to any known sources of radiation. Cosmologists soon interpreted this finding as the long-sought relic radiation from the big bang.[13] Unbeknownst to Penzias and Wilson, some physicists had already predicted this as a consequence of the hot big bang but not the steady-state model. As a result, Penzias and Wilson's discovery sounded the death knell for the steady state.

Ralph Alpher, George Gamow, and Robert Herman presented the first modern picture of the big bang's residual radiation in the late 1940s. They understood that the expanding universe models implied a much denser and hotter time in the distant past. If we went back far enough, we would find a time when the density of matter prevented the free flow of photons, when matter and radiation were in "thermal equilibrium." The temperature and pressure were then high enough to ionize most of the hydrogen, the most abundant element. Prior to that, photons did not travel far before being scattered by free electrons.

In a sense, the radiation "decoupled" from the matter after the protons and electrons recombined. Most photons produced during the decoupling era (CMB photons)—when the gas in the universe was about 3,000 degrees Kelvin (5,000 degrees Fahrenheit)—traveled unimpeded as the universe continued to expand. Since then, the expansion of space and time has stretched the photons, redshifting them about a thousand times. These cosmic microwave background photons are the oldest in existence. They have been in transit for most of cosmic history. Thus, the CMB sky is called the "surface of last scatter," since it represents the last time these photons strongly interacted with matter before setting off on their long journey.

Current models indicate that the CMB photons were liberated about 380,000 years after the beginning. That's how long it took for the universe to expand and cool enough for protons to combine with electrons to form neutral hydrogen. This cosmic stretch has preserved the shape of the spectrum of the cosmic microwave background radiation—a critical clue to its origin. The universe today is far too transparent to photons to have produced the background radiation we observe. It's like sunlight (produced by a 10,300 degree Fahrenheit object) streaming through our atmosphere (near 60 degrees Fahrenheit). These two are way out of equilibrium, and so the gas in the atmosphere cannot be the source of the sunlight. The present background radiation points back to a time when the universe was much denser and hotter.

The steady-state models failed to predict not only the background radiation but also observations showing reverse galaxy evolution with distance. That is, when we look farther and farther into space, we are really looking back in time. And as we peer deeper and deeper into the past, we see galaxies in much earlier stages—just what we would expect, if the big bang theory is correct.[14] In the famous images taken by the Hubble Space Telescope, called Hubble Deep Fields, we see galaxies whose redshifts place them some 13 billion light-years away. And the James Webb Space Telescope has detected even more distant galaxies. This means we are not seeing distant galaxies in their "present" state, but as they existed some 13 billion years ago. Similarly, the background radiation delivers information about the newborn universe. John Barrow and Frank Tipler explain why this matters for science. "The background radiation," they write, "has turned out to be a sort of cosmic 'Rosetta stone' on which is inscribed the record of the Universe's past history in space and time. By interpreting the spectral structure of the radiation, we can learn of violent events in the Universe's distant past."[15]

At the same time, we learn something of the expansion of the universe, since changing space and time modified the background photons on their way to us. Cosmologists especially value the subtle fluctuations in the intensity, and hence temperature, of the background radiation across the sky. Caused by sound waves traveling through the ionized gas just prior to decoupling, these fluctuations are a rich source of information about the early universe. There were regions of gas that were hotter and denser than other regions. We now see these as variations in the surface brightness of the background radiation on the sky. (See plate 20.) In the same way, we learn about the Sun's interior by studying oscillations on its surface, except that the background radiation is a snapshot in time. Cosmologists can extract at least ten properties of the universe, cosmological parameters, from these variations.[16]

But since they don't reveal every detail equally well, these data points don't make other observations obsolete. For example, Type Ia supernovae are still vital because they probe so much of the observable universe. We need three crucial numbers to grasp the universe as a whole: the Hubble constant, the matter energy density, and the cosmological constant, also called the vacuum energy density. The Hubble constant is the present rate of expansion of the universe, which astronomers determine by measuring the distances and redshifts of galaxies. The matter energy density is the total amount of matter and energy in the universe (not including dark or vacuum energy), which we discern from its light emissions and gravitational effects. The vacuum energy is more mysterious, since it contains energy that, at very large scales, counteracts gravity. Astronomers have detected its effects by observing distant standard candles. We'll discuss this below.

For discerning the latter two, the data gleaned from Type Ia supernovae are nearly perfect complements of certain data derived from the background radiation.[17]. Neither is redundant, since together, the two sets of

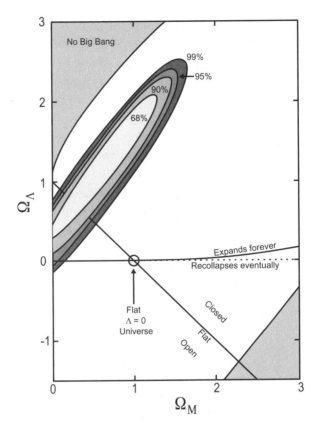

Figure 9.3: Two of the most important cosmological parameters for understanding the geometry of the universe are the matter-energy density (Ω_M) and the cosmological constant energy density (Ω_Λ). The region of the diagram constrained by Type Ia supernovae data is at right angles to the region constrained by the microwave background data. Because of this, the two sets of data complement each other in determining the values of these two parameters.

data narrow the possible values of these two important properties. Notice that their intersection is only a small region on the plot. This permits the most efficient use of our measurements and demonstrates that all visible matter helps us learn about the universe.[18] And the distribution of galaxies fills in gaps in information left from these two sources of information.

COSMIC STEW

In the 1940s, no one knew the origin of the chemical elements. As big bang cosmology took hold, theorists soon recognized that the universe had cooled too quickly to form anything beyond lithium. Hydrogen and helium predominate in the universe at large. But whence came all the other elements? In time, astrophysicists concluded that the elements heavier than lithium were produced inside stars well after the big bang. By the late 1950s, they worked out many details of this theory, called stellar nucleosynthesis.

More recently, the observed abundance of helium compared to hydrogen in metal-poor galaxies, the deuterium in Lyman-alpha clouds (see below), and lithium in nearby metal-poor stars has confirmed the expectations from big bang models.[19] In a sense, the abundance of the light element isotopes in the universe is a type of telescope that can peer just beyond the time when the background radiation formed. Particle physics and cosmology merge just after the big bang. So, three independent findings all support the big bang theory: Hubble-Lemaître's galaxy redshift relation, the cosmic background radiation, and the relative abundances of the light element isotopes.[20]

MORE TESTS OF
COSMOLOGICAL EXPANSION

Discovering the background radiation and measuring the abundances of light elements added more support to the big bang theory. Still, they did not show that the redshifts are due to space itself expanding. Fortunately, our universe provides still more tests for the theory, and Earthlings have access to them.

The late Allan Sandage, a cosmologist who worked at the Observatories of the Carnegie Institution of Washington in Pasadena, was perhaps best known for his decades-long pursuit to estimate the Hubble constant. Another of his pursuits was an independent test of cosmic expansion. Richard Tolman proposed the first such test in 1930 with a crucial prediction,[21] which Sandage confirmed with observations of early-type galaxies.[22] At the same time, Sandage debunked the "tired light" hypothesis, which posits that redshifts result from photons losing energy over vast distances. He showed that galaxies do not change with redshift in the way proponents of the tired-light hypothesis predicted.

Four other tests have confirmed that the universe is expanding. One is that the changes in brightness (light curves) of the distant Type Ia supernovae appear broader compared to nearby ones. This means that the light from a distant supernova seems to wax and wane more slowly than a nearby one. According to the big bang standard model, this is due to time dilation. This time dilation was first confirmed in 1995 from supernova light curves.[23] In 2023, astronomers reported that distant quasars also show this effect.[24]

The second test that confirms the model involves comparing redshifts of many galaxies and quasars to where they are on the sky.[25]

The last two tests involve the cosmic microwave background radiation: one confirmed the change in its temperature with redshift,[26] and the other the shape of its spectrum.[27] With one exception, we can only do these tests because we can see other galaxies.

This is a mere glimpse at the many ways access beyond our galaxy has allowed cosmologists to reject many models of the universe and uncover supporting evidence for others.[28] What a loss to our inquisitive spirit if we couldn't choose among static, steady-state, and big bang models. That our time and place in the universe is so well suited not only for life but also for observation has, in one of the most important questions about the universe, made all the difference.

QUASARS AND INTERVENING MATTER: A RECORD OF COSMIC HISTORY

Quasars are another useful probe of the distant universe since they report about an epoch shortly after matter and radiation decoupled. These powerful beacons are the most luminous and distant objects in the visible universe. Quasars are believed to be galaxies in their early stages, when their central black holes were growing rapidly by accreting gas. Before disappearing into the black hole, the gas forms a very hot, bright accretion disk around it. It's this accretion disk that makes quasars so bright even from vast distances.

Quasars tell us about themselves as well as about intervening matter. As the light from a distant quasar traverses space and time, gas clouds along the way absorb some of it, impressing absorption lines in its spectrum. The specific wavelength of the absorption depends on the amount of cosmic expansion between the background quasar, the absorbing gas, and, to a lesser degree, on the relative velocities between them. Quasars emit light most strongly in the ultraviolet part of the spectrum. The intervening neutral hydrogen gas absorbs most strongly at a distinctive spot in the electromagnetic spectrum called Lyman-alpha. In laboratories, its wavelength is 1,216 Ångstroms—the far ultraviolet. But looking into the heavens from our vantage, each intervening gas cloud or galaxy has a different redshift and thus produces a Lyman-alpha absorption line at a *different* wavelength. Together, the intervening gas clouds and galaxies produce a Lyman-alpha "forest" of absorption lines.[29] This cosmic forest, like ancient petrified forests on Earth, presents a time-ordered fossil record of cosmic history. Conveniently, cosmic expansion redshifts the ultraviolet forests into the region of the spectrum where the Earth's atmosphere becomes transparent (starting at 3,000 Ångstroms). Only this happy circumstance allows astronomers to study them from ground-based observatories. A contracting universe would shift the absorption lines into the far ultraviolet end of the spectrum, making them invisible from the ground.[30]

THE COSMIC HABITABLE AGE

We've discussed our planetary system's habitable zone and the larger-scale galactic habitable zone. But there's a still larger scale framework, which we can call the cosmic habitable age (CHA). When pondering the global properties of the universe, age is more basic than location. And not all ages of the universe are equally habitable.[31] This is obvious in the very early universe prior to decoupling. At that epoch the universe was a dense, hot plasma of elementary particles and light nuclei, a dismally hostile place for life of any sort. But the beginning is not the only no-man's-land. If we think of everything a place needs to support life, especially complex life, then only a relatively tiny slice of cosmic history is habitable.

The life-essential elements heavier than helium weren't present in the universe until they were made in the first stars and ejected from their interiors. The first generation of stars began seeding their environment a few hundred million years later. The life-essential elements concentrated more quickly in the larger galaxies, especially in their inner regions. So even if some stars had Earth-size planets only a few billion years after the beginning, they would have been stranded in the most perilous neighborhoods. Unlike today, the early universe was poor in heavy elements and rich in high-energy quasars, star births, gamma ray bursts, and supernovae, which bathed early-forming planets with lethal levels of gamma ray, X-ray, and particle radiation.

Much of what we wrote in chapter 8 about chemical evolution in the Milky Way applies to the broader universe, with some key differences. Our galaxy is more massive than most other galaxies, so it has accumulated

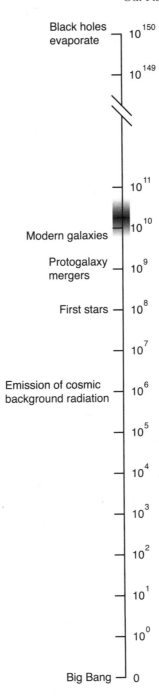

Figure 9.4: Important events in the history of the universe on the way to habitability. The Cosmic Habitable Age (shaded) is a tiny segment of the total span of time that cosmologists often consider (up to 10^{150} years). It coincides with the best time to discover the geometry of the universe.

heavy elements more quickly. As a result, planets probably formed around stars in the Milky Way earlier than they did in most other galaxies.[32]

The universe, then, has been getting more habitable. But this trend won't continue forever. We also mentioned in chapter 8 that the long-lived radioisotopes so crucial to Earth's geology are declining, compared to stable heavy elements. This global decline limits the future birth times for life-friendly planets. Eons from now, some smaller galaxies may grow as metal rich as the Sun, but by then, these radioisotopes will be far less abundant. That means their planets will produce less internal heat. A typical Earth-size planet forming at that time would not have a lively 4½-billion-year geology as the present Earth does.

Yes, a larger planet could compensate for this. The average rocky planet would need to be more massive in the future as the metals keep building up. But that comes with costly trade-offs. As we've argued, many processes on a terrestrial planet depend on its size—asteroid and comet impact rate, surface relief, ocean depth, internal heat retention, atmosphere loss rate, time to oxygenate the atmosphere, and the stability of the planet's orbits. A larger planet could buy some extra time but would throw up big road-blocks for life.

We can't be precise about the maximum future birth time of a habitable planet. We can note, however, that the three most vital radioisotopes for heating the Earth—potassium 40, uranium 238, and thorium 232—have half-lives of 1.3, 4.5, and 14 billion years. That's about the lifetime of sunlike stars on the main sequence—their active phase of life. One of these corresponds to the age of the universe, and another to the age of the Earth. In any event, the long-lived radioisotopes, like the long-lived stars, delay their release of energy. On this evidence alone, we can estimate that the remaining birth time of habitable planets is probably less than ten billion years. This is a tiny slice of time compared to timescales common in cosmology—up to 10^{150} years into the future.

Although the buildup of metals has made the universe more habitable, this trend cannot continue forever. We noted in previous chapters that metal-rich stars are more likely to host giant planets. Nevertheless, most of the giant planets around these stars have orbits that are hostile to habitable planets. It's beginning to look like a star needs to have just the right amount of metals to serve as a host for a habitable system.[33] This translates into a narrow range of ages of the universe for building habitable planets.

If we combine all the relevant properties of the universe that vary over time—declining star formation rate, declining high-energy radiation, increasing metals, declining radioisotopes—we begin to get a picture of the cosmic habitable age. The extremes as hostile to life. The very early universe lacked galaxies and planets altogether. And even after they formed, there were still not enough metals to build habitable planets and organisms in safe places. In the distant future, the universe will be dominated by black holes, neutron stars, white dwarfs, and red dwarfs. Any terrestrial planets still in close orbits of these stars will be tidally locked and probably tectonically dead.

The cosmic habitable age may be even narrower since the ways it's defined vary from galaxy to galaxy. As a result, many galaxies may have had no habitable planets for much of their history. For example, low-mass galaxies that don't form many stars may not amass enough heavy elements to build habitable systems for another five or ten billion years. By then, long-lived radioisotopes may be too dilute to sustain plate tectonics. Some small galaxies might already have stopped forming stars. If so, then perhaps only galaxies at least as massive as the Milky Way with the right histories of star formation enjoy the cosmic habitable age. Perhaps the 3.7-billion-year period of life on Earth is the only cosmic period, give or take a few billion years, that's compatible with our existence.[34]

A PRIVILEGED TIME AND PLACE

Astronomers entered the twentieth century not comprehending the Milky Way or spiral nebulae, the source of energy in stars, or the origin of the elements or the cosmos. Is the Milky Way the whole show, they wondered, with the spiral nebulae small objects within it? Does gravity power the stars, or is the answer some new source of energy? Were all the chemical elements formed inside stars? Is the universe static and eternal, or is it changing and finite? Why is the night sky dark?

Today, we have good, if general, answers to these questions. Hardly a month passes without news of an important discovery. As we write this, the James Webb Space Telescope has spent the last year collecting data on the most distant galaxies.[35] We have seen in preceding chapters how the universe has revealed itself to us in a few brief centuries on scales ranging from the Earth's surface to the nearby galaxies. Our vision now extends to the universe at large—in both space and time. As we peer into the vast distance, we now know we are reaching back to an epoch close to the beginning.

We marvel at the ingenuity that has allowed us to decode such information. But we shouldn't ignore the prerequisites for such ingenuity. Our place in the Milky Way gives us a view of the distant universe and of various and sundry nearby stars. Without these local clues, distant galaxies would be beyond our ken.

In fact, for scientific discovery, time may matter as much as place at the largest scales. If we imagine improbably hardy residents of the early universe—say, a billion years after the beginning—they would have had a front-row seat at a dazzling firework display of nearby supernovae and quasars, their central black hole engines fed by gas falling into their deep gravity wells. By studying the Hubble Deep Fields, we get a glimpse of what these hardy denizens of the early universe would have witnessed at night. The images reveal a young universe filled with crowded—and thus distorted—galaxies. Partly as a result, the intense heat from the many massive stars and supernovae bequeaths to the galactic dust a bright and sometimes beautiful glow. But that means that when most of the stars in the Milky Way galaxy formed, hot dust veiled the view of the distant universe. If they could have existed, early cosmic residents might have enjoyed the local show. But they would not have seen far beyond it.

The universe would have looked much smaller then. The distance observers can see at a given age of the universe is called the particle horizon. It's as large as the distance light has traveled since the big bang, given the intervening expansion of the very fabric of space. It limits what we can directly glean from the universe. The particle horizon swells as the universe ages, giving astronomers an ever-larger sample of the physical world.

Will the universe keep expanding at the same rate, slow down, or speed up? That depends on the competing effects of the matter/energy density and the dark energy density. Imagine gravity working to gather everything in, and dark energy working to do just the opposite. So, which one now predominates at the cosmic scale? According to the best measurements of Type Ia supernovae light curves, dark energy now has the edge at cosmic scales. As a result, the universe's expansion is accelerating. Until about six billion years ago, gravity still ruled the roost, and the whole show was slowing down.[36] If we're rightly interpreting the messages from Type Ia supernovae, then the universe should keep expanding and even accelerating.

Today, the universe presents astronomers with quite a large sample of the sources that anyone could, in principle, observe—about 40 percent.[37]

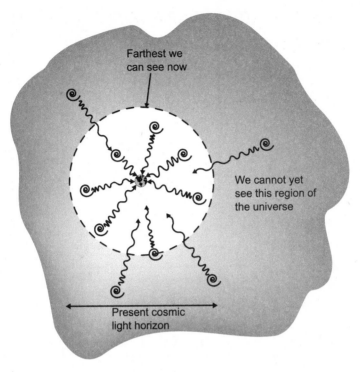

Figure 9.5: The distance we can see at any one time, called the *particle horizon*, is set by the speed of light and the age of the universe. As the universe ages, the particle horizon will continue to expand, revealing a larger volume. But billions of years from now, the effects of the *event horizon* will begin to reduce the volume of the universe we can see.

But as the cosmos expands ever faster, objects will appear at ever greater redshifts and gradually fade from view. This dying of the light will touch the most distant objects first—since they are receding the fastest. It will be like watching an object pass through the event horizon of a black hole. Indeed, this cosmic space-time limit of vision is called an event horizon. We can't yet see its effects, since it's beyond the particle horizon. But in perhaps twenty to thirty billion years, the farthest reaches of the universe, such as the background radiation, will go gently into the cosmic night, and no astronomer's rage will retrieve it.[38]

Even before that darkness descends, however, the echo of the cosmic starting gun will become ever harder to measure, since it gets fainter as the universe expands and redshifts it. This began in the decoupling era and will continue. Moreover, right now we have an advantage: many sources of radiation—cosmic rays, the microwave background, the galaxy's starlight, synchrotron emission from its magnetic fields, and warm dust—*happen*

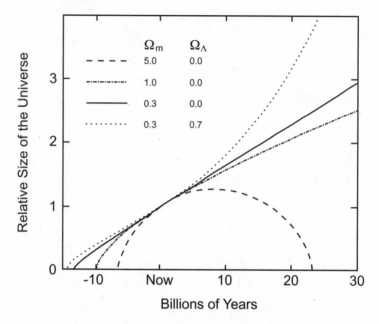

Figure 9.6: Four possible histories of the universe shown in terms of the matter-energy density (Ω_M) and the cosmological constant energy density (Ω_Λ). A closed universe (dashed) has enough matter-energy density to cause it to collapse. A flat, or critical, universe (dashed-dotted) is just barely able to avoid collapse and expands forever, but at an ever-decreasing rate. An open universe (solid) expands forever and experiences less deceleration than a flat universe. An open universe with a cosmological constant (dotted) accelerates indefinitely. This last case best matches what we know about the universe.

to be enough alike to allow us to do both galactic and cosmic astronomy. "Possibly we just have to accept this as a coincidence," said mathematician and cosmologist Michael Rowan-Robinson, "as we have to accept the similar apparent sizes of sun and moon."[39]

We owe this coincidence to the fact that we live during the most life-friendly age of the cosmos. And, at this most habitable moment, the distinct sources of radiation in a large galaxy like ours are enough like the background radiation so we can access them, while differing enough to allow us to tease them apart. If the background radiation were much stronger than the local sources, galactic astronomy would suffer; if the local sources were much stronger, cosmology would suffer. As it is, we can disentangle the cosmic background and the galactic sources of radiation. But this happy convergence will not persist.

Rowan-Robinson is right to compare it to the match between the Sun and Moon in our sky, but not, we think, to shrug it off as a coincidence.

CONTAMINANTS TO THE RESCUE

Early in the modern debates in cosmology, the similarities between the cosmic and local sources of radiation suggested to some astronomers, like Fred Hoyle, that the background radiation could have a local origin. So, why are astronomers now so sure that they have detected the *cosmic* microwave background radiation, rather than a galactic imposter?

After all, it might have been thwarted by foreground contaminants such as dust between the planets, warm dust between the stars, and radiation emitted by free electrons spiraling along the galaxy's magnetic field lines.[40] As it happens, however, each of these contaminants (that is, to the CMBR) has a unique spectral signature that helps us tell them apart from each other, and from the background radiation itself.

We say *help* rather than *allows* because that alone isn't enough. Astronomers require a second lucky break to sift the proverbial wheat from the chaff. That lucky break is our local setting.

Since the background radiation is cosmic, it's expected to be mostly isotropic. That means that it should look pretty much the same whichever way we look. In contrast, most local sources of radiation are highly anisotropic. This fact plus the fact that we live in a highly flattened galaxy far from its hub provides a contrast that allows us to tease out the galactic foreground contaminants from the background radiation. (See plate 20.)

Had we been much closer to the galactic center, trying to detect the background radiation would have been, well, hellish. All the local radiation contaminants not only get stronger but also appear more isotropic closer to the galactic center, where the bulge dominates the disk. The very worst place for observers to discover and study this evidence of the big bang (or any other cosmic phenomenon) is at the center of the galaxy, since they wouldn't know whether they were detecting local or cosmic radiation. They would get a close view of the giant black hole at the galactic center—like a fly getting a brief close-up of a windshield on a racing car. But even bracketing off the risks to life and limb of living so close to the galaxy's central black hole, the front-row view of it would be a bad trade since it's not unique; there are many smaller and less hazardous black holes strewn throughout the Milky Way galaxy. The background radiation, in contrast, is unique. We can also be glad that the Milky Way's nucleus is currently inactive.

Similarly, studying the background radiation is much easier between spiral arms, where there is less dust than inside a spiral arm.

In general, then, those very places in the galaxy most threatening to complex life are also the poorest places to measure this echo of the big bang (and any other cosmic source).

Most other types of galaxies would offer poorer views of the background radiation than we get. The worst would be the spherical ellipticals and their smaller cousins, the globular clusters. Observers in such systems would see nearly uniformly distributed sources of radiation except those living near their peripheries. Irregular and active galaxies would also be fairly low-rent districts. Active galaxies are buzzing star factories, and so have high infrared fluxes from supernova-heated dust. And like adding fog to rain, supernovae themselves and galaxies with active nuclei spew high-energy particles, producing intense radio emissions. You might suppose this would be easy to correct for, since it comes from a single direction, but charged particles permeate a galaxy along its magnetic field lines, creating the illusion that they come from all directions.

Like everything else, foreground contamination changes with time. Before the first stars were born, the universe contained no metals to speak of, and, hence, no dust. Shortly after the first stars seeded the interstellar gas with metals, dust began to form. The frequent supernova explosions warmed it, making it glow brightly. There were fewer stars to contaminate the foreground, but particle emissions would have been more intense, and the background emissions would peak closer to the optical region of the spectrum, where more of it would be absorbed by interstellar dust.

With all these competing factors, it's hard to say if the background radiation was easier to measure in the early universe than it is now. Today, dust and particle emissions probably interfere less, but as stars and metals accumulate in our galaxy and the universe continues to expand, it will become ever tougher to measure. Our educated guess is that, apart from the epoch just before stars began forming, background radiation is more accessible now than it was in the past or will be in the future.

After the background radiation, distant galaxies and quasars will be the next objects to fade from view. Future observers would not have quasars to probe matter over vast distances. The galaxies will be much farther apart.[41] In about 150 billion years, galaxies presently beyond thirty million light-years will fade as they approach the event horizon.[42] Before that happens, these nearby galaxies will be farther away from Earth than the farthest known quasars are from us now. One hundred and fifty billion years, keep in mind, is a tiny fraction of the timescales cosmologists often ponder.

We live in a time when there are still plenty of stars forming. However, as fewer and fewer stars are produced, there will be fewer massive stars, and thus, fewer Classical Cepheids and supernovae—all valuable measuring tools for astronomers. Millisecond pulsars await a similar fate. Typically, pulsars spin down in a few hundred million years.[43] Since they were once massive stars, millisecond pulsars—the best cosmic clock in the business—will track the declining rate of star formation. Type Ia supernovae will keep popping off for a few billion years after stars cease to form, but by then these valuable measuring rods also will be less common.

Over time, large galaxies tend to gobble up their satellites, as the Milky Way is currently doing to the Sagittarius Galaxy. As a result, the future will have fewer of these galactic groupies hanging around the large galaxies. Future observers would have nothing like our Magellanic Clouds, which serve as middle rungs in the cosmological distance ladder, nor will any satellite galaxy do. To contain Classical Cepheids, some stars must still be forming. How long would Henrietta Leavitt's discovery have been postponed without something like the Magellanic Clouds?

The distant future will contain fewer objects relatively uncontaminated by the buildup of elements inside stars. This will make it harder to discern the primordial abundances of the light element isotopes. Today, we still have a few samples, such as nearby old stars in our galaxy's halo and ionized nebulae in low-mass nearby galaxies. But, as the universe ages, stars will keep processing gas and shifting the number of light elements in the universe away from the early amounts early in cosmic history. For their part, the old, metal-poor stars will continue to die away.[44] As quasars fade, studying the intervening Lyman-alpha clouds (a useful source of data on the original deuterium abundance) will prove ever harder. Much of what we can now discover will disappear beyond reach.

What if the first observational cosmologists had not been born until, say, A.D. 10 billion? Well, they probably wouldn't have the period-luminosity relation for Classical Cepheids at their disposal, because there probably wouldn't be anything like the Magellanic Clouds orbiting the Milky Way galaxy. With few stars forming, Classical Cepheids would probably be rare in the Milky Way galaxy, too. These observers might know about the less luminous Type II Cepheids and maybe even about their period-luminosity relation, but these standard candles would be faint and thus harder to detect in other galaxies. Most galaxies (and galaxy clusters) would be farther away. Nearby galaxies would be more diffused and their redshifts greater. Scientists might at some point discover the Hubble Law and set out to search for the elusive background radiation, but by then, it would be much weaker than the foreground contaminants. Cosmology would be a more ponderous endeavor. Some profound discoveries might never be made.

Closer to home, there are several direct ties between Earth's surface and the broader universe. For example, we mentioned in chapter 3 that radio astronomers measure the motions of the continental plates by observing distant quasars. Cosmic ray particles hitting the Earth's atmosphere produce carbon 14 (as well as beryllium 10 and chlorine 36), which is very useful for dating the remains of once living things. Of course, Earth's atmosphere does blur the images of distant objects in our sky, but a cosmic coincidence helps us see distant galaxies. Nearby objects look smaller as they get farther away—as you would expect. Paradoxically, at a certain distance, about one redshift, the effect reverses and objects appear larger. The farther the distance, the greater this effect. It's as if space itself serves as a magnifying glass. The minimum apparent size of a Milky Way–like galaxy is just about the same size as the blurring effect of our atmosphere.[45] As a result, our atmosphere is not nearly the impediment for viewing the most distant galaxies that one might expect.

In sum, over the next few billion years most galaxies will be farther apart, the particle horizon farther away, the background radiation dimmer, cosmological standard candles and pulsars rarer, and quasars fainter. Fewer samples preserving the original ratios of the light elements will be available. Most profoundly, once the effects of the event horizon start to become apparent in about a couple of Hubble times (a Hubble time is roughly the present age of the universe), some key diagnostics about the universe will slowly disappear. Observers living in the far future would enjoy a more distant particle horizon, but at the cost of most of the astrophysical tools used today. We already have access to plenty of the maximum theoretical number of radiant sources—about 40 percent.

All told, we're living in the best overall age of the universe to do cosmology. In the deep past or far-flung future, many discoveries made in the past century, such as the big bang itself, could not be made.[46] Of course, if we're correct about the cosmic habitable age, then it's unlikely that observers could exist at such a time, so they would be spared the burden of their ignorance.[47]

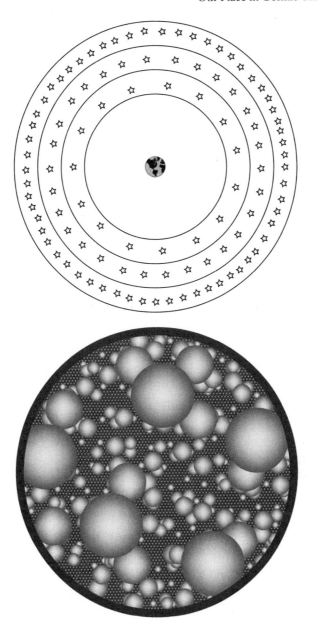

Figure 9.7: (top) Although Kepler had first described it, Olbers wrote more clearly on the paradox of the darkness of the night sky. Consider a spherical shell of stars around the Earth. Compared to a closer, equally thick shell, each star in a more distant shell will be fainter, but it will contain more stars, so each successive shell will be as bright as the interior shell. Therefore, if the universe is infinite in age and extent, why is the night sky so dark? (bottom) Hypothetical high-powered telescopic view in an infinite universe. In such a universe every line of sight will encounter the surface of a star. Every patch of the sky will be as bright as the surface of the Sun. Such a universe would be hostile to life and scientific discovery.

PARADOX SOLVED

In 1826, the German astronomer Heinrich Wilhelm Mathäus Olbers gazed at the heavens and asked a deceptively modest question: "Why is the night sky dark?"[48] Generations of astronomers learned to call his question "Olbers' paradox." The dark sky seemed to be a paradox because an eternal universe uniformly filled with an infinite number of stars—just the universe assumed by many scientists—should produce a uniform and intensely bright sky, day and night.

Many proposed solutions to the paradox failed. For instance, some claimed that intervening gas and dust would block much of the starlight. But, with enough time, even such inert matter would begin to glow hot and bright with the energy absorbed from an infinite swarm of stars.

The modern understanding of a universe with a finite past, initiated by Hubble's and Lemaître's discoveries early in the last century, has allowed us to solve this long-standing conundrum. There's no paradox if

Figure 9.8: Heinrich Wilhelm Mathäus Olbers (1758–1840).

the universe is neither eternal nor infinite in the requisite sense. In fact, a dark night sky is itself evidence for a beginning.

Besides answering a popular scientific riddle, this fact clearly matters for both life and discovery. Imagine the obstacles posed by a sky as bright as the surface of the Sun. It's precisely because the universe is *not* infinite and eternal that we can discover so much about it, despite its vastness. We can distinguish the variety of information transmitted to us from the heavens, and lest we forget the obvious: life in a static and eternal universe bathed in intense radiation would be unlikely to prosper.

The discovery of a changing universe with a finite past astonished many scientists at the dawn of the twentieth century. It transformed how we see not only the universe but also our very existence.

CHAPTER 10

A UNIVERSE FINE-TUNED FOR LIFE AND DISCOVERY

There is for me powerful evidence that there is something going on behind it all. . . . It seems as though somebody has fine-tuned nature's numbers to make the Universe. . . . The impression of design is overwhelming.

Paul Davies[1]

Imagine you're kidnapped by "Q" from *Star Trek*, a member of a group of uber-intelligent and obnoxious aliens who can manipulate space and time without breaking a sweat. For kicks, Q transports you back to the big bang and takes you into a spacious room with a large console on one side, adorned with scores of dials like those on a Master padlock. Walking up to it, you notice that every knob is inscribed with numbered lines. Above each knob is a title like "Gravitational Force," "Weak Nuclear Force," or "Electron Mass."

You ask Q what the machine is, and after some snide comments about the feebleness of the human mind, he tells you that it's a Universe-Creating Machine. According to Q, the great collective Q continuum used it to create our universe. The machine has a viewing screen that allows the Q to preview what different settings will produce before they press Start. Q explains that the dials must all be set precisely, or the machine will spit out a worthless piece of junk, as shown on its preview screen, like a universe

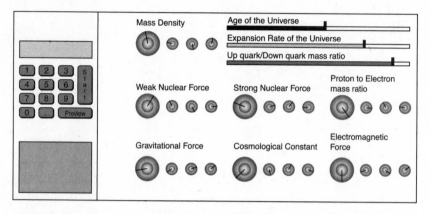

Figure 10.1: Imagine the fine-tuning in terms of a Universe-Creating Machine with many dials, each of which sets the value of a fundamental law, constant, or initial condition. Only if each dial is set to the right combination will the machine produce a habitable universe. Most of the settings produce uninhabitable universes.

that collapses within a few seconds into a single black hole or exists interminably as a lifeless hydrogenated soup.

"How precise do the knobs have to be set?" you ask.

After some uncharacteristic hemming and hawing, Q tells you that, so far, they've found only one combination that produces a universe even mildly habitable—namely, our own.

"So," you ask, "do you mean that there are only two habitable universes, the one you're from, and ours that you created?"

"Uhm, no," Q says, busying himself with a pointless inspection of the knobs. "There's just this one."

"Wait a minute," you say. "What sort of bootstrapping magic allowed you to create the universe you yourself exist in?"

Q glances around the room, pulls at his collar, clears his throat. "Well, we didn't really find the right combination ourselves. The machine, it doesn't exactly belong to us. We found it . . . with the dials already set. The machine had done its work before we arrived. Ever since then, we've been looking for another set of dial combinations to create another habitable universe; but the infernal thing! So far, we haven't found one. We're certain that other habitable universes are possible, though, so we're still looking."

This fanciful story illustrates a very real and very startling discovery made by modern cosmologists: the universe's properties, as described by its initial conditions and constants, appear fine-tuned for the existence of life—and in many cases, to an almost unimaginable degree of precision.

Physicists sometimes refer to this fine-tuning as "anthropic coincidences." Take Newton's law of universal gravitation, $F=Gm_1m_2/r^2$. It describes how two bodies interact. G is the constant gravitational force, which has a specific value in our universe. Both the law and G itself are fine-tuned for life, meaning that if they were slightly different, the universe could not contain any kind of life. If gravity were a bit weaker, the expanding universe would have dispersed matter, and galaxies and planets could not form. If it were a bit stronger, the universe would have re-collapsed, retreating into oblivion. In either case, the universe would lack the ordered complexity that organisms require.

Physicists often refer to the value of one force—say, gravity—relative to other forces, like electromagnetism. In this case, the ratio of gravity to electromagnetism must be just so if complex life as we know it is to exist. If we were just to pick the values of these two forces at random, we would almost never find a combination compatible with life or anything like it.

Given the prevailing biases in science since the mid-nineteenth century, discovering that the universe is fine-tuned was a *surprise*. If generated blindly, the settings for the initial conditions and constants for a universe are far more likely to render it uninhabitable. Astronomically more likely.

And yet here we are.

Previous chapters focused on local fine-tuning: those features of our setting within the universe conducive to life and discovery, but most discussions of fine-tuning in physics, astrophysics, and cosmology focus on features of the cosmos. Scientists have filled many pages describing cases of such fine-tuning.[2] Let's consider just a few.

FINE-TUNING IN CHEMISTRY

We can trace the idea of fine-tuning in chemistry to Lawrence Henderson's 1913 work, *The Fitness of the Environment*,[3] which showed how carbon and water suit living organisms, actual or theoretical. Slight changes in their properties would disrupt the fitness of the environment for life, as we noted in chapter 2.[4] They're also vital for science. Consider one of many examples: concentrated mineral ores are easy to mine only because water is a great solvent (as we discussed in chapter 3). Our technology depends on it. That's why chemists have spent most of the last two centuries working with water. If it could not dissolve so many substances, chemistry would have developed at a snail's pace, perhaps never reaching the status of a science. At the same time, life wouldn't be possible for this and other reasons. This and many other features of the chemistry of our universe make life possible while, at the same time, helping chemists and other scientists make a range of astonishing discoveries about the natural world.

CARBON AND OXYGEN, ACT II

Such fine-tuning for both life and discovery goes well beyond chemistry. In 1952–1953, Fred Hoyle discovered one of the most celebrated examples of fine-tuning in physics. He was contemplating how carbon and oxygen could be forged in the hot interiors of red giant stars. Given the amount of carbon and oxygen in the universe, Hoyle predicted that carbon 12 must have a very specific nuclear energy resonance.[5] A nuclear resonance is a range of energies that makes it much easier for a nucleus to interact with another particle—for example, the capture of a proton or a neutron. An energy resonance in a nucleus will speed up reactions if the colliding particles have just the velocities. Resonances tend to be very narrow, so even very slight changes in their properties would lead to enormous changes in the reaction rates. This may seem obscure but think of a wineglass shattering when playing just the right acoustic note. That's a resonance.

The relevant nuclear reactions occur in the stage of a star's life during so-called helium-shell burning. Fortunately for us, many stars have reached this stage, seeding our galaxy with a healthy dose of carbon and oxygen. During this advanced stage, helium nuclei, also called alpha particles, which abound in a star's deep interior, can collide, forming unstable beryllium 8 nuclei. They remain bound just long enough (10^{-16} seconds) to collide with another alpha particle to form carbon 12. However, this won't forge much carbon 12 without a resonance.

It was the lack of a known resonance at the right energy level that led Hoyle to make his famous prediction. Since the universe contains plenty of carbon, Hoyle deduced that such a resonance must exist.

In fact, the abundances of carbon *and* oxygen depend on a few other coincidences, including a lack of a resonance in oxygen at the typical alpha particle energy in stars prevents them from using up their carbon to make oxygen. Thankfully, the closest oxygen energy resonance is just a little bit too low. However, if that were the whole story, the universe would still have squandered most of its oxygen well before any star

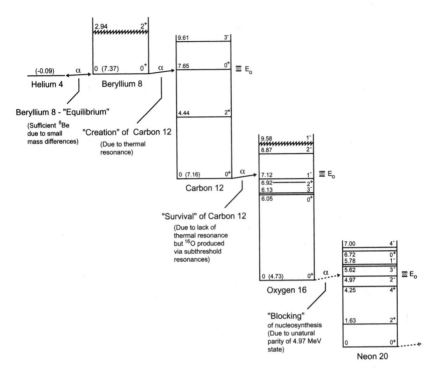

Figure 10.2: The primary nuclear reactions involved in producing nuclei heavier than helium in red giant stars. Four remarkable coincidences give the universe comparable amounts of carbon and oxygen. Energy levels (in MeV) and spin-parity (J^π) are shown for each nucleus. During "helium-burning," light elements are built up by acquiring α particles (helium nuclei), starting with two α particles combining to form the very short-lived beryllium-8 nucleus. A "coincidence" is present at each of the four α-capture steps from beryllium to neon. The most famous is the presence of a nuclear energy resonance in the carbon nucleus at 7.65 MeV above the ground state. This resonance greatly improves the probability that the very short-lived beryllium nucleus will capture an energetic α particle.

system had time even to think about hosting life. Fortunately, there's more to the story. Certain conservation laws prevent easy capture of alpha particles by oxygen 16 to form neon 20, even though a resonance exists in neon 20 at just the right place. Otherwise, little oxygen would remain.

As a result of this nest of coincidences, stars produce carbon and oxygen in comparable amounts. A change in the light quark mass[6] by more than about half a percent or 7.5 percent in the strength of the electromagnetic force (the force between charged particles) would yield a world with either too much carbon compared to oxygen or vice versa, and thus little if any chance for life.[7]

THE REST OF THE FANTASTIC FOUR

In 1913, Henderson singled out the life-friendly properties of carbon and water in chemistry, which involves the electromagnetic force. This is one of the four "fundamental forces." The strengths of these forces—the gravitational, strong nuclear, weak nuclear, and electromagnetic—affect pretty much everything. And like those dials on the Universe-Creating Machine, each one must fall within a narrow range to render a life-friendly universe.

We've covered the electromagnetic force, so let's look briefly at the remaining three.

THE STRONG NUCLEAR FORCE

The strong nuclear force, often called the nuclear force, holds protons and neutrons together in the nuclei of atoms. In such close quarters, it's strong enough to overcome the electromagnetic force and bind the otherwise repulsive, positively charged protons together. It's as short a range as it is strong, extending no farther than atomic nuclei, but despite its short range, changing the strong nuclear force would have wide-ranging effects disastrous for life.

This force plays a leading role in the periodic table of elements. Like carbon and oxygen production in stars, heavier elements depend on the values of the nuclear and electromagnetic forces. Because there are no stable elements with atomic weights of five and eight, carbon 12 can only be built in stars following the pathway described above. Stars build the heavier elements with the ashes of carbon and oxygen. Irrespective of the carbon bottleneck, the periodic table of elements would look different with a changed strong nuclear force.[8] If it were 50 percent weaker, there would be fewer stable elements[9]—less than a third as many.[10] It would lack iron and molybdenum, for instance, which are crucial for life.

Now, the more complex organisms require about twenty-seven chemical elements, while our universe has more than ninety. From this one might infer that our universe has far more than it needs to for life. But in a universe with a shorter periodic table, one or more of the isotopes of the abundant light elements, such as hydrogen, carbon, nitrogen, and oxygen, would likely be radioactive. If any of their main isotopes were even slightly unstable, with half-lives measured in tens of billions of years or longer, the radiation from their decay would pose a serious threat to organisms.[11] In our universe, potassium 40 is probably the most dangerous light radioactive isotope needed for life. As a result, its abundance must be balanced on a razor's edge: enough to help drive a planet's plate tectonics and volcanism but not so much that it harms life on the surface.[12] And, as you have probably guessed, it is balanced on that razor's edge.

If the strong nuclear force were a bit less strong, each element would have fewer stable isotopes. We've already discussed how this diverse mix of elements and their stable isotopes is crucial for scientists exploring the Earth and the universe. So, a large periodic table isn't simply a dirty trick on teenage science students, saddling them with a sprawling array of facts to memorize. It makes life possible while also making the universe much more helpful to science.[13]

THE WEAK FORCE

Several processes relevant to life are especially sensitive to the strength of the weak force.[14] For instance, the weak force governs how protons convert to neutrons and vice versa, and how neutrinos interact with other particles. The event that triggers a massive star to explode as a supernova is the collapse of the core, which itself is triggered by the electrons combining with protons to form neutrons and neutrinos. The weak force mediates this transformation. As the neutronium core collapses, the overlying layers are left without support. Like Wile E. Coyote in the Looney Tunes cartoons who runs off a cliff and hesitates before he starts falling, the overlying atoms accelerate toward the core. The falling matter crashes onto the hard surface of the core and bounces back, like a tennis ball off a brick wall. The bounced matter then collides with the still-infalling layers, creating a shock front. If not for the neutrinos depositing some of their energy on the way out, the shock would peter out and matter would fail to escape. The weak force allows some stars to return their metal-enriched outer layers to the galaxy. Without it, there would not be enough of certain heavy elements for life.

The weak force is also critical for producing primordial helium-4 soon after the big bang, in a cosmic cauldron hot and dense enough for brief nuclear reactions. Slight tweaks in the expansion speed of the cosmos or in nuclear physics would lead to a quite different end. The early big bang produced about 25 percent helium 4 by mass. Although stars have been cycling the interstellar gas for about thirteen billion years, the fraction of helium in the universe has only increased by a few percent. All stars that have ever formed in our universe have started with similar amounts of helium.

Helium stars are like flashbulbs, burning brightly and quickly. Hydrogen stars, in contrast, are more like wax candles. Our hydrogen-burning Sun consumes its nuclear fuel more than one hundred times more slowly than a pure helium star of similar mass. A pure helium star wouldn't last nearly long enough for life to develop on any of its planets, not that it would matter, since a planet orbiting it would have no water or organic compounds. This would rule out life whatever the time scale.

It's not even clear that stars could form from contracting clouds of gas in a universe of pure helium. Unlike hydrogen, helium does not form molecules. They are the best way to cool dense interstellar clouds, which then contract to form stars.

GRAVITY

It's the same story with gravity. Think of its role in stars. A star is in a state of temporary balance between gravity and pressure from hot gas — which, in turn, depends on the electromagnetic force. A star forms from a parcel of gas when gravity overcomes the pressure and turbulence and causes the gas to coalesce and contract. As the gas concentrates, it heats up until the nuclei begin to fuse, releasing radiation, which itself heats the gas.

What would happen if the force of gravity were a million times stronger? "The number of atoms needed to make a star [a gravitationally bound fusion reactor] would be a billion times less . . . in this hypothetical strong-gravity world," explains Martin Rees, Britain's Astronomer Royale, "stellar lifetimes would be a million times shorter. Instead of living for ten billion years, a typical star would live for about 10,000 years. A mini-Sun would burn faster and would have exhausted its energy before even the first steps in organic evolution had got under way."[15] Such a star would be about one-thousandth as bright, three times as hot on the surface, and one-twentieth as dense as the Sun. For life, such a mini-Sun would be a mere flash in the pan.

Most stars transport energy from their interiors through radiation and convection. Think of the Sun's warmth on your skin on a clear day. That's radiative energy transport. Now, think about boiling water in a pan on your stovetop. That's convective energy transport. In the Sun, radiation transports energy most of the way, but convection mostly takes over for its outer 20 percent. Cosmologist Brandon Carter first noticed the coincidence that midrange-mass stars are near the dividing line between convective and radiative energy transport. This dividing line is a razor's edge, a teetering balance between gravity and electromagnetism.[16] If it were shifted one way or the other, the main-sequence stars would be either all blue or all red. Either way, stars on the main sequence with the Sun's surface temperature and luminosity would be rare or nonexistent.

This would surely be a loss for Martha Stewart and other lovers of yellow. But would such a universe, though less well adorned for the Fourth of July, be less habitable than ours? The neighborhood around red stars would be less welcoming for life, for some reasons we gave in chapter 7. For instance, it would take much longer for oxygen to build up in a planet's atmosphere. A bigger problem if our universe had all red stars would be that the red stars burn all their hydrogen to helium, then helium to carbon, and so on. In our universe, the late stages of massive giant stars develop a layered onion structure resulting from all the previous nuclear burning stages. This process preserves a rich assortment of elements, which they disperse when they go supernova.[17]

Moderately bluer stars might still support life, but very blue stars would not, since they spew too much harmful ultraviolet radiation. We have already shown that the Sun's spectrum is far more useful to science than bluer or redder stars.

What about planets? A stronger gravity for an Earth-mass planet would make everything heavier on the surface. It would also boost the planet's self-compression, making everything even heavier. Martin Rees notes that on such a strong-gravity planet, organisms could not grow very large.[18] Such a planet would also endure more frequent and powerful impacts from asteroids and comets. A planet with less mass could avoid these problems if it had a surface gravity comparable to the Earth's. But a smaller planet would lose its internal heat faster,[19] possibly preventing long-lived plate tectonics and volcanism.

Tinkering with gravity would also change things at larger size scales. For example, the expansion of the universe must be carefully balanced with the deceleration caused by gravity. Too fast and the atoms would disperse before stars and galaxies could form; too slow, and the universe would collapse before stars and galaxies could form.[20] The density fluctuations of the matter in the universe when the cosmic microwave background was formed also must be just so for gravity to coalesce it into galaxies later and for us to be able to detect them.[21] Thus, our ability today to measure the fluctuations in the cosmic microwave background radiation is bound to the habitability of the universe. Had these fluctuations been much smaller, we wouldn't be here.

THE COSMIC TUG-OF-WAR

Recall that in the 1920s astronomers discovered that the farther a galaxy the faster it is moving away from us. The Hubble-Lemaître Law implies a simple linear expansion of the space between the galaxies. Then, in 1998, astronomers found that distant galaxies are not only receding but are accelerating away from us. The simplest explanation is that there is a nonzero cosmological constant in Einstein's general relativity cosmological equation, denoted by the capital form of the Greek letter lambda—Λ. The energy associated with it is called dark energy.

On the largest scales, there's a cosmic tug-of-war between gravity working to pull everything together, and the dark energy working to spread everything out. By observing Type Ia supernovae, astronomers have determined that today the push of dark energy has recently overtaken the pull of gravity at cosmological scales. There's only one special time in the history of the universe when the dark and matter energy densities are the same, and we're living near it. What if the dark energy had become dominant several few billion years earlier? No galaxies. A bit earlier still? No stars.

The leading theory for dark energy is that it results from the energy density of the vacuum. All the various particles have associated fields that contribute to the vacuum energy, and they add up to a value much greater than the observed value—10^{120} times greater! This is why, until the Type Ia supernovae results showed that L is not zero, most cosmologists had assumed that some unknown law of physics required that it be exactly zero (zero being a special but simple value). The effective value of L is the sum of Einstein's L and the vacuum energy from all the fundamental particle fields. Lewis and Barnes commented on the cosmological constant fine-tuning problem:

> It's actually several problems. Each quantum field—electron, quark, photon, neutrino, etc.—adds a ludicrously large contribution to the vacuum energy of the universe. . . . Particle physics probably won't help. Particle physics processes—those described by quantum field theory—depend only on energy *differences*. We can change the absolute values of all the energies in particle physics and all the interactions remain the same. It is only gravity that cares about absolute energies. Thus, particle physics is largely blind to its effect on cosmology, and thereby life.[22]

Physicist Lawrence Krauss describes the consequences of this discovery:

> Our current understanding of gravity and quantum mechanics says that empty space should have about 120 orders of magnitude more energy than the amount we measure it to have. That is 1 with 120 zeroes after it! How to reduce the amount it has by such a huge magnitude, without making it precisely zero, is a complete mystery. Among physicists, this is considered the worst fine-tuning problem in physics.[23]

But this is not even the fine-tuning of the dark energy for life. It turns out that the dark energy density must not be less than -10^{-90} or greater than 10^{-90} (in units of the Planck energy density). In the first case, the universe re-collapses after one second, and in the second case, structure formation ceases after one second.[24] This is thirty orders of magnitude larger than the observed value.

Why does this matter? Over-tuning by such a large factor compared to what life needs is highly improbable from chance alone, even if you grant that we are self-selected from a multiverse from habitability requirements (see chapters 15 and 16). Something else must be going on.

INITIAL ENTROPY

We still haven't mentioned the most impressive instance of fine-tuning — that distinction belongs to the initial entropy of the universe. In everyday parlance, entropy is a measure of the amount of disorder in a closed system, which goes up as energy flows through it. More precisely, the more ways there are of distributing the components or energy "microstates" of a system, the higher the entropy. So, for example, increasing the volume of a gas in a container will increase its entropy because there are more ways of distributing the molecules.

In the case of the universe as a whole, where gravity dominates at large scales, the lowest entropy state would be a perfectly smooth spacetime. The highest entropy corresponds to a black hole. The initial entropy must be small enough so that gravity can cause matter to clump and form structures like galaxies and stars, which still have much lower entropies than black holes.

It turns out that the initial entropy of our universe had to be insanely small to reach the value we observe today. Roger Penrose estimated that the initial entropy of the universe was fine-tuned to one part in 10 raised to the power 10^{123}.[25]

MULTI-TUNING

In some analyses of the fine-tuning of the initial conditions and constants of nature, only one is adjusted at a time to make the problems more tractable. This would correspond to changing one dial at a time on our Universe-Creating Machine, while leaving the other dials unchanged. Even taken individually, each of these examples of fine-tuning is impressive, but in the real universe, the values of all the constants and initial conditions must be satisfied simultaneously to have a life-friendly universe. In addition, the setting for a particular constant must be just right for multiple reasons, making the level of fine-tuning even more challenging.

So, for instance, the strong nuclear force tolerates a narrow range of possible settings that will allow stars to produce about the same amounts of carbon and oxygen. However, an overlapping, though not identical, narrow range of settings is required of it if beryllium 8 is to remain bound at least 10^{-16} seconds, keeping the deuteron bound, to allow a minimum periodic table for life, and to keep the light-abundant isotopes stable. When we vary another force, say electromagnetism, the permitted region for life is smaller still. The range for the nuclear force strength is narrow, as is the electromagnetic force strength. The range within which both forces are satisfied is smaller still, though now it is a two-dimensional plot. It's like having to hit not just the tiny bull's-eye on the far side of the field, but having to hit the tiny slice of the tiny bull's-eye that slightly overlaps with a second bull's-eye beside it. Add the required range for the weak force strength and the bull's-eye becomes smaller still, and so on for the other constants.[26] Add other requirements for life—e.g., the chemistry of water and carbon—and it becomes even smaller. Add additional parameters necessary for advanced and then technological life, and it becomes smaller still.

Imagine a set of equations, each describing a different constraint on the laws of nature to permit life.[27, 28] While physicists can't do the complete calculation, it's clear that it's absurdly unlikely that changing several constants at the same time would leave any point where all the parameters still overlapped and therefore allowed for a habitable universe.[29]

While we think of laws and constants as governing the cosmos, as we've seen, changing one or more of them will in these universal variables have profound consequences on particular objects within the universe. Fiddling with the relative strengths of gravity and electromagnetism, for instance, affects not only the cosmos but also galaxies, stars, and planets.

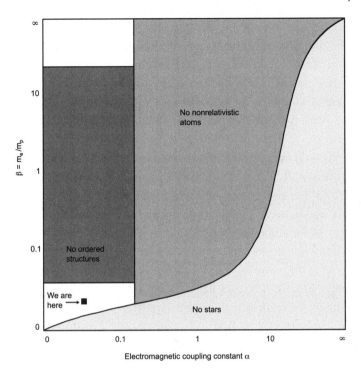

Figure 10.3: Multiple-tuning plot for the electron to proton mass ratio (ß) and the electromagnetic coupling constant (α, also called the fine structure constant). Universes with ordered structures occupy a tiny region. They require that ß be much less than one; otherwise, the nuclei in atoms would have unstable locations. Large values of ß appear consistent with ordered structures, because the electron is assumed to take the place of the nucleus, but this is probably impossible for elements more complex than hydrogen. α must also be much smaller than one to keep electrons in atoms from being relativistic. The third major exclusion region in the diagram shows where stars cannot exist. (The axes are scaled by the arc tangent of the logarithm of ß and α.)

The strong and weak nuclear forces determine the composition of the universe and, thus, the properties of galaxies, stars, planets, molecules, atoms, and people. As a result, we ultimately can't divorce the chemistry of life from planetary geophysics or stellar astrophysics.

FINE-TUNING FOR LIFE AND DISCOVERY

The laws, constants, and initial conditions of the universe must also be finely tuned to allow for scientific discovery. This reaches all the way to matter at the smallest scale. For example, the discrete energy states at the quantum level allows astronomers to extract detailed information from the spectra of bodies that emit light. As we showed in chapter 7, every element and molecule leave a telltale fingerprint in the electro-magnetic spectrum. This is why we can learn so much about the universe from studying the light of stars and galaxies. As the astronomer E. A. Milne once noted,[30] if the laws of nature had produced a non-quantum micro-realm where energy was distributed smoothly rather than bit by bit—like planets in orbit about stars—spectroscopy would have been far less useful.

The fact that every fundamental particle has a universal mass also matters for science. It allows astronomers to apply what they discover in labs on Earth to the most distant and early parts of the universe. Since the elements, and hence life, depend on distinct quantum states and the mass constancy of fundamental particles, it seems that life and discovery are yoked all the way down.

CARBON AND OXYGEN, ACT III

Yes, we know we've already discussed carbon and water—two main characters in the story of life. Both rely on exquisite fine-tuning in physics. And, as we noted above, chemistry and high technology depend on water's role as a solvent. But carbon and oxygen each make the universe much easier to measure. The isotope carbon 14, for instance, helps us date once-living things, since its half-life is conveniently close to the time it takes buried organic matter to decompose. Carbon 14 also testifies to the recent history of the solar system, since it's production in the atmosphere is sensitive to radiation from the galaxy, the Sun's activity cycles, and the Earth's magnetic field. Silicon carbide pre-solar grains in primitive meteorites are the main sources of samples from previous generations of stars.

The information in the grains share only a few features with the spectra of present-day stars. The carbon 12 to carbon 13 isotope ratio is one of the shared features.[31] Without these isotopes, meteorites would be much less useful records of galactic nucleosynthesis history. Finally, carbon atoms in their neutral, ionized and molecular ([C I], [C II], CO) forms, seen in distant intergalactic gas clouds against background quasars, are the best probes of the temperature of the microwave background at various redshifts.[32]

Oxygen also has some endearing qualities for astronomers. An oxygen-rich atmosphere, unlike many alternatives, is transparent to optical light, giving us a clear view of the broader universe. And carbon monoxide is the best tracer of dense molecular gas between stars and around young stars.

DISCOVERABILITY

Paul Davies, perhaps more clearly than anyone, has pointed to the features of our universe that allow us to discover the laws of nature.[33] Discovering a law has much to do with its simplicity. The inverse-square laws of gravity and electric fields helped lead to the early discoveries of the universal Law of Gravitation and Maxwell's Equations. The trek from Kepler to Newton to Einstein was aided, then, not only by our local environment (chapter 6) but also by the fact that gravity takes a simple mathematical form (more on this below). At the same time, the laws of physics are not so simple that they prohibit the variety and complexity needed for life.

For Kepler to formulate his laws, he needed large bodies that could be captured by classical laws that allow for distinct positions and movements. Paul Davies has argued that we should not assume that any universe popping out of a quantum state would later have these properties.[34] "It is amazing," Harlow Shapley once marveled, "what grand thoughts and great speculations we can logically develop on this planet—thoughts about the chemistry of the *whole* universe—when we have such a tiny sample here at hand."[35]

Paul Davies contends that the success of science depends on the locality and linearity of the physical laws. Concerning locality, he writes:

> It is often said that nature is a unity, that the world is an interconnected whole. In one sense this is true. But it is also the case that we can frame a very detailed understanding of individual parts of the whole without needing to know everything. Indeed, science would not be possible at all if we couldn't proceed in bite-sized stages.[36]

In other words, we can discover truths that hold throughout the universe and extrapolate them in both time and space, by studying a representative local sample. If it were otherwise, we would be limited to what we could see directly and locally.

Discovering the laws of nature also depends on their stability.[37] (In fact, the idea of a law of nature implies this.) Likewise, the laws and fundamental constants of nature must remain constant long enough to provide the kind of stability life requires through the building of nested layers of complexity. The most basic physical units we know of, quarks, must remain constant to form larger units, protons and neutrons, which form atoms, and so on, all the way to stars, planets, and in some sense, people. The lower levels of complexity provide the structure and carry the information of life. There is still much we don't know about how the levels relate, but clearly, at each level, structures must remain stable over vast stretches of space and time.[38]

And our universe does not merely contain complex structures; it also contains elaborately nested layers of higher and higher complexity. Think of complex carbon atoms, within still more complex sugars and nucleotides, within more complex DNA molecules, within complex cell nuclei, within complex neurons, within the complex human brain, all of which are integrated in a human body. None of this could hold in either a chaotic universe or a homogeneous one made up of, say, hydrogen atoms or quarks. And surprisingly, our universe allows such higher order complexity alongside quantum indeterminacy and nonlinear interactions—which, unless precisely fine-tuned, would dissolve into chaos.

ARE THE CONSTANTS REALLY . . . CONSTANT?

But are the physical laws and constants truly constant? To date, cosmologists and physicists have only set tight upper limits on possible variations in the fine structure constant, the gravitational constant, the speed of light, and the electron-to-proton mass ratio.[39] These limits are many orders of magnitude more stringent than life requires. For science, that's a good thing; since we can apply the physical laws, we discover locally to the rest of the universe.

In these two ways the constancy of the constants is like the very low initial entropy of the universe. The volume of low-entropy space we observe is much greater than it needs to be for life. Roger Penrose famously calculated that if we need a low entropy region to be only 10 percent of the universe for life's needs, then the chance of the rest of the universe having as low an entropy is 1 part in $10^{10^{123}}$.[40] In fact, Penrose was being very generous. At most, life only needs a low entropy bubble big enough to contain a galaxy the size of the Milky Way; the universe of galaxies we see around us is extravagantly larger than it needs to be for us to exist.

"In theory at least," observes philosopher of science Stephen Meyer, "we could have lived in a universe in which only our local galactic environment exhibited such low entropy. The rest of the universe could have been characterized by . . . vast spaces filled with black holes . . . Since the rest of the universe is not dominated by black holes, we can observe and investigate it and learn about its history by observing other galaxies and stars beyond our Milky Way."[41]

Philosopher Robin Collins considers the low entropy of the observable universe the premier example of fine-tuning for discovery:

> The existence of ECAs [embodied conscious agents] only requires a relatively small region of order, and hence only a relatively small region of low entropy in the universe. Yet, low entropy is required wherever there are stars and galaxies. Consequently, the universe must have relatively low entropy over vaster regions than what is required for life in order for ECAs to be able to observe other galaxies. The latter is in turn crucial for discovery in astrophysics and cosmology—for example, without other galaxies we would not know that the universe came into existence via a big bang. So, although this large-scale low entropy of the universe is often cited as a case of fine-tuning for life, it is really a case of extreme fine-tuning for discovery.[42]

Of course, although nature's laws are stable, simple, and linear, they do take more complicated forms. However, they usually do so only in those regions of the universe far removed from our everyday experiences: deep inside atoms, in the cores of stars, or near a black hole, where even angels fear to tread.

Yet even in these exotic locations, nature is still a helpful tutor. In a complex field, such as quantum mechanics, for instance, we can describe a lot with the simple Schrödinger Equation. Eugene Wigner famously spoke of the "unreasonable effectiveness of mathematics in natural science."[43] It's unreasonable only if we assume that the universe is not the product of reason.

HIERARCHICAL CLUSTERING

The delicate balance of the forces and constants determines how matter is distributed. Matter is spread out in a handy balance between uniformity and diversity, homogeneity and clumpiness. "Why does our universe have the overall uniformity that makes cosmology tractable," Martin Rees asks in his book *Just Six Numbers*, "while nonetheless allowing the formation of galaxies, clusters and superclusters?"[44]

This pattern is another nonnegotiable for both life and discovery. If matter were evenly distributed, which might have happened if space had expanded faster early on, then no galaxies, stars, planets, and rocks would have formed—and so no life. The situation would be no sunnier at the other extreme, if an imbalance of forces caused all matter to clump together into black holes.

No life, but also no discovery even if there were life. The cosmological distance ladder we described earlier works because matter has just the right amount of clumpiness over many size scales. This hierarchical clustering of matter favors objects on size scales of moons, planets, stars, star clusters, satellite galaxies, galaxies, and clusters of galaxies. These are the rungs on the ladder that have allowed astronomers to climb from the Earth's surface to the edge of the visible universe. If matter did not clump on size scales between small planets and galaxies, there would be no stars, and the universe would be far less friendly not just to life but to science.

A universe with clustered matter is also more transparent than a less clustered one. Consider the humble comet. Why does a comet like Hale-Bopp appear so bright in the night sky? The solid nucleus of a typical comet may be only a few tens of kilometers across, but a comet brightens as it approaches the Sun because parts of it disintegrate into myriad tiny dust particles, which reflect far more light. If the universe were a similarly diffuse sea of solid particles, radiation would have trouble making it even one light-year before it got hassled to death in the cosmic dust bowl.

In such a dusty, nebular universe, it would be tough to measure the distance of anything. As we noted in chapter 7, astronomers can measure the positions of stars in part because they are much smaller than the distances that separate them. This gives us a striking contrast in the night sky. The diffuse nebulae in galaxies provide much less contrast. If we had found ourselves inside such a nebula, or in a nebular universe, the sky would be a confused mess.

Clumping helped Kepler derive his simple laws of planetary motion, because a planet's mass is typically only a small fraction of its star's mass. When Newton derived his Law of Gravity a few decades after Kepler formulated his laws, he found imperfections in Kepler's models of our solar system. Kepler had assumed that the Sun is exactly at one focus of an ellipse. In fact, it wobbles in reaction to the bodies around it. He also assumed, incorrectly, that the square of the orbital period of a planet is exactly proportional to the cube of its semimajor axis. Fortunately, the planets' masses are low enough, and they are far enough apart, that their mutual gravity is slight. As a result, Kepler's estimates, and his elegant and discoverable laws of planetary motion, were good enough to neatly explain Tycho Brahe's naked-eye observations. Thanks to the hierarchical clustering of objects in our solar system, Kepler's equations were good enough to allow astronomy to progress until these minor complications could be dealt with, and then, happily, resolved with Newton's Laws. If our universe didn't clump most of its matter in a series of generously separated size scales, then the size difference between stars and planets probably would have been too small for Kepler to succeed.

This hierarchical clustering has also allowed physicists to distinguish the fundamental forces. Robin Collins argues — following physicist Anthony Zee — that it's only because these four forces — gravity, electromagnetism, the weak force, and the strong nuclear force — vary widely in strengths and scales of effect that we can disentangle and study one force at a time. For instance, in studying objects at the scale from planets to galaxies, the dominant force is by far gravity.[45] At the level of chemistry and atomic structure, however, scientists can focus exclusively on electromagnetism. Zee notes that because of this "we can learn about Nature in increments. We can understand the atom without understanding the atomic nucleus. . . . Physical reality does not have to be understood all at once."[46]

HIERARCHY AND SIMPLICITY

One final example. We discussed the importance of size scales between quarks and the universe as a whole, with each step along the scale functioning as a crucial rung in the ladder of discovery. The same holds for laws at different levels of depth and precision, which allow scientists to ascend over time to new, more complex conceptual levels. Think of the shift from Kepler's laws of planetary motion to Newton's theory of gravity to Einstein's general relativity. The latter builds on the former. We have already looked at the move from Kepler's laws to Newton's theory of gravity. Consider now the step from Newton to Einstein.

Newton's picture of the world is quite different from Einstein's. At bottom, Newton describes particles tugging at each other over distances. Einstein describes curved space-time. We can translate each theory into the other — at the cost of simplicity. But both laws — Einstein's and Newton's — are simple within their own frameworks. This is just an instance of a wider pattern. For some reason, the ultimate laws of physics can be apprehended, even if not exhausted, by simple laws at each conceptual level, even for those that are later judged inadequate, such as Newtonian space-time. As a result, each conceptual level serves as a rung to get us to the next level. If the theoretical laws could not be simple and yet relatively precise at each level, we would likely fail to discover them for that level, and hence could not progress from level to level toward the more encompassing laws of physics.

Robin Collins argues that general relativity may have been beyond our ken if the Newtonian theory of gravity had not already been in place; even as it was, developing general relativity took a true act of genius. Developing Newton's law of gravity also demanded an act of genius and required not only that the laws of gravity be simple,[47] but that Newton's law reduce to simple rules of planetary motion—namely Kepler's three laws. Even with simple laws of planetary motion, it took Kepler fifteen years of trial and error to discover them. The universe is an excellent tutor. It has not been so demanding as to ensure failure, but instead has presented us with worthy challenges that we can meet with diligence.

OUR PLACE IN THE COSMOS

So, the physical universe has a hierarchy of levels of complexity and diverse sizes of clumps within it. Of course, we're the tiny clumps that we care most about. And no discussion of the cosmos is complete without a reproachful sermon about our puny size compared to the universe. Scientists and science writers often imply that our size is a mark of disrepute. Astronomer Stuart Clark probably spoke for many when he said, "Astronomy leads us to believe that the Universe is so vast that we, on planet Earth, are nothing more than an insignificant mote."[48]

Actually, over the extreme range of size scales from quarks to the observable universe, the range from humans to the Earth is smack in the middle on a logarithmic scale. Why does that matter? As it happens, our middling size maximizes what we can observe, both large and small.

If we were only a few orders of magnitude smaller—such as the size of an ant—the realm beyond the Earth's surface might be largely out of view. The size of the pupil determines the eye's maximum resolution. A creature with an eye 0.1 millimeters in diameter could not quite resolve the full Moon.

Humans, in contrast, can discern details sixty times finer than that. Astronomers can greatly improve this limit by building telescopes. As with the pupil, the diameter of a telescope's primary light collecting lens or mirror, called an objective, determines how well it can resolve. Our size makes it easy for us to grind objectives about 12 inches in diameter. Amateur astronomers often make mirrors this size in their garages. The resolution provided by a 12-inch mirror is over one hundred times

Plate 1: The famous *Earthrise* photo taken by Bill Anders on December 24, 1968. Anders took the picture with the Moon situated to the right since he understood them to be orbiting the Moon at its equator. Moments later, Frank Borman took a black-and-white photograph with the lunar horizon at the bottom. Popular reprints of the color Anders photo usually turn it 90 degrees so that it is oriented like the Borman photo. The image has become a moving if ambiguous symbol of Earth's place in the universe.

Plate 2: Earth as a "Pale Blue Dot." *Voyager 1* took this image in 1990 while looking back at the solar system from a distance of some four billion miles. Earth is immersed in one of the rays of sunlight scattered in the camera's optics, resulting from the small angle between Earth and Sun.

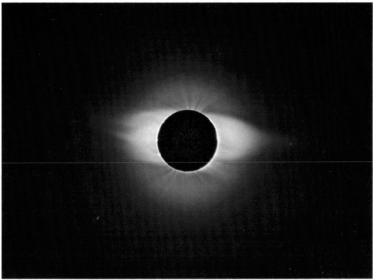

Plate 3: Two total solar eclipses. The first is a photo of the October 24, 1995, total solar eclipse from Neem Ka Thana, India. Six photos taken over a range of shutter speeds were combined to reveal detail in the bright and faint regions of the Sun's atmosphere. Notice the pink chromosphere hugging the dark lunar disk. The long coronal streamers were plainly visible to the unaided eye. The second is a composite image of ten pictures of the June 21, 2001, eclipse, taken from Chisamba, Zambia. The less flattened appearance of the corona compared to the earlier eclipse is caused by the changing solar magnetic field over the sunspot cycle.

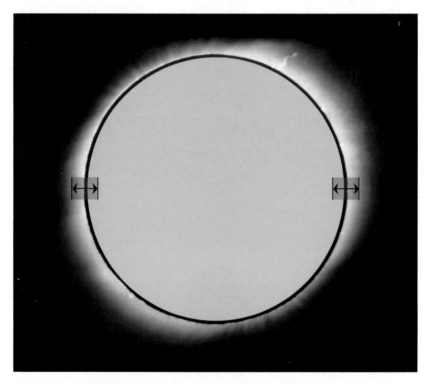

Plate 4: Colorized image of the October 24, 1995, total solar eclipse. The yellow disk, representing the occulted solar photosphere, is superimposed over the image. The red markers indicate the range of apparent sizes of the Moon. Notice the black outer edge of the Moon's disk, which just covers the Sun's bright photosphere, revealing the thin chromosphere and prominences.

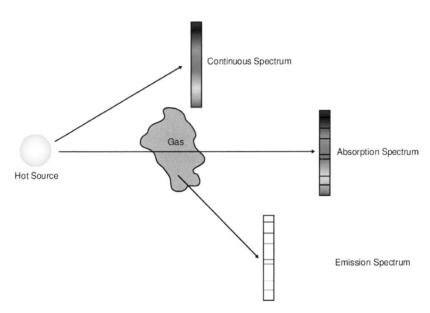

Continuous Spectrum

Gas

Hot Source

Absorption Spectrum

Emission Spectrum

Plate 5: Gustav Kirchhoff first recognized the basic types of spectra. A hot dense object emits a continuous spectrum. If a cooler, low-density gas is placed between a continuous source and the observer, atoms in the gas remove light at specific wavelengths, producing an absorption spectrum. The same gas observed from a different direction (without the hot source behind it) produces an emission spectrum.

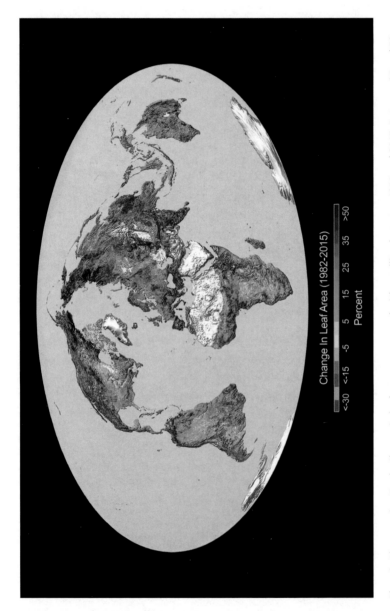

Change In Leaf Area (1982-2015)

<30   <-15   -5   5   15   25   35   >50
Percent

Plate 6: The greening Earth. Change in leaf area as measured with satellites between 1982 and 2015. Between 25 and 50 percent of the vegetated land experienced significant greening during this time. The Earth has literally become greener, mostly from the increased carbon dioxide in the atmosphere.

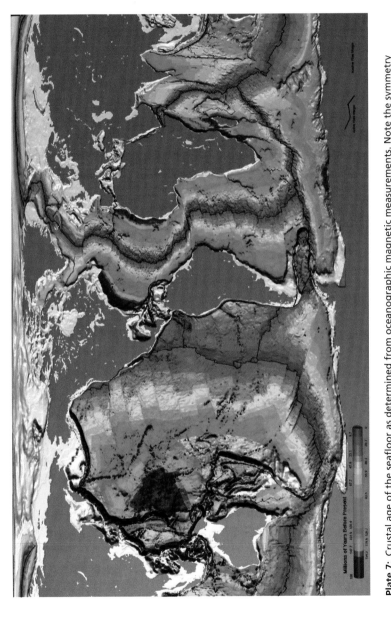

Plate 7: Crustal age of the seafloor as determined from oceanographic magnetic measurements. Note the symmetry of the ages on opposite sides of the mid-ocean ridges. This map was constructed mostly from oceanographic magnetic measurements.

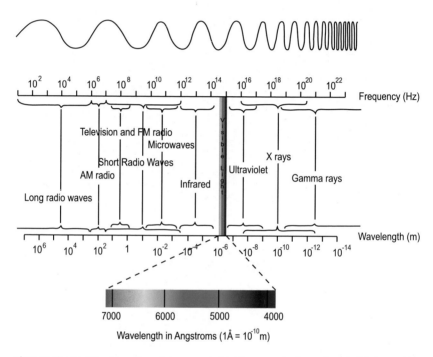

Plate 8: The Earth's atmosphere is transparent to the narrow sliver of optical light and to radio waves. Shorter wavelengths are visible as blue and longer ones as red. (This image is called a "spectrum.")

Plate 9: The Earth's transparent atmosphere allows us to see objects beyond Earth with remarkable clarity. The atmosphere is also transparent to the radio region of the electromagnetic spectrum. A radio telescope antenna at Fick Observatory in central Iowa is shown against the backdrop of the northern constellations.

Plate 10: Hale-Bopp as viewed from the Washington Cascade mountains April 1, 1997. Hale-Bopp was a large comet, which would have caused mass extinctions if it had hit the Earth.

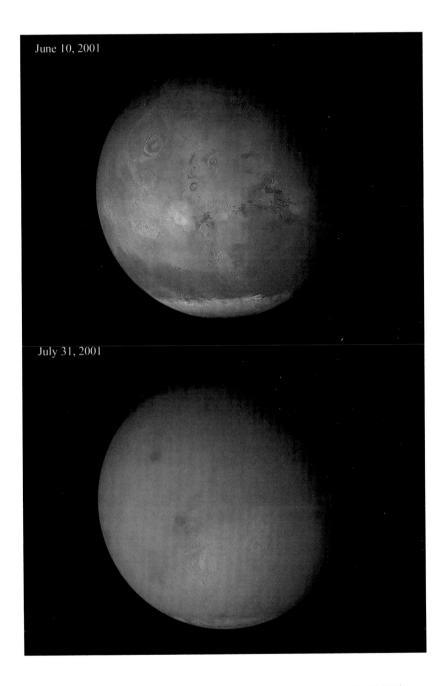

June 10, 2001

July 31, 2001

Plate 11: Mars Global Surveyor images of Mars just before a global dust storm (top) and shortly after it completely engulfed the planet (bottom). The storm continued to intensify through September. A single dust storm can hide Mars's face from our view. A planet-wide Martian dust storm prevented the Mariner 9 probe from mapping its surface for a few days in 1973. An even stronger dust storm greeted the Mars Odyssey probe upon its arrival in October 2001.

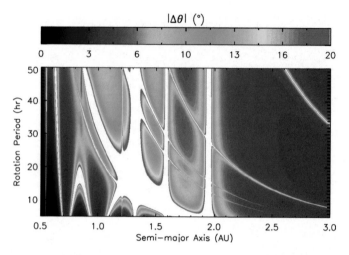

Plate 12: The range over which a planet's obliquity (axis tilt) varies depends on several factors, including the planet's internal structure, rotation period, and distance from its host star, as well as the locations of other planets, and the presence of a large moon. Shown in the figure are the results of numerical experiments on the expected obliquity variations of Mars with various combinations of rotation period and distance from the Sun; Mars's actual period is 24.6 hours and is 1.52 AUs from the Sun. White areas correspond to variations greater than 20 degrees. Stability comparable to Earth's (about 2 degrees) is rare for a Mars twin inside about 2.5 AUs.

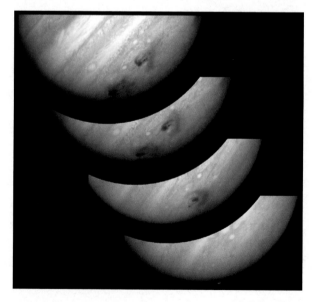

Plate 13: Jupiter takes some hits for the solar system. This composite shows several impacts of comet Shoemaker-Levy 9 on Jupiter. The images from lower right to upper left show: the impact plume July 18, 1994 (about five minutes after impact); the fresh impact site 1.5 hours after impact; the impact site three days after the first G impact and 1.3 days after the L impact; and further change of the G and L sites from winds and an additional impact (S) in the G vicinity, 5 days after the G impact. Within a few years, all traces of the impacts had disappeared.

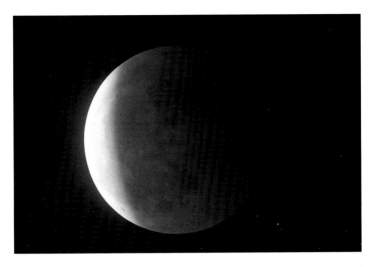

Plate 14: Partial phase of the total lunar eclipse of May 15, 2003. Notice the curved shape of the Earth's shadow. Aristotle cited observations of lunar eclipses as evidence that the Earth is round.

Vega

Arcturus

Antares

Plate 15: Low-resolution spectra of three bright stars. Antares is a very cool red supergiant. Molecules in its atmosphere produce the broad dark bands in its spectrum. The warmer red giant star, Arcturus, is slightly metal-poor. Absorption lines are faintly visible in its spectrum. Vega is a hot main sequence star (a dwarf). Hydrogen atoms in its atmosphere cause the prominent absorption lines in its spectrum.

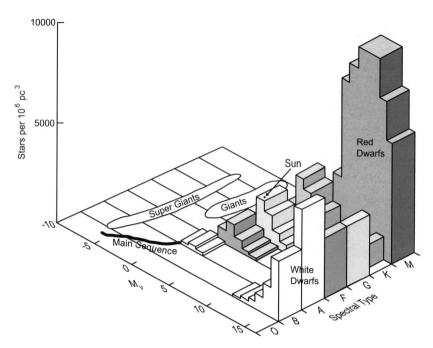

Plate 16: Space densities of nearby stars over a range of spectral types. Most stars are dim red dwarfs, which are unlikely to support complex life. The Sun is one of the relatively rare luminous main sequence stars.

Plate 17: Hubble Space Telescope image of the aligned galaxies NGC 2207 and IC 2163. This and other similar pairs of spiral galaxies have allowed astronomers to verify that spiral arms are very dusty, while interarm regions are transparent in the outer disk. But even the interarm regions become very dusty toward the center of the galaxy.

Plate 18: The galactic habitable zone. The inner disk is more dangerous, and the outer disk lacks enough heavy elements to build Earth-size planets. The circumstellar habitable zone is shown in the lower left.

Hubble Deep Field
Hubble Space Telescope · WFPC2

PRC96-01a · ST ScI OPO · January 15, 1995 · R. Williams (ST ScI), NASA

Plate 19: The first and most memorable "Hubble Deep Field," imaged by the Hubble Space Telescope in 1995. This image reveals myriad distant galaxies as they existed billions of years in the past.

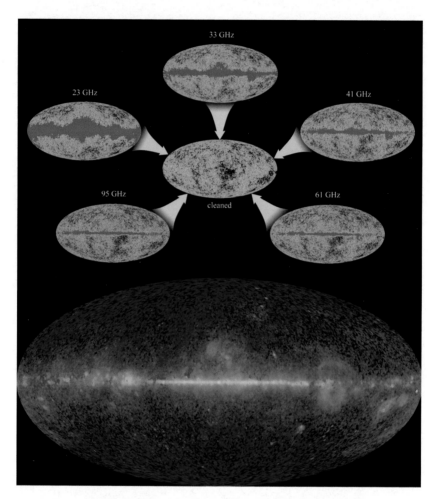

Plate 20: (top) Full-sky maps at five frequencies in the microwave region of the spectrum (red is bright, blue is faint). The horizontal band running across each map is the galactic plane. The final "cleaned" map with the foreground contaminants removed is shown in the center. The remaining variations in brightness are due to the cosmic background radiation. (bottom) The 41 GHz map with the galactic foreground contaminants color-coded: synchrotron radiation is red, free-free radiation (interacting charged particles) is green, and thermal dust is blue.

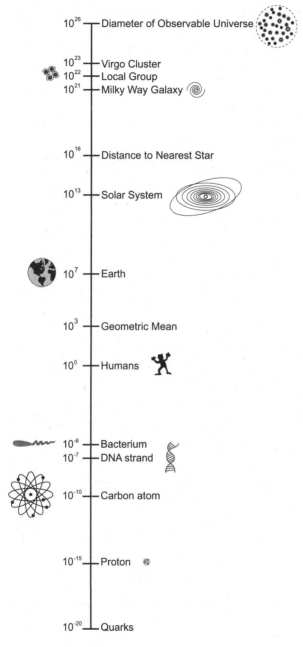

Figure 10.4: Size scales in the universe. Although Earth and human beings are often considered "insignificant" compared to the vastness of the cosmos, the Earth-human size range actually includes the geometric mean of extremes of sizes in the universe. We are near the middle of the size scale ranging from quarks to the observable universe, when figured on a logarithmic scale (in powers of ten). For scientific discovery of both large- and small-scale structures, this midway size is near optimum. (Approximate sizes and distances are indicated in meters.)

better than the naked eye. That's comparable to the blurring effect of the Earth's atmosphere on distant celestial sources. A skilled ant-sized being might construct a telescope objective maybe half a millimeter in diameter, which would be just enough resolution to observe the Moon, Sun, and maybe a few bright planets and stars. But the many fainter stars would be forever beyond its view.

Ironically, a tiny being would also have a hard time detecting the smallest building blocks of matter. Physics labs built to measure matter at the smallest scales tend to be about a hundred times larger than we are. That's because it takes big machines using a lot of energy to do the job—namely, accelerating particles with giant electromagnets to near the speed of light.

We're near the upper end of the animal size scale. Larger animals tend to be less dexterous than we are, but without gaining a critical edge in vision. And the largest ones live in the oceans, where they can enjoy the extra support from water—which is hardly ideal for viewing the distant universe.

Of course, we can always imagine some larger ethereal life-form. Think of the ghostly electromagnetic beings in Fred Hoyle's *The Black Cloud* or the imaginary floaters in Jupiter's atmosphere from Carl Sagan's *Cosmos*. Even if such creatures could exist, however, they could not control their environment as we can.

In his books *Nature's Destiny*, *Fire-Maker*, and *The Miracle of Man*, biologist Michael Denton describes how our size is well adapted for technology—especially for controlling fire. If we were ant-sized, or smaller, fire might be out of our reach. Note that fires never seem to be smaller than a few centimeters. Fire is a great, simple source of energy because it's hot but nonexplosive. That's because hydrocarbons and oxygen, in turn, are relatively inert. So, both life and fire—a crucial step for high technology—need an environment with plenty of these two elements.[49]

Over the last ten chapters, we've moved from Earth's geophysics and atmosphere to the beginning of cosmic time and the forces and constants that apply throughout the universe. Over and over, we've seen a pattern: the rare conditions required for life also provide excellent overall conditions for discovering the universe around us. At some point, this pattern should lead us to re-evaluate contrary assumptions about the universe and our purpose (or rather, our supposed lack of purpose) on this pale blue dot.

So, let's get to it.

SECTION 3

IMPLICATIONS

CHAPTER 11

THE REVISIONIST HISTORY OF THE COPERNICAN REVOLUTION

The great Copernican cliché is premised upon an uncritical equation of geocentrism with anthropocentrism.

Dennis Danielson[1]

B y now, the result of the preceding should be clear: Our local setting is rare and is far better suited for both life and scientific discovery than any other known place. Further, in our universe these two features go together. That is, those rare places best fit for complex and intelligent observers also provide the best overall settings for making diverse and wide-ranging scientific discoveries. We will argue later that this pattern has profound meaning.

Standing in our way, however, is the popular claim that our place in the cosmos is wholly unremarkable. The partisans of this view claim it's rooted in the spirit of modern science, in the revolutionary work of Galileo, Kepler, and, especially, Copernicus. They also lay claim to the four hundred years of scientific discovery that have followed. In truth, however, this so-called Copernican principle has a doubtful pedigree. Indeed, Copernicus himself rejected it, and for good reason.

THE OFFICIAL STORY

The revisionist history propping up the misnamed Copernican principle is, like all great myths, a mixture of truth and falsehood rather than unalloyed fiction. One of its champions, Nathan Myrhvold, distills it in a *Slate* article, aptly titled "Mars to Humanity: Get Over Yourself":

> Ptolemy (second century) was the first and boldest in a long succession of spin doctors for the primacy of human beings. The whole universe, he postulated, rotated around us, with the Earth sitting at the center of Heaven itself. Any marketing consultant will tell you that positioning is everything, and center-of-the-universe is hard to beat. A Polish astronomer named Copernicus (1473–1543) rudely pointed out: Sorry, Earthlings, we spin around the Sun, not vice versa. . . . Giordano Bruno, a sort of sixteenth-century Carl Sagan, popularized these concepts . . . saying, among other things, that "innumerable suns exist. Innumerable earths revolve around those suns. Living beings inhabit these worlds. . . ." Bruno's crime, like Galileo's, was to undermine the uniqueness of our planet, and by doing so, to threaten the intellectual security of the religious dictatorships of his time. . . . Over time, advances in astronomy have relentlessly reinforced the utter insignificance of Earth on a celestial scale.[2]

Myrhvold's picture of duplicitous spin doctors is tendentious. But many still regard his description as basically right. After all, ancient superstition put Earth at the center of a small, human-centered universe. The benighted masses thought the Earth was flat,[3] while the educated elites, following Ptolemy and Aristotle, knew that it was a sphere. They wrongly assumed, however, that not just the Moon, but also the Sun, planets, and stars revolved around it.

Copernicus, we're told, demoted us by showing that ours was a sun-centered universe, with the Earth both spinning on its axis and revolving around the Sun like the other planets. "If Copernicus taught us the lesson that we are not at the center of things," claims Robert Kirshner, "our present picture of the universe rubs it in."[4]

Bruce Jakosky, in *The Search for Life on Other Planets*, further informs us, "Because of this tremendous change in world view, Copernicus' views were not embraced by the Church: the history of his persecution is well known."[5] Never mind that Copernicus wasn't persecuted but died the same year his book was published (1543), not at the oil-soaked stake, but peacefully and of natural causes. These facts muddy the popcorn movie simplicity of the Official Story, with its cast of intrepid, steely-eyed scientific heroes on the one hand, and its one-dimensional villain priests on the other.

The popcorn movie continues from Copernicus's persecution with a bravura medley of fact and fiction: the savior Copernicus leaves his even less fortunate disciples, like Bruno, the first martyr, and Galileo, the first saint, to suffer more hideous ends. In time, however, the brave and unflagging march of science overwhelms the darkness and idiocy of religion—swelling and triumphant musical score followed by cheers and the film's credits. The test audience loves it; everyone goes home smug in the knowledge that modern man is superior to the superstitious fools of a dead and defeated past.

The Copernican Revolution, we're led to believe, was the opening battle in the ongoing war between Science and Religion, with Copernicus as the enduring symbol of science's unflinching fidelity to the facts.

Figure 11.1: This famous image is often mistaken for a seventeenth-century woodcut. In fact, its earliest known appearance is in Camille Flammarion's *L'Atmosphere: Météorologie Populaire* (Paris, 1888), 163. Flammarion and many others have used it to claim that the medievals believed the Earth was flat.

Textbooks and science writers on the subject display varying degrees of reductiveness and aversion to detail. But with a few exceptions, the central message is the same: Religious superstition maintained the myth that we are the cherished center of the universe. Modern science rescued us from this delusion. As astronomer Stuart Clark puts it, "Astronomy leads us to believe that the Universe is so vast that we, on planet Earth, are nothing more than an insignificant mote."[6]

Strangely, some even cast the Copernican Revolution in the role of moral teacher. It "will not have done its work," philosopher Bertrand Russell once said, "until it has taught men more modesty than is to be found among those who think Man sufficient evidence of Cosmic Purpose."[7]

The subtext, of course, is that one can be scientific only to the extent that one is nonreligious, and vice versa. To be religious, in the sense intended here, is to believe that there is something special or intentional about the world and our place in it. Science, on this view, has a special definition as well. Rather than a search for the truth about nature based on evidence, systematic study, and the like, science becomes applied materialism. That is, the conviction that the material world is all there is, and that chance and blind necessity explain, indeed must explain, everything.

Toward this end, the official storyline comes close to reversing the key historical points. Prominent in that storyline is the link between our "central" location and our status in the scheme of things. As planetary scientist Stuart Ross Taylor puts it:

> Copernicus was right after all. The idea that the Sun, rather than the Earth, was at the centre of the universe caused a profound change in the view of our place in the world. It created the philosophical climate in which we live. It is not clear that everyone has come to grips with the idea, for we still cherish the idea that we are special and that the entire universe was designed for us.[8]

The spin on the meaning of the cosmic "center," however compelling some may find it, is deeply misleading. Historians of science have protested it for decades; but so far, their protests have not trickled down to the masses or the textbook writers.

Given a chance to speak, the protesting historians will explain that the ancient cosmology was a sophisticated synthesis. It drew on the physics of the Greek philosopher Aristotle (384–322 BC) and the careful observations and mathematical models of the Greek Egyptian Ptolemy (circa 100–175 AD) and other astronomers. The universe, they thought, was a set of nested, concentric spheres that encircled our spherical, terrestrial globe. On each sphere was affixed a different celestial body. This model wasn't far-fetched. It nicely explained the movement of the Moon, Sun, planets, and background stars when no one had telescopes. The crystalline spheres were thought to connect so that the movement of the outer, stellar sphere of the stars moved the inner spheres that housed the planets, Sun, and Moon.

This model explained the east-to-west movements of the Sun across its daily path and of Moon at night, the celestial sphere encircling the poles, and the perplexing paths of the known planets. It explained the seeming stability of the Earth itself.

Figure 11.2: This cosmic section from Peter Apian's popular astronomy textbook (1540), shows the Aristotelian/Ptolemaic world picture that most medieval people (at least those with even a modicum of formal education) took for granted: a spherical Earth in the undignified middle, surrounded by elemental spheres and spheres for each of the planets, Sun, Moon, fixed stars and other astronomical periods, and finally, as Apian puts it, the "empyrean heaven or habitation of God and of all the elect."

It was also backed up by plausible arguments. For example, if the Earth were moving, one would expect a stiff east wind, and that an arrow shot straight in the air would come down west of the archer. The pre-Copernican cosmos may appear naive to modern minds, but it stood above other models because it took account of these and other observations.[9] In this sense, it reflected what we now see as a virtue of science, namely, openness to evidence from the natural world.

Contrary to the textbook story, neither Aristotle nor Ptolemy thought the Earth was a large part of the universe. Aristotle thought it of "no great size" compared to the heavenly spheres.[10] Likewise, in Ptolemy's master-work, the *Almagest*, he says, "The Earth has a ratio of a point to the heavens."[11] Both of them based this view on observations of Earth's relation to the stars, from which they surmised that the stellar sphere was a vast distance from the Earth. Copernicus based one of his arguments for Earth's rotation on this shared assumption. "How astonishing," he exclaimed, "if within the space of twenty-four hours the vast universe should rotate rather than its least point!"[12]

More to the point, *they did not see the center of the universe as a place of honor*, any more than we think of the center of the Earth as being such. And the Earth was certainly not thought to be "sitting at the center of Heaven itself," as Myrhvold puts it. Quite the opposite. The sublunar domain was the mutable, corruptible, base, and heavy stuff of the cosmos. Things were thought to fall to Earth because of their heaviness. The modernist spin on geocentrism has it wrong. In our contemporary sense of the words, the Earth in pre-Copernican cosmology was the *bottom* of the universe rather than its center.

In contrast, Aristotle saw the heavens as unchanging. Whereas the regions below the Moon were made up of the four mutable elements of earth, water, air, and fire, the heavens were composed of a fifth element, a quintessence or ether. Heavenly bodies were perfect spheres, moving in perfect circles. From this it followed that the laws governing the heavenly realms were quite different from the laws governing the sublunary realm.

When Christian theology was later added to the mix, the center or bottom of the universe became, quite literally, *hell*. Dante's *Divine Comedy* immortalized this vision, taking the reader from the Earth's surface through the nine circles of hell, which mirror, and hence reverse, the nine celestial spheres above. Man, composed of both earth and spirit, occupied an intermediate state as a sort of micro-cosmos.[13] He could ascend to the heavenly realm, or descend to the realm of evil, death, and decay.

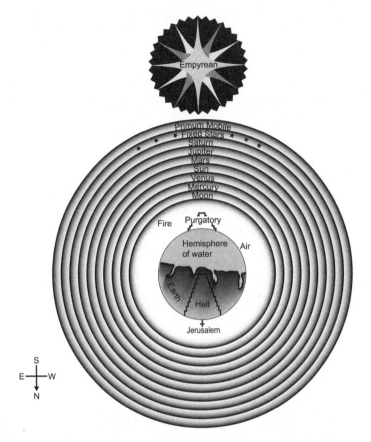

Figure 11.3: The Aristotelian/Ptolemaic universe according to Dante. Contrary to the common stereotype, the pre-Copernican cosmology did not view the "center" of the universe as the most important place. For Aristotle, Earth was cosmic sump, where earth, air, fire, and water mix to cause decay and death. The spheres of the moon, planets, and stars were the realm of the eternal and immutable. In *The Divine Comedy*, Dante describes the levels of hell mirroring the celestial spheres, with Satan's throne at the very center of the Earth. Not surprisingly, against this backdrop, Copernicus, Galileo and others could argue that the heliocentric view elevated the status of Earth.

Other purely spiritual beings filled the wider creation. God dwelt beyond the outer empyrean sphere as the Unmoved Mover of everything else. Metaphysically, reality in the medieval scheme is God-centered, not man-centered. Thus, Augustine argued that God did not create the world "for man" or out of some necessary compulsion, but simply "because he wanted to."[14] It's false, then, to say that the pre-Copernicans elevated the Earth and man, while Copernicus relegated us to a trifling backwater.[15]

WAR AND RUMORS OF WAR

Overlaid on this bad history is the larger myth of warfare between science and religion.[16] We're often told that religious belief hindered science, but many careful historians have argued, rightly, that science grew out of a Christian and theistic milieu—and not by coincidence.

People in every culture could observe the world around them. Only one culture gave birth to modern science. Why? The Judeo-Christian tradition, contrary to modern stereotype, was decisive. It enriched and transformed the pagan Greek tradition when the two began to interact after the birth of Christianity. As science historian Reijer Hooykaas puts it, "Metaphorically speaking, whereas the bodily ingredients of science may have been Greek, its vitamins and hormones were biblical."[17]

Science inherited the biblical idea that linear time is real and not an illusion. In other words, cosmic history goes somewhere rather than merely in circles.[18] In this way it distinguished itself from the cyclical view of time held by Eastern religions, and, to some extent, by the pagan Greeks. Also crucial was the distinction between Creator and creation, which had at least two profound results.

First, in contrast to much Greek thought, the biblical authors saw matter as good and manual labor, ennobling.[19] Second, man was to respect nature as God's good creation, but not revere it as a god or a divine offspring. As Christopher Kaiser puts it, "In total contradiction to pagan religion, nature is not a deity to be feared and worshipped, but a work of God to be admired, studied and managed."[20] This break with Greek thought paved the way for scientists to get their hands dirty (sometimes literally) in the work of experiments. This allowed them to make fresh discoveries about nature and develop technology.[21]

Another key biblical idea is the view that God freely created the world.[22] That means nature is *contingent*. It might not have existed, or it might have been different. As a result, the facts of nature must be discovered rather than deduced from the principles of logic or mathematics.[23]

Balancing the view that God is free is the belief that he is good and rational rather than fickle or malicious, as the pagan gods are so often depicted. Jews and Christians expected nature to be orderly and even lawful in its workings—not deceptive. The birth of modern science drew from this careful balance of contingency and order. Science emerged in a culture that held the rich inheritance of Greek and Judeo-Christian ideas in careful balance.[24]

Finally, since they believed that God is one and that human beings are created in God's image, Christians and Jews could expect nature to have unity (to be a *universe*)[25] and to be accessible to the human mind.[26] These ideas, mingled with the best of Greek thought, were the seedbed from which natural science slowly grew. Science draws on the interest of theology.[27]

CHRISTIAN THEOLOGY AND ANCIENT COSMOLOGY

To discuss the pagan Greeks does not, of course, give us a full and fair view of their gifts to modern science. Plato wanted to kick the poets out of his ideal republic because they so often depicted the gods as amoral, fickle, and irrational. Both Plato and Aristotle groped their way toward a unifying cosmic principle. This encouraged them to see a unified cosmos rather than a confused chaos. But science didn't fully develop until Western civilization had been leavened for many centuries with Christianity. We're used to attributing modern discoveries and inventions—from the seventeenth century to the present—to the so-called scientific method, which tends to ignore the soil from which modern science sprang. The writings of Plato, Aristotle, the Neoplatonists, and the ancient materialists such as Democritus were part of that soil. The work of Muslim scholars during the Middle Ages also played a role.[28]

To avoid the textbook myths, however, it's crucial to understand how Christianity relates to Aristotle's world picture. For centuries, the Christian West had no official picture of the cosmos, in part because the biblical texts and imagery lack that sort of detail. Over time, however, the exposure of Western scholars to Aristotle's thought—mostly transmitted from the Muslim world—led to an integration of the Aristotelian/Ptolemaic cosmology with Christian theology.

At first, this was not a happy marriage. In 1277, the bishop of Paris, prompted by Pope John XXI, issued a series of decrees denouncing "radical" Aristotelianism.[29] The main problem was that some disciples of Aristotle denied God's freedom in creating the world. Aristotle viewed the cosmos as eternal, while the Christian saw it as the free act of God who spoke it into being. Nevertheless, Thomas Aquinas's use of Aristotle soon won the day and became the basis of thought in Europe's universities.[30]

For Aristotle, everything was purposeful in the sense that it was directed toward some end or goal. Everything moves, finally, because of the movement of the Unmoved First Cause, which is also the Final Cause toward which everything tends. Once properly chastised and baptized, such a way of thinking provided a plausible argument for the existence and actions of God from the features of the natural world.

At the same time, Christian theology did not require Aristotle's view of the cosmos. The faith had prospered without it; it would do so again.[31] Christian doctrine did not require teleology in Aristotle's sense—even though there's much to commend it. Eastern Christianity never adopted it, and many later non-Aristotelians, like Galileo, Kepler, Newton, and Boyle, continued to see nature as purposeful by pointing to the orderly configuration of natural objects.[32] By the time of Boyle in the seventeenth century, nature was seen more as an interlocking machine, an artifact of intelligence rather than a guided organism whose parts tended toward some internally motivated end.

And last but not least, the figures who proposed the new world picture were religious. Copernicus, for instance, was a Catholic canon who worked for the local bishop. Astronomy was his hobby![33] Indeed, many of the founders of modern science were not mere cultural Christians; they were devout believers who saw their scientific work as a way of glorifying God.[34]

THE (REAL) COPERNICAN REVOLUTION

Copernicus lived during what we now call the Renaissance and the early Reformation (1473–1543). This was a time of enormous upheaval, which helped foment Copernicus's revolutionary ideas. Besides his intellectual context, he was aware of an empirical problem: the motion of the planets never quite fit in Aristotle's scheme. Explaining the weird back-and-forth of the planets across the sky required appealing to inelegant "epicycles" and "equants"—think circles riding on circles.[35]

In his book On the Revolution of the Heavenly Spheres, Copernicus proposed that Earth rotates around its axis and revolves around the Sun along with the other planets. This scheme promised to resolve some of the convolutions in the planetary orbits, but it didn't dissolve the need for epicycles and other ad hockery. Copernicus lacked observations precise enough to verify his scheme when he proposed it.[36] Still, he made a staggering imaginative leap.[37] Unlike most who came before him, he focused on the anomalous evidence of the planets, and suggested a solution that to

Figure 11.4: Nicholaus Copernicus (1473–1543).

many seemed to contradict the evidence of the senses. History would show that on this point, he had chosen correctly.

He also hinted at a vision of an even larger cosmos, one that culminated in the infinite universe of Newton. But Copernicus wasn't the cool, detached scientist of modern lore, and he didn't see his theory as demoting man's status.[38] Quite the contrary, he was a man of the Renaissance. As a result, his argument was suffused with anthropocentric (man-centered) premises.[39]

The Neoplatonic strain of Renaissance humanism motivated his search for mathematical simplicity and harmony in nature. It instilled in him an aesthetic sense that the inelegant motion of the planets was a problem to be solved. "A felt necessity was the mother of Copernicus' invention," says Thomas Kuhn. "But the feeling of necessity was a new one."[40] Neoplatonism also inspired his reverence for the Sun and allowed him to reinterpret the status of the physical center from that of the earlier

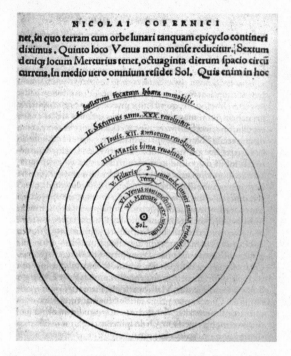

Figure 11.5: This heliocentric cosmic cross section, from Copernicus's *On the Revolution of the Heavenly Spheres* (1543), shows the Earth with its Moon circling the Sun, along with the other then-known planets.

Aristotelian cosmology. Otherwise, heliocentrism would have *demoted* the Sun, by putting it at the bottom of the universe. As science historian Dennis Danielson puts it, Copernicus "needed to renovate the [cosmic] basement, because the basement needed renovating."[41]

So, Copernicus enthuses:

> In the middle of all sits Sun enthroned. In this most beautiful temple could we place this luminary in any better position from which he can illuminate the whole at once? He is rightly called the Lamp, the Mind, the Ruler of the Universe: Hermes Trismegistus names him the Visible God; Sophocles' Electra calls him All-seeing. So the Sun sits as upon a royal throne ruling his children the planets which circle round him.[42]

Notice what's happening here. It's subtle. In the medieval scheme, centrality was not an honor since Earth was a cosmic sump. Copernicus, however, boldly reinterprets the center, so that the center for the Sun is a place of honor. At the same time, since Earth reflects the light of the Sun, it, too, receives a boost in status.

Philosophy doesn't tell the whole story, of course. Copernicus grounded his insights in observations, thus departing from the Neoplatonic distinction between the worlds of illusory sense appearances on the one hand, and intelligible reality on the other.[43] His was a Christian Neoplatonism. The material world itself, as the creation of an omnipotent God, could reflect the precision of mathematics, something unthinkable to the strict Neoplatonist. Without this revision of pure Platonism, the inelegant movement of the planets would not have struck Copernicus as a problem to be solved. This Neoplatonism, tempered by Christianity, was a guide, then, for Copernicus and for other scientists who followed him, including Galileo and Kepler.[44]

Despite his innovation, Copernicus retained the perfectly circular, celestial spheres, along with other elements of the old cosmology. "The

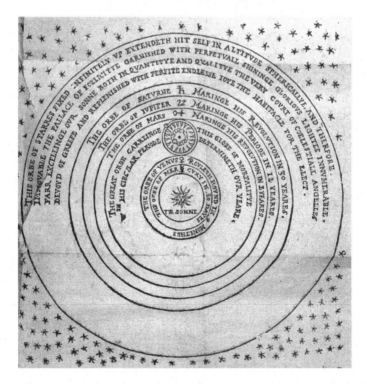

Figure 11.6: An early English Copernican cosmic section by Thomas Digges (1576). Although Digges's Copernican model was otherwise conservative, he's usually considered the first to break through the finite outer wall of the cosmos and posit an infinite expanse of stars.

Figure 11.7: Tycho Brahe (1546–1601).

significance of *De Revolutionibus*," Thomas Kuhn observes, "lies . . . less in what it says itself than in what it caused others to say."[45]

One of Copernicus's successors was the anti-Copernican Tycho Brahe (1546–1601) .[46] Brahe, a Danish naked-eye astronomer, proposed a model that made the Earth the fixed center of the sphere of stars that defined the outer edge of the universe, but with the Sun, and Moon, and the other planets revolving around the Sun.[47] His enduring legacy, though, was his careful observational record. It allowed his Copernican under-study, Johannes Kepler (1571–1630), to resolve the problems that beset Copernicus's own scheme. The appearance of a "new star"—a nova—in 1572, and Brahe's observations of comets, previously mistaken for events in the atmosphere, helped hasten the demise of two other Aristotelian convictions: the immutability of the heavens and the concentric heavenly spheres.

Kepler's access to Brahe's data allowed Kepler to correct the often-flawed ancient records that Copernicus had inherited. A deeply religious Lutheran,[48] Kepler internalized the logic that Copernicus had only par-tially embraced. He had a mystical attraction to harmonic regularities that inspired him to search for simple laws of celestial motion. At the same

Figure 11.8: Frontispiece of 1660 book by Jesuit Athanasius Kircher, depicting the Tychonic system. In Tycho's system, the Sun continued to circle around the Earth, as in the Ptolemaic scheme, but the other planets circle around the Sun, as in Copernicus's scheme. While Tycho's system is often considered a timid compromise between Ptolemy and Copernicus, it signaled a radical break with the traditional idea of solid celestial spheres.

Figure 11.9: Frontispiece of *Almagestum novum*, by Giovanni Battista Riccioli, 1651. In this richly detailed image, Urania, the Muse of astronomy, is holding a scale weighing the Tychonic and Copernican systems. Tycho's system outweighs the Copernican one. For many years, the Tychonic system was the chief alternative to Copernicanism. Notice that the Ptolemaic system in the bottom right corner has clearly been rejected, and is not even weighed in the scales.

Figure 11.10: Johannes Kepler (1571–1630).

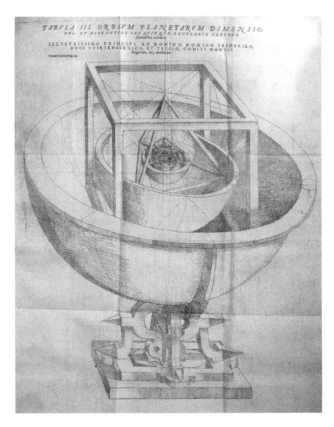

Figure 11.11: This "geometric planetarium" by Johannes Kepler (1596) illustrates Kepler's mystical attraction to mathematical harmonies in nature.

time, his realism and his access to reliable data led Kepler to propose that the planets orbited in ellipses that vary in speed rather than in perfect circles. These ideas would have irked strict Aristotelians, who preferred the perfection of the circle.[49]

Kepler even anticipated our argument—that the needs for life and scientific discovery correlate—by noting that, for astronomy, it's much better to be somewhere other than the fixed center of the universe:

> Thus it is apparent that it was not proper for man, the inhabitant of this universe and its destined observer, to live in its inwards as though he were in a sealed room. Under those conditions he would never have succeeded in contemplating the heavenly bodies, which are so remote. On the contrary, by the annual revolution of the earth, his homestead, he is whirled about and transported in this most ample edifice, so that he can examine and with utmost accuracy measure the individual members of the house. Something of the same sort is imitated by the art of geometry in measuring inaccessible objects. For unless the surveyor moves from one location to another, and takes his bearing at both places, he cannot achieve the desired measurement.[50]

The Italian scientist Galileo Galilei (1564–1642) added the telescope.[51] Peering through his portal to the heavens, Galileo resolved a dizzying panoply of stars invisible to the naked eye. He discovered the four major moons of Jupiter—which we call the Galilean moons—and saw the phases of Venus. Both discoveries showed that not all bodies are centered on the Earth. He also spied craters, valleys, and mountains on the Moon and found spots on the Sun. Such blemishes contradicted the belief that all heavenly bodies are pristine spheres. But they fit neatly with the new model of the heavens, which regarded the Sun's, Moon's, and planets' bodies subject to the same natural laws we find on Earth.

So, to sum up, far from demoting the status of Earth, Copernicus, Galileo, and Kepler saw the new scheme as exalting it. Galileo spoke of earthshine, in which Earth reflected the light and glory of the Sun more perfectly than did the Moon. He thought Earth's new position removed it from the place of dishonor it had been thought to occupy and placed it in the heavens. As he put it in his book *Sidereus Nuncius*:

Figure 11.12: Galileo Galilei (1564–1642), from the 1613 frontispiece of his discourse on sunspots.

Many arguments will be provided to demonstrate a very strong reflection of the sun's light from the earth—this for the benefit of those who assert, principally on the grounds that it has neither motion nor light, that the earth must be excluded from the dance of the stars. For I will prove that the earth does have motion, that it surpasses the moon in brightness, and that it is not the sump where the universe's filth and ephemera collect.[52]

THE GALILEO AFFAIR

Like the story of Copernicus, the popular story of Galileo's run-ins with the Church is a farrago of fact and fiction. Galileo was forced to recant his views before the Inquisition in 1633. That part is true. But the episode

Figure 11.13: The figures on this frontispiece of Galileo's *Dialogue Concerning the Two Chief World Systems, Ptolemaic and Copernican* (1632), which led to Galileo's infamous and widely misunderstood trial before the Inquisition, are often misidentified as the characters in the dialogue. In fact, it shows Copernicus (right) holding a Sun-centered model, in dialogue with Aristotle and Ptolemy (left) holding an Earth-centered armillary sphere. Galileo chose an easy target by making the Ptolemaic system the foil against Copernicanism. He ignored the Tychonic system, which was the leading alternative to Copernicanism at the time.

can't be reduced, as it is in textbooks, to a simple conflict between science and religion. We could just as well view it as a conflict between two competing scientific models—the Aristotelian and the Copernican. In 1633, the evidence couldn't prove either one. That would not come until 1727, when astronomer James Bradley discovered the aberration of starlight, proving Earth's motion around the Sun.

The conflict involved many twists and turns, with one-dimensional heroes and villains in short supply. The intense challenge of Protestantism loomed much larger than it had a century earlier when Copernicus proposed his theory. And then there was Galileo's temperament. Never the diplomat, he insisted that the Church endorse his views rather than allow them gradually to gain acceptance as evidence accumulated. He also mocked Pope Urban VIII in his 1632 *Dialogue Concerning the Two Chief World Systems*, assigning to Urban the name Simplicio.[53] Seriously.

At the same time, he angered the Aristotelians, who ruled the universities and whose careers were built on the older world picture. He further irritated them by publishing his ideas in the Italian vernacular, rather than in the scholarly Latin.

Church diplomats sought a compromise that they hoped would keep everyone happy. They even got Galileo to agree that while he could teach Copernicus's theory, he wouldn't teach that it was literally true. An awkward compromise, to be sure, but it further shows that the actors in this drama were complex and, frankly, more interesting than the cardboard characters of the popular myth.

Alas, Galileo went back on his promise and was reproved as a result.[54] Yet far from being thrown in prison or tortured, he was sentenced to house arrest in his own villa—with a servant—and received a Church pension for the rest of his life. Not an ideal outcome, of course, but also far from the treatment one would expect from a religious monomaniac. Recall that at the time there were major witch hunts in places like Germany and New England.[55]

Also keep in mind that Copernicus and Galileo got some key details wrong. The orbits of the planets weren't perfect circles. For instance, the Sun is neither stationary nor the center of the universe, and Galileo's primary evidence for Earth's motion, the tides, turned out to be wrong.[56] So, if the Church had endorsed Copernicus's theory as Galileo wanted, it would have gotten some things wrong.

The above are a just a few of the details that make the incident poor evidence for an eternal war between science and religion.[57]

Dominican monk Giordano Bruno's execution in Rome in 1600 is another misunderstood event. Though often connected to the Galileo Affair, it has little to do with our story. His execution is a dark spot in Church history. But his Copernican views were incidental, since he also defended pantheism and an eternal universe with an infinite number of occupied worlds. These were the claims—neither a part of Copernicus's theory—that troubled Church leaders. Bruno wasn't even a scientist. Nevertheless, writers often call him into service to prove that religion sought to destroy every brave and clearheaded scientist it could find.[58] In fact, Bruno was not executed for his Copernican views—such as they were—but for his heretical views on the Trinity, the Incarnation, and other doctrines.[59]

Finally, bear in mind that much of the resistance to Copernicus's Sun-centered model was because it left many key puzzles unanswered. In particular, it failed to answer how the planets and stars moved and cohered without the celestial spheres. Solving this puzzle required radical new concepts. One was the switch from Aristotle's belief in projectile motion, in which a moving object must be acted upon directly to keep moving, to the modern concept of inertia,[60] in which a moving object keeps moving unless slowed by a force, such as wind drag.

A related insight was Newton's mathematical view of gravity, in which bodies act on each other from a distance without direct contact. Without the earlier dichotomy between the terrestrial and celestial realms, Newton could describe, even if he could not explain, both the fall of objects to the Earth and the revolution of the planets around the Sun. But Newton's insights wouldn't arrive for nearly half a century after Galileo's death.

Newton is yet another figure distorted by the science-vs.-Christianity template. He was no mere deist, as is often claimed. Much like Thomas Aquinas, he viewed natural laws as God's ordinary ways of acting in the world, and miracles, His extraordinary ways of doing so.[61] God acted by sustaining the motion of celestial spheres, and by setting up the initial orbits of the planets.[62] Newton even seems to have viewed gravity itself as God's constant action.[63] That's not the hands-off God of deism.

Still, the ancient materialism revived during the Renaissance (bits of which were present already in Galileo and others[64]) and the skepticism of the later Enlightenment led many to view God as ever more distant and uninvolved. The concept of nature's Lawgiver slowly disappeared, and with it Newton's concept of natural law.[65] In its place stood an atomistic and impersonal concept of natural law. Newton's successors left a picture

Figure 11.14: Isaac Newton (1642–1727).

of nature as an eternal, infinite, and self-sustaining clockwork machine. Others, having abandoned Aristotle's teleology and Newton's design arguments, threw the baby out with the bathwater, and banished any notion of design from nature altogether.

But when it comes to natural science, nature has a say. The limits of materialism to fully explain nature soon started to show.

WHERE ARE WE AND WHY DOES IT MATTER?

The above is but a quick sketch of the origins of modern science, but it's enough to make clear that the story most of us learn is shot through with mistakes. The centrality of Earth in pre-Copernican cosmology meant something quite different to the pre-Copernicans, Copernicus, Galileo, and Kepler, than it does in the textbook orthodoxy we've all learned.[66] "The great Copernican cliché," Dennis Danielson explains," is premised upon an uncritical equation of *geo*centrism with *anthro*pocentrism."[67]

The official story gives the false impression that Copernicus started a trend, such that removing the Earth from the center of the universe was the first step in science establishing that, in the scheme of things, we don't matter. By sleight of hand, the Copernican principle transforms a series of metaphysically ambiguous discoveries into the Grand Narrative of Materialism.

That said, there are issues lurking in the neighborhood, namely, the question of design and purpose in nature, and our own significance. While distinct one from the other, these issues are usually part of a package. The Book of Genesis says men and women are created along with the heavens, Earth, and animals. Though formed from the dust of the ground, humans are specially endowed to perceive the starry heavens above and the moral law within—to paraphrase Immanuel Kant. Man is made in God's image, free to choose the good, or deny it. Christianity adds to this the Incarnation, whereby God became man to reconcile man and creation to himself.

The conviction that we're here for a purpose is not unique to the "religions of the book"—Christianity, Judaism, and Islam. For instance, the Roman philosopher and statesman Cicero argued—even more strongly than Jews and Christians would—that since the gods and man alone share the gift of reason and are aware of the passing of time and seasons, "all things in this universe of ours have been created and prepared for us humans to enjoy."[68]

One might argue that the insights that started with Copernicus do not prove but do *suggest* something different. Forget the muddle about the center of the universe. What about all those other planets? What about the sight of moons encircling many of those planets? What do we make of the fact that the Sun is just one of hundreds of billions of stars in the Milky Way, which is one of hundreds of billions of galaxies in a very large, very old universe? These facts aren't trivial. Even without false stereotypes and bad arguments, they leave many with a gnawing sense of dread and isolation. "When I consider the short duration of my life," wrote the *Christian* Blaise Pascal in 1670, "swallowed up in an eternity before and after, the little space I fill engulfed in the infinite immensity of spaces whereof I know nothing, and which know nothing of me, I am terrified. The eternal silence of these infinite spaces frightens me."[69]

In the middle of the last century, astronomer Harlow Shapley transformed this into a scientific rule, which Carl Sagan later popularized. Shapley called this rule the Copernican principle. It's now often called the principle of mediocrity, but does the rule hold up to scientific scrutiny? We take up that question in the next chapter.

CHAPTER 12

THE COPERNICAN PRINCIPLE

*Because of the reflection of sunlight . . . the Earth seems to be sitting
in a beam of light, as if there were some special significance to this
small world. But it's just an accident of geometry and optics. . . . Our
posturings, our imagined self-importance, the delusion that we have
some privileged position in the Universe, are challenged by this point of
pale light. Our planet is a lonely speck in the great enveloping cosmic dark. In
our obscurity, in all this vastness, there is no hint that help will come
from elsewhere to save us from ourselves.*

Carl Sagan[1]

Every job has its risks. Coal miners can contract lung disease.
Transcribers, if they don't take frequent breaks, can develop car-
pal tunnel. Crop dusters, if they let their minds drift, can have
short careers.

Other professions have less obvious risks. For instance, some scientists risk drifting into far-flung speculations and confusing observation
with assumptions. The Russian physicist Lev Landau once observed,
"Cosmologists are often wrong, but never in doubt."[2] Their talk about
such esoteric matters as multiple universes and time travel may bring to
mind Mark Twain's quip in *Life on the Mississippi*: "There is something
fascinating about science. One gets such wholesale returns of conjecture
out of such a trifling investment of fact."

But the risk comes with the territory. Astronomers and cosmologists deal with very large-scale phenomena. At times, they must ponder the history of the entire universe over billions of years. To manage, they need guiding rules with wide scope. These rules may be hard to prove, but that doesn't mean scientists should disavow them. It just means they have to beware of the dangers of their profession.

In the early twentieth century, for instance, Albert Einstein used the so-called cosmological principle to expand the reach of his general theory of relativity. The principle was simply this: We should assume that at very large scales the universe is homogeneous and isotropic—that is, matter is evenly distributed, and the universe looks the same in every direction. Nothing in Einstein's theory implied that the universe conforms to this principle; but the principle allowed him to apply his theory to the universe as a whole.[3] And although the cosmological principle has needed ongoing qualifications, it has allowed cosmologists to produce mathematical models of the universe that fit our best observations.

Sometime in the twentieth century, however, Einstein's principle came to be identified with the Copernican principle,[4] also known as the principle of mediocrity or principle of indifference. In its modest form, the principle states that we should assume that there's nothing special or exceptional about the time or place of Earth in the cosmos.[5] This seemed plausible, since, by definition, there are more usual than unusual places to be.[6] Besides, it need not be just an assumption, since one can treat it as a scientific hypothesis, make predictions, and compare those predictions with the evidence.

But as we've noted, the principle of mediocrity has a closely related but more expansive metaphysical form, which says: we're not here for a purpose, and the cosmos isn't arranged with us in mind. Our status is as trivial as our location. This denial of purpose is usually accompanied by materialism, the view that the blind material world is all there is and that it exists for no purpose.

Although an eccentric view for most of Western history, materialism has always had adherents. In its early form among some ancient Greek philosophers, it amounted to a conviction that the universe emerged from an infinite and eternal chaos, without purpose. These thinkers supposed that, given enough time, space, and matter, anything that can happen will happen. Even some theists like Descartes seemed to prefer this view.[7] Still, only in the modern age has such a denial of purpose in nature enjoyed official status among the cultural elite.[8] To openly question it among the fashionable is to all but guarantee an awkward pause in the conversation.

The metaphysical Copernican principle often guides scientific research. By itself, of course, it's not irrational or unscientific to allow your philosophical views to guide your research. If it were, no one would be scientific since it's unavoidable. Besides, as we've already seen, the history of science is filled with examples of philosophical ideas that led scientists to new discoveries. The problem is not when scientists allow their views to guide their scientific study, but when those views blind them to uncooperative evidence.

What's great about natural science is that it provides a way to publicly test what we believe against the natural world, and to sift out individual motives and opinions. We can do this by testing assumptions against careful observations. What, for instance, would count against the Copernican principle? That's easy to answer: if human beings, the Earth, or our local environment, were unusual or unique in important ways. If the cosmos seemed specially fitted for our existence, or the existence of life, then that would also count against the Copernican principle. Conversely, evidence that confirmed the mediocrity of our surroundings would count in its favor.

This means that if it's to be more than an untestable dogma, we should test the Copernican principle, much as Einstein proposed tests for his general theory of relativity. Such an approach may seem like common sense, but some of its advocates mistake the principle for science itself. Take astronomer Mario Livio, for example. He has defended—correctly in our view—the concept of beauty as a key ingredient of good scientific theories, even going so far as to propose the "cosmological aesthetic principle." But in a startling bait-and-switch, he identifies beauty as conformity to the Copernican principle. To be beautiful, he says, a theory "should be based on symmetry, simplicity (reductionism), and the generalized Copernican principle."[9] In other words, a theory is beautiful to the degree that it avoids taking us into account or noticing that we're exceptional.

However we judge Livio's quirky concept of beauty—and it is quirky—clearly he's trying to acquire by definition what must be won by evidence and argument. Scientists don't have to assume the Copernican principle, and it's not grounded in basic rules of reasoning. Further, to make it a dogma violates the scientific virtue of openness to evidence. We can't determine the nature of reality by imposing a definition of science that restricts what we can ask or notice. And we shouldn't arbitrarily protect our assumptions.

In any case, the central issue is not whether the Copernican principle is beautiful, but whether it's *true*. And one way to determine this is to test its predictions.

PREDICTIONS OF THE
COPERNICAN PRINCIPLE

In practice, these predictions tend to be unstated, which protects them from scrutiny. All the more reason, then, to make them explicit.

We all take some of its implicit predictions for granted. For instance, we think that the same laws of physics and chemistry govern both the heavens and the Earth. We also assume that nature's laws are uniform, so that the law of gravity doesn't differ from the Earth to the Moon to a star on the other side of the Milky Way. Moreover, there are trillions of stars, and we expect that many of them have planets encircling them. We understand that, at least in these ways, Earth is not unique. This understanding is the firm legacy of the Copernican Revolution. If we stopped here, the Copernican principle might appear to be well founded. But on closer inspection, we find that many of its predictions turn out to be false.

The principle manifests itself in cosmology, physics, and biology. But we'll hold those issues for the following chapters. Here let's consider, one at a time, the predictions of the Copernican principle in the domain of astronomers, who study objects as small as meteorites and individual planets, and as large as galaxy clusters.

Copernican Principle Prediction 1:
Earth isn't exceptionally suited for life in our solar system. Other
planets in the solar system probably harbor life as well.
This was one of the earliest expectations of modern astronomers. When they had only scant evidence, many respected scientists expected to find intelligent life on other planets in our solar system. Kepler conjectured that the structures on the Moon were built by intelligent beings. More recently, Giovanni Schiaparelli (1835–1910) described Martian "channels," which to Percival Lowell (1855–1916) suggested the presence of a Martian civilization. Translating, or mistranslating, Schiaparelli's "channels" as "canals," Lowell founded his own observatory in Flagstaff, Arizona, and dedicated his time to gathering evidence to support his belief.

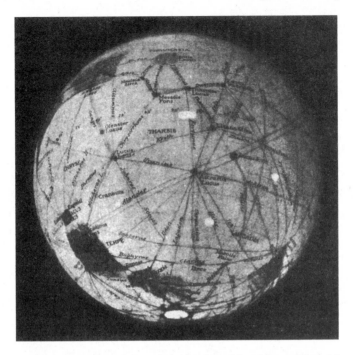

Figure 12.1: Percival Lowell's drawings of Martian "canals" and "oases" from *Mars as the Abode of Life* (1908). As telescope resolution improved, it became clear that Lowell's canals were a combination of optical effects, wishful thinking, and naturally occurring structures.

Lowell matters because of his influence and because he explicitly linked the idea of Martian life to his opposition to anthropocentrism, thus embodying the spirit of the Copernican principle. "That we are the sum and substance of the capabilities of the cosmos is something so preposterous as to be exquisitely comic," he wrote. Man "merely typifies in an imperfect way what is going on elsewhere, and what, to a mathematical certainty, is in some corners of the cosmos indefinitely excelled."[10] According to Carl Sagan, Lowell's enthusiasm "turned on all the eight-year olds who came after him, and who eventually turned into the present generation of astronomers."[11]

But the Mariner, Viking, and Sojourner missions to Mars revealed a forbidding world, and dashed hopes of finding advanced Martian life. Similar hopes have been dashed as our knowledge of the other planets has grown. Expectations nursed by the Copernican principle have thus proven deeply misguided.

To be clear, the belief that Mars once harbored life lives on, more recently in the excited announcement of the discovery of microscopic magnetite crystals in the Martian meteorite ALH84001 and the discovery of vast water-ice fields under the Martian surface. At the same time, there's a flickering hope that life exists on one or more moons orbiting Jupiter and Saturn, such as Europa and Enceladus, where liquid water likely exists below their surfaces. Conjectures abound regarding the exotic creatures that may dwell in the deep, icy crevices of those moons.

However, we have no compelling evidence for even primitive life on Mars or any of these moons, and we have great evidence for skepticism. Much of the optimism ignores the myriad ways Earth is uniquely well suited among the bodies in our solar system for life, as detailed in chapters 2–6. No other place in our solar system comes close to providing the properties that make Earth habitable. If anything, our planetary neighbors underscore how stringent the needs are for building a habitable planet, even for the other planets in a system with one inhabited planet.

The point is worth repeating because it's so often forgotten or ignored. From the seventeenth to the twentieth century, many expected to find intelligent, even superior life on the Moon as well as on Mars and other planets in the solar system. This expectation required direct contrary evidence to overturn it. Now, at the beginning of the twenty-first century, the hope of finding intelligent life elsewhere in our solar system is dead, and

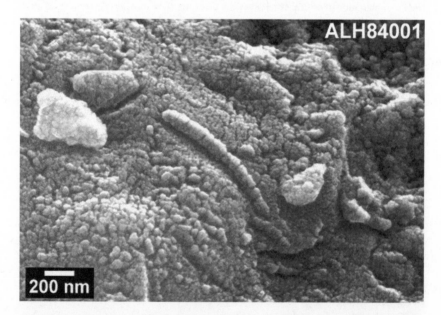

Figure 12.2: Electron micrograph of sausage-shaped structure in the carbonate minerals in the 4.5-pound Martian meteorite, Allen Hills 84001 (ALH84001). This image and other data related to ALH84001 were cited in 1996 as evidence for ancient life on Mars. Today, this evidence remains highly controversial.

Figure 12.3: Illustration of Martians for a nonfiction article by H. G. Wells, "The Things that Live on Mars," in *Cosmopolitan Magazine*, March 1908. The article appeared during the controversy over Lowell's alleged canals on Mars when it was widely believed that the planet was home to intelligent life. In the following century, expectations have diminished as we have learned more about the Red Planet.

the search for primitive life has moved from the planets to a few obscure outlying moons. ET advocates now behave as if finding the Europan equivalent of slime mold would be as profound as finding intelligent Martians. Such claims are a serious downgrade.

Add to this pattern the evidence of chapter 6: some of the planets once said to diminish the Earth's status now serve as its guardians. Finally, recall that the rare properties of Earth discussed in chapters 2–6 have been crucial in all manner of scientific discoveries here, from the nature of gravity to the internal structure of our planet. These facts about Earth's superiority both for living and observing count as a sobering contradiction of the Copernican principle.

Let's try another prediction on for size—in this case, as it applies to our host star.

Copernican Principle Prediction 2:
Our Sun is an unremarkable star.

Textbooks and science writers have repeated this claim so often that it has become entrenched dogma. This isn't the result of conscious deception, but of the power of the Copernican principle to shape expectations. We often see what we expect to see. However, as discussed in chapter 7, we now know that our Sun has all sorts of anomalous, life- and science-friendly features. These include its luminosity, variability, metallicity, and galactic orbit. So, our Sun doesn't fit this prediction well at all. Even among the minority of stars in its class, our Sun is exceptional in several key ways, again contrary to the Copernican principle.

Now, let's see how the principle fares when applied to our solar system.

Copernican Principle Prediction 3:
Our solar system is a typical planetary system.

This prediction may yet prove partly true, but it's had a bad run for the past several decades. Still, it inspired whole research programs in astronomy and planetary science. In 1995, Swiss astronomers Michel Mayor and Didier Queloz discovered the first planet orbiting a sunlike star, 51 Pegasi, in the autumn constellation Pegasus. At the time, many theorists had predicted that other planetary systems would resemble our own. In particular, they predicted that small terrestrial planets would have fairly circular orbits close to their host star, with large gas giants farther out, also in circular orbits.

But no. Although 51 Pegasi is a gas giant like Saturn or Jupiter, it whips around its host star every 4.2 days, only one-eighth the distance from its star that Mercury is from our Sun. It's not what the theorists expected. "It was very strange to consider the attitude of people facing something completely in disagreement with theory," Mayor noted after the discovery. Some astronomers, he reported, even said things like, "Oh, this is not a planet, because you cannot form Jupiter-like planets close to their stars."[12] The discoveries of thousands of other extrasolar planets since then have continued to contradict conventional wisdom.

This wasn't entirely the fruit of the Copernican principle. The idea of a water-ice condensation line, which we find from Jupiter and beyond, gave astronomers some reason to expect a similar pattern in other systems. But for many, the expectations went well beyond mere faith in a sturdy theory. Many of their models left little room for flexibility. As a result, fed on the Copernican principle, they were shocked when gas giant exoplanets parked well inside the ice condensation line started turning up.

We're just starting to have enough hard data to say how atypical our solar system is. The time required and the primary search method employed for detecting Jupiter analogues—radial velocity—are biased toward systems with massive planets with short orbital periods. To detect a Jupiter analogue around a sunlike star, astronomers must monitor it for at least twelve years—the period of Jupiter's orbit. Such data are just now coming in.

Recent studies show that 42 percent of systems with close-in small planets—that is, two to thirty times Earth's mass—have cold, outlying Jupiters. Nine out of 100 stars host giant planets between 8 and 32 AU.[13] In addition, almost all Jupiter analogues have far more eccentric orbits than Jupiter does—as we discussed in chapter 5. In the next few years, astronomers will have a better estimate of the fraction of planetary systems similar to our own. SETI enthusiasts are holding out hope that this silent data will yet vindicate the Copernican principle. It's more likely that we'll find that our solar system is atypically friendly to life.

We don't oppose the Copernican principle. We simply treat its predictions as open questions that could prove true or false. After all, this seems to be the track record. Maybe most other planetary systems have their gas giants in their outer regions, the terrestrial ones in the inner, with pretty circular orbits, just as with our solar system. Or maybe not. Let's look and then follow the evidence where it leads, rather than treating the principle as dogma. When that happens, it inspires fail-safe predictions, like the epicycles in the Ptolemaic model of the solar system, serve mainly to protect the dogma from refutation.

The next prediction looks very much like that.

Copernican Principle Prediction 4:
Even if our solar system proves atypical, lots of planetary
systems are consistent with life. Many variables, like the
number and types of planets and moons, have little to do with
a system's habitability.
This prediction has the endearing quality of being impossible to falsify, since a determined theorist can always claim that an exotic mystery critter hides in some bizarre, far-flung solar system and then tries to shift the burden of proof to the skeptic. What's more, this belief provides an escape hatch if predictions 2 and 3 above turn out to be false.

The evidence from chapters 1–6, however, weighs heavily against it. Our large, well-placed moon, circular planetary orbits, a properly placed asteroid belt with felicitous properties, the early bombardment of these asteroids on the Earth, the outlying gas giants that sweep the solar system of sterilizing comets and asteroids later on: complex life here depends on all these factors and many more. Moreover, any sort of life will need sophisticated information-processing systems. These, in turn, need a delicate balance of order and flexibility, which, as we've seen, needs a precise interlocking system of planets to get anywhere near habitability. As a result, star systems with living planets are likely to be rare, relative to lifeless ones. This claim fits everything we know. And, unlike the alternative, it's falsifiable. Find just one radically different system with thriving, native, complex life, and our argument falls apart. (More on this point later.)

Ironically, despite the claims made for the Copernican principle, this prediction—that life doesn't need a planetary system like ours—slowed the progress of science. After all, it led astronomers to downplay or overlook the key role played by comets, asteroids, moons, and outlying planets in the story of life. Similarly, it may have discouraged astronomers from giving due credit to the strictures of our system's habitable zone. If earlier astronomers had done so, they might have put much less energy into quixotic searches for ETs in our solar system that turned out to be wild-goose chases. They could have focused instead on how the details of our solar system mattered for life on Earth. Scientists are now starting to explore this topic.

If you're following our argument, you can probably guess the next prediction of the Copernican principle.

Copernican Principle Prediction 5:
Our Sun's location in the Milky Way is not that important.

This expectation inspires the belief that most stars, wherever they are, are receptive hosts for habitable planets, including, for instance, stars in globular clusters or near the galactic center. In 1974, at the Arecibo radio telescope in Puerto Rico, Frank Drake and other astronomers transmitted a radio message to the globular cluster M13, the Great Cluster of Hercules.[14] With perhaps 300,000 old, densely packed stars, this was a very unlikely place for planets in general, let alone habitable ones. While they may not have stated it, Drake and his colleagues were clearly assuming prediction five. They aimed their telescope at a small target with lots of stars because they assumed this would improve their chances of contacting an intelligent race. But the number of stars doesn't matter if the stars in that patch of cosmic real estate are unfit for life.

When telling the history of Milky Way astronomy, writers often tell a story parallel to the Copernican stereotype we criticized in chapter 11. Because of errors due in part to cosmic dust that obscured the light from distant stars, earlier astronomers William Herschel and Jacobus Kapteyn had placed our solar system near the center of our Milky Way galaxy. Later, Harlow Shapley, using evidence from globular clusters, determined that we are really thousands of light-years from the galactic center and near a spiral arm.[15] As you might expect, advocates of the Copernican principle claim this confirms their theory. "Just as Copernicus had removed the Earth from the center of the solar system," one science writer wrote in 1995, "so Shapley would yank the Sun from the center of the Milky Way and put it in the celestial equivalent of a suburb."[16]

As we now know, however, the galactic center, like the lowest circle of Dante's hell at the center of the Earth, is the *last* place we'd want to be. Happily, our solar system is within what may be a quite narrow galactic habitable zone, and far from dusty, light-polluted regions, permitting a great overall view of both nearby stars and the distant universe.

But what about our galaxy as a whole?

Copernican Principle Prediction 6:
Our galaxy is unexceptional. Life could just as easily exist in old, small, elliptical, and irregular galaxies.

In a 1998 study of the Milky Way, astronomers S. P. Goodwin, J. Gribbin, and M. A. Hendry argued that our galaxy is slightly smaller in diameter than its closest large neighbor, Andromeda (M31), and is, "at most, an averagely sized spiral galaxy." They concluded that their finding provides further support for the "principle of terrestrial mediocrity" (a.k.a. the Copernican principle), by which they mean "that there is nothing special about where and when we live and observe from. We seem to live on an ordinary planet orbiting an ordinary star, and it is natural to infer that the solar system resides in an ordinary galaxy."[17]

Yes, the Milky Way probably is less massive and smaller than M31.[18] But does this mean the Milky Way is ordinary? As we noted in chapter 8, the Milky Way is atypical in some ways compared to otherwise similar spirals.[19] More to the point, one might ask why mere galactic size or mass would matter when seeking support for that principle. As we've seen, the fact that some stars are larger than the Sun hardly counts against the Sun's life-friendly properties. Similarly, when we consider the features of a spiral galaxy most valuable for life, such as mass and true brightness—which correlate with metal content—we find that the Milky Way is exceptional.

For life, the age and galactic Hubble type of a galaxy matter a lot. All galaxies in the early history of the universe, and low-mass galaxies forming now, are metal poor. Also, young galaxies are dangerous places, making them unlikely habitats for life. Similar problems attach to globular clusters and irregular galaxies. We now have reason to suppose that large, spiral galaxies like the Milky Way (formed at about the same time) are more habitable than galaxies of different ages and types. The metal content of a galaxy depends on both its age and mass. Without enough metals, there aren't enough materials to build habitable planetary systems. Our very massive and luminous spiral galaxy is an especially suitable home for a habitable planetary system, while providing an optimal platform for viewing and discovering both our own galaxy and the wider universe. The Milky Way ranks among the 2 percent most luminous galaxies in our cosmic neighborhood.

One might think the Copernican principle had nowhere else to go, no bigger field in which to be wrong, but one remains: the universe as a whole.

CHAPTER 13

THE ANTHROPIC DISCLAIMER

The next battle of the Copernican revolution is thrust upon us. Just as our planet has no special status within our solar system, and our solar system has no special location within the universe, our universe has no special status within the vast cosmic mélange of universes that comprise our multiverse.

Fred Adams and Greg Laughlin[1]

In recent decades, the Copernican principle has fallen on hard times, so much so that what is known as the anthropic principle has been brought in to shore it up. The hard times stem from two of the greatest scientific advances of the twentieth century: that the universe is fine-tuned for life and that it is finite in age. Each of these discoveries dealt a damaging blow to the Copernican principle.

THE COPERNICAN PRINCIPLE IN COSMOLOGY

There was a protracted debate about the age of the universe in Europe starting in the mid-seventeenth century that later included the United States. By the nineteenth century, however, orthodox scientific opinion held it to be both infinite and eternal. This view provided an easy way to avoid questions about the origin of matter, space, time, and natural laws. Newton unwittingly lit the spark. First, positing an infinite universe answered a vexing question, namely: Why did gravity not cause the parts of the universe to collapse in on each other? If it were infinitely extended, there would be no center point toward which everything would converge.

Second, Newton viewed the universe as the "divine sensorium"—the medium through which the infinite God acted in the world. To be adequate to this task, the cosmos, he surmised, also needed to be infinite.

Of course, later thinkers, who added eternity to infinity, did not share Newton's theology. So, these became another prediction of the Copernican principle, but this time applied to the entire universe.

Copernican Principle Prediction 7:
The universe is infinite in space and matter, and eternal.

This assumption persisted into the twentieth century, until Edwin Hubble detected the redshift of distant galaxies and inferred the expansion of the universe. His findings provided a way to reconcile classical Newtonian physics with a finite universe. More importantly, they confirmed an implication of Einstein's general theory of relativity highlighted by the Belgian physicist Georges Lemaître and the Russian physicist Alexander Friedmann.

These findings played a key role in the development of big bang cosmology, which implied that the universe, including cosmic time itself, had a beginning. There could be no greater contrast between assumption and observation. The difference between a temporally finite universe, implied by the new cosmology, and the eternal universe assumed by two centuries of scientists, dwarfs all other shortcomings of the Copernican principle. This is because it's hard to resist the thought that anything that begins to exist must have some outside cause to bring it into existence.[2] This is not the sort of evidence partisans of the Copernican principle—who, as we noted earlier, tend to deny cosmic purpose—anticipated or welcomed.[3]

So, as we discussed in chapter 9, until contrary evidence made it untenable, Einstein resisted the fact that his own theory implied of a cosmic beginning, and scientists such as Fred Hoyle, Hermann Bondi, and Thomas Gold stumped for a steady-state model to preserve an eternal universe without a beginning.[4] According to this theory, as the universe expanded, new matter would come into existence and fill the emerging gaps. In this way, the theory accounted for evidence of an expanding universe but avoided awkward questions about a cosmic starting gun.

This virtue alone won it some adherents. In 1961, for instance, Denis Sciama used the Copernican principle to choose the steady-state model over the big bang model. A key goal of science, he argued, "should be to show that no feature of the universe is accidental."[5] By that, he meant that the purpose of science should be to show that everything can be reduced to an impersonal natural law—as opposed to unique event like a beginning. In this case, at least, the principle misguided him. As we saw, the discovery of the Cosmic Background Radiation, a remnant of the big bang, and the explanatory power of the nucleosynthesis of the light elements led to the demise of the steady-state model.

Still in pursuit of an eternal past, some scientists have suggested an "oscillating universe" model. The idea is that our universe is just one episode in an eternal cycle of big bangs, expansions, and collapses. But this proposal also had problems. Philosopher of science Stephen Meyer summarizes the two important ones: "As physicist Alan Guth showed, our knowledge of entropy suggests that the energy available to do the work would decrease with each successive cycle. . . . Thus, presumably the universe would have reached a nullifying equilibrium long ago if it had indeed existed for an infinite amount of time. Further, recent measurements suggest that the universe has only a fraction . . . of the mass required to create a gravitational contraction in the first place."[6]

We can add to these problems a third, discussed in chapter 9: the evidence that the universe is not only still expanding, but that that expansion is speeding up. If one were hoping to smuggle infinite time with an exotic dash of Far Eastern reincarnation into modern cosmology, this recent observation adds insult to injury. If this evidence prevails, it will render the idea of a cosmic re-collapse doubly unthinkable, and further confirm the idea of a cosmic beginning.[7]

With nowhere else to go, our chastened but still bloated Copernican principle lumbers into the lab and tries its hand at physics. Here we get one final prediction:

Copernican Principle Prediction 8:
The laws of physics are not specially arranged for the existence of complex or intelligent life.

The Copernican principle stumbles here, too. It turns out that the universe's laws, constants, and initial conditions are intricately fine-tuned for life. If these features of the universe were slightly different, nothing even roughly resembling our habitable cosmos would have existed.

Although it's easy to get lost in a swamp of inconceivable numbers and obscure philosophical nitpicking, the idea of fine-tuning can be grasped intuitively. Recall the story of the Universe-Creating Machine from chapter 10. Alternately, imagine an enormous white wall extending in every direction, up to the clouds and out to the horizon. This represents a range of possible universes, with their respective laws, constants, and initial conditions. Now, imagine the wall is covered with countless black, red, and green dots. Around the outer boundary and receding beyond the horizon are black dots, which represent different chaotic and disorderly hypothetical universes, ones that don't follow regular laws. And just as there are vastly more ways the atoms in a block of marble could be arranged to yield a disheveled pile of rubble than to form perfect square bricks arranged as a giant arched bridge, so too these possible chaotic universes vastly outnumber the possible worlds with stable laws and constants. Thus, we find on our enormous wall, a much smaller cluster of red dots nearer in and surrounded by the vast sea of black dots. These red dots represent possible worlds with stable laws and constants, but which can't host life, intelligent or otherwise.

At first, you think you're close to the gargantuan wall and that the patch of red dots is purely red. But after walking several hours toward the wall, you realize that the red patch is itself many miles high and wide, and inside it is a tiny patch of green dots. These dots represent possible worlds whose settings would yield habitable worlds. Within this small patch of green dots is a tiny, glowing green circle labeled "The Actual Universe."

Although we take it for granted that our cosmos is a cosmos rather than a chaos, such order would be surprising if the basic features of our universe were thrown to the winds of chance.

To summarize: Almost the entire wall will be black. A much smaller proportion of the wall will be red, and a tiny fraction inside the red will be green. The question is: Why, among all those possible worlds, does the glowing green one exist?

Before tackling that more fundamental question, we should address one possible point of confusion. Why the little green circle instead of just a single glowing green point in the illustration? Well, saying our universe's laws, constants, and initial conditions take very narrow values required for life does not mean that no other universe of any sort could be habitable by any sort of organism.[8] To return to our universe-creating machine, it means that if we were to fiddle with the values that hold here in our universe—if, for instance, we were to begin adjusting the dial for the value of the gravitational constant in either direction—the universe might remain habitable, at first, but would very quickly become hostile to life of any sort as the adjustment became more than trivial. In other words, one quickly

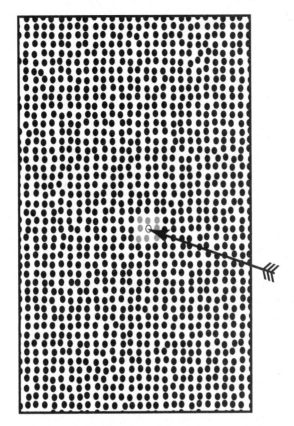

Figure 13.1: Of the many possible universes within our "universe neighborhood," only those with laws and constants very much like our own would be compatible with complex life.

wanders beyond the little green patch and into the vast expanse of red. And whether a green spot or dot might appear in some far-off section of the wall doesn't affect the fact that our universe falls within a vast sea of red dots that dwarfs the tiny patch of possible universes with settings close enough to ours to remain habitable.

This one glowing green spot that was selected out of a vast otherwise populated area with red dots, and beyond this, a still vaster sea of black dots, cries out for an explanation. And this is why so many physicists and astronomers have concluded that the universe is designed. Physicist Paul Davies, for instance, reversed his earlier views, saying, "The impression of design is overwhelming."[9]

The late astrophysicist Fred Hoyle, one of the founders of the steady-state model and long an intransigent atheist, put it even more emphatically:

"A commonsense interpretation of the facts suggests that a superintellect has monkeyed with physics, as well as chemistry and biology, and that there are no blind forces worth speaking about in nature."[10]

Some scientists, such as the late Stephen Hawking, hope the problem will be resolved by collapsing the fundamental forces and everything they entail into a single grand unified theory.[11] Given such a theory, the forces that now seem fine-tuned relative to each other will be found to be the inevitable outcome of some single overarching law. While this might resolve the appearance of fine-tuning between *independent* variables, it would inherit the larger problem it was supposed to fix. For any such unified theory would posit some particular value, or number, or formula. While all the actual laws of physics might follow from it, the higher-level formula would not. The fine-tuning would simply get moved up one level. In fact, it would get worse. Instead of multiple variables, there would be a single, *grand* one, from which the array of sub-laws would produce our habitable universe. It would be like a billiard player on a table with countless balls who sinks every ball in one shot.

Others argue that no matter how much our universe may seem fine-tuned, we can't get a probability out of it because there's no coherent way to assign probabilities to all the possible dots on the wall.[12] The wall is infinite in extent, so any portion of the wall should have the same probability as any geometrically equal part of the wall. It has the same problem as the number line: you can always place a number between any two different numbers you pick out. Because the wall—representing possible universes—lacks a uniform probability distribution, we can't assign any probability to any part.

This argument assumes a limited view of probability. By any reasonable assessment, the number and extent of uninhabitable universes overwhelms the number and extent of habitable universes. And there are approaches to probability theory that can reflect this fact, underscoring the relative rarity of habitable universes and thus their fine-tuning. (See the notes for more details.[13])

But even without the comfort that a precise number might provide, everyone recognizes that the dizzying swarm of so-called coincidences is fishy and needs explaining, or at least explaining away. Clearly, the many scientists who will go to unsurpassed speculative lengths to salvage the Copernican principle recognize this need for an explanation (as we will see below).

As before, a story might help clarify. Return again to the Universe-Creating Machine. Perhaps Q, your host, had withheld one crucial detail. Although satisfied at first with his truthfulness, you decide to take a closer look at the mysterious machine. In addition to the many dials, you now notice a camouflaged cover. Under it, you find a hidden keypad and display screen with one giant number on it—say, 91215225.79141425. Embarrassed once again, Q confesses that this one number, the Grand Unified Combination, controls the whole show. With the keypad, one can change the number, *without limit*, in either direction on either side of the decimal. Unlike the individual dials, neither the number itself, nor its number of digits, is fixed. But any change to that Grand Unified Combination affects all those combination dials designating the different physical values. Q, it turns out, has spent his time trying to change this number, rather than bothering with the individual dials. He's still convinced that there are other numbers that will produce habitable universes; he just hasn't found any yet.[14]

Now, the fact that there is only one combination that designates the values of all the forces, constants, and parameters means that these aren't independent. And since you're dealing with a potentially infinite range of options, you can't assign a probability to the total. You could only consider the (many) numbers near the original combination that have been tried so far and run a probability on those. Would this prevent you from concluding that the original Grand Combination had been intentionally set? Surely not.

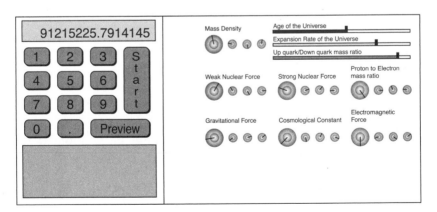

Figure 13.2: Even if a single unifying law determined all the cosmological parameters, and even if the range of possible universes were infinite, we could still recognize that the universe is fine-tuned. Such a universe would be like a universe produced with a digital Universe-Creating Machine, which requires a very specific numerical combination to produce a habitable universe.

THE ANTHROPIC PRINCIPLE AND MANY-WORLDS HYPOTHESES TO THE RESCUE

So, on the surface, this series of false predictions looks like a stunning defeat for the Copernican principle, especially since many of these large number coincidences are so large that they trivialize even the prospects for habitable settings within our large, ancient universe. But the Copernican principle, having degenerated into dogma, is highly adaptable. In this case, it survives by enlisting a contrary concept, the anthropic principle.

The anthropic principle has been the subject of many books and articles since the 1970s, and theorists have formulated it in a variety of ways, ranging from the trivial and tautological to the obscure and outlandish. The most widely discussed version is the weak anthropic principle touched on in chapters 7 and 8. It states that we can expect to observe local conditions necessary for our existence as observers—however improbable they are.[15]

In a sense, the weak anthropic principle is a check on an overly zealous use of the Copernican principle.[16] Because of a selection effect, we will see only those local conditions that allow us to exist. This is handy, since it helps explain why, if the Copernican principle is true, we don't find ourselves in a far more common slice of the universe, such as intergalactic space hundreds of trillions of years in the distant future.[17]

The Copernican principle, then, explains all those ways in which our setting is commonplace; the anthropic principle accounts for the exceptions. Using both these principles together, however, is a bit like the railway stationmaster who claimed that all the trains were running on time. When passengers complained that their trains were hours late, he qualified his assertion: "Actually, what I meant to say was that the trains all run on time—except when they don't."[18] As with the epicycles patched onto the geocentric model of the solar system to make it fit weird reversals of the planets in the sky, the anthropic principle is a sort of reusable epicycle, invoked to prevent the Copernican principle from dying the death of a thousand qualifying cuts. In reality, while the weak anthropic principle may go some small way toward explaining our strangely habitable setting without recourse to cosmic purpose, it does little to salvage the Copernican principle. The cuts remain and the principle continues to bleed out.

The strong anthropic principle (as some define it) applies this same reasoning to the laws, constants, and initial conditions of the universe as a whole: We can expect to find ourselves in a universe compatible with our existence.[19] True enough. But what are we to make of this appeal to the obvious?

Some think the anthropic principle explains the fine-tuning. It doesn't. It just states a *necessary condition* for our observing the universe. What cries out for explanation is *not* that we observe a habitable universe, but why the universe—the one that that exists—is habitable.

Consider the popular story of the firing squad.[20] Imagine an important American intelligence officer, captured by the Nazi SS during World War II, who is sentenced to death by firing squad. The SS assign fifty of Germany's finest sharpshooters to the task. They line him up against a wall and take their positions three meters away. After they fire, however, the officer finds that every sharpshooter has missed him. Instead, their fifty bullets have made a perfect outline of his body on the wall behind him.

The officer looks around. "I suppose I shouldn't be surprised to see this," he says. "If the sharpshooters hadn't missed, I wouldn't be here to observe it." The last part is true enough, and yet we know that response is crazy since the far more likely explanation is that the non-execution was rigged. Perhaps the sharpshooters had been ordered to

Figure 13.3: Confusing a necessary condition with an explanation.

miss, or they had colluded with each other for some unknown reason. Shrugging one's shoulders and concluding that it's a chance occurrence is just dense.

Sensitive advocates of the anthropic principle have realized that more is needed. What is needed, they argue, is recourse to other universes. But not just any set of universes will do. For instance, following the so-called "Everett Interpretation" of quantum physics, some physicists suggest that other universes exist in the sense that different lines of universes split off due to the collapse of wave functions at the quantum level. At every collapse, a different branch splits off and goes its own way. Such a process would leave many universes in its wake.

Similarly, some inflationary cosmological models postulate a vast array of distinct universes or discrete domains, which were produced very soon after the big bang. After expanding exponentially, each such domain slowed down to the sort of modest expansion we now detect in our universe.[21] Whatever the virtues of these theories, however, they won't do the job, and not merely because we can't observe these other universes or domains. The problem is more basic than that: both theories presuppose the very fine-tuned laws that need explaining.

What is needed, many suppose, is a full-blown world ensemble, as postulated by a many-worlds hypothesis (MWH), with a more or less exhaustive (that is, infinite), random distribution of different properties in separate, causally disconnected, universes.[22] In such a scenario, countless universes (a "multiverse" or "world ensemble") exist. The only ones observed will be those capable of having observers. In just those universes, observers will look around, and the more benighted ones will marvel that their universe is so well suited for them. In this way, and only in this way, does the anthropic principle have any hope of saving the Copernican principle from a stinging defeat.[23]

In other words, if all this were true, we might observe a habitable universe merely because of a selection effect. Its improbability would be apparent, not real. We would be like a lone dandelion growing in a thin crack in the middle of a sprawling asphalt parking lot. The dandelion might look around and be surprised it was planted in a very narrow area where it could grow and prosper. Wow, what are the chances! If it knows nothing about how dandelions reproduce, it might think that someone planted it in the crack. But it might just be subject to a selection effect. Dandelions release lots of seeds into the wind. One of them was bound to land in this dirt-filled crack in the middle of the big parking lot.

Figure 13.4: This image by Thomas Wright (1750) illustrated Wright's "Original Theory of the Universe," and in particular, his ideas on the structure of the Milky Way. To modern readers, however, it might suggest the more modern and more speculative idea that our universe is only one member of a multiverse consisting of many different "universes."

Without more information, the dandelion should assume that its location is the result of impersonal processes such as wind, not of a quirky landscape engineer.

But there's a whopping difference between this story and our situation. First, it's not obvious that an *actual* infinite set of anything, including universes, is even possible.[24] Second, we know that there are gobs of dandelion seeds floating around in the wind every spring, but we have no independent evidence to think other universes exist, except that fine-tuning contradicts the Copernican principle and lots of people don't like what that might suggest.[25] Astronomers Fred Adams and Greg Laughlin are explicit on this point:

The seeming coincidence that the universe has the requisite special properties that allow for life suddenly seems much less mirac-
ulous if we adopt the point of view that our universe, the region of
space-time that we are connected to, is but one of countless other
universes. In other words, our universe is but one small part of a
multiverse, a large ensemble of universes, each with its own variations of physical law. In this case, the entire collection of universes
would fully sample the many different possible variations of the
laws of physics. . . . With the concept of the multiverse in place,
the next battle of the Copernican revolution is thrust upon us. Just
as our planet has no special status within our solar system, and
our solar system has no special location within the universe, *our
universe has no special status within the vast cosmic mélange of universes that comprise our multiverse.*[26]

Notice that they don't say that the next *task* of the Copernican revolution is to find evidence for this multiverse. They say "battle," which suggests an ideological crusade rather than a scientific quest. We have come
full circle to the infinite chaos with its chance pockets of order posited
by certain Greek rivals of Plato and Aristotle. This multiverse argument,
more fanciful than most science fiction stories, simply presupposes the
Copernican principle, along with the grand mythology that has built
up around the Copernican Revolution. But why do that? Not only do
we have good reason to doubt the Copernican principle, but using it

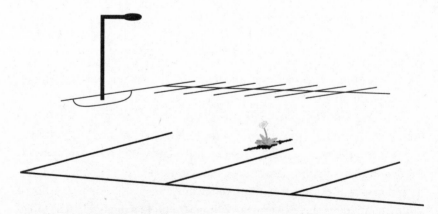

Figure 13.5: The danger of a selection effect. The dandelion, if it is ignorant of how
dandelions reproduce, might mistake its location in a relatively rare hospitable crack in a
parking lot as evidence that it was intentionally planted there.

to posit other universes is bad faith since we would never accept such reasoning elsewhere.

Recall the American officer who has survived the Nazi firing squad. Imagine if he concluded, "Well, apparently there must be millions and millions of similar executions going on all around the globe. Given so many tries, it's inevitable that some group of sharpshooters will all miss their target. I'm just the lucky one." No one would or should accept this. First, there's no independent evidence of these other events, which makes the explanation ad hoc. Second, these other firing squads are not causally connected to the one in question, so they can't be included in any assessment of probability.

Third, if allowed, such reasoning would destroy our ability to make practical judgments. Indeed, it would undermine reason itself. Using the same method, the firing squad survivor could chalk up every event in his life to chance,[27] no matter how strong the evidence of a setup. *No amount of evidence for apparent design could ever count as evidence of actual design.* But surely, natural science should be the search for the best explanation based on physical evidence, rather than merely a search for the best materialistic explanation—reason itself be damned! If so, then what sense is there in adopting a principle that blinds us to an entire class of evidence?

Many-worlds hypotheses are diverse, and any adequate treatment would require multiple chapters, such as those found in *The Return of the God Hypothesis* by Stephen Meyer.[28] But most of these hypotheses suffer from the same dilemma. If the alternative universes are causally disconnected from our own, then merely postulating them isn't a causal explanation. So, they don't even compete with, say, intelligent design—which is a possible causal explanation for the fine-tuning of *this* universe—or a habitable universe like this universe. If, on the other hand, the many-worlds theorist does offer some causal mechanism, some equivalent to a Universe-Creating Machine, then he has merely shifted the problem back one level, since the machine itself would need fine-tuning.[29]

One could just assert that the multiverse exists. After all, we must all stop the regress of explanation somewhere. But this strategy exposes the many-worlds hypothesis as merely the bitter end of the Copernican principle, with little more basis than the strength of its motivating premise. The principle's devotees are free to look for theories to justify a multiverse. But that doesn't obligate those of us who aren't so committed. And given the persistent failure of the Copernican principle's testable predictions, we're wise not to doubt its most speculative consequence.[30]

THE FAILURE OF THE COPERNICAN PRINCIPLE

In sum, when we can test the Copernican principle against the evidence, it tends to fail.[31] And when it does not fail, it's because it retreats to a strong-hold that makes it unfalsifiable. Although such speculative rearguard defenses may persist, the evidence still points away from the Copernican principle, and toward a single, expanding, fine-tuned universe with a finite past. This cosmos, moreover, has changed profoundly over time. And we occupy not only an exceptional perch within that universe, but also a special moment in cosmic history. While we and our environs are not the physical center of the universe, we are special in the ways that really matter. For life and scientific discovery, we enjoy pride of place.

These findings cast a different light on the story of discovery from Copernicus to the present. The existence of other planets, stars, galaxies, and the like is unambiguous (though limited) support for the Copernican principle only if these aren't relevant to our own existence. But as we've seen, they are. Our existence depends on such local variables as a large stabilizing moon, plate tectonics, intricate biological and nonbiological feedback and greenhouse effects, a carefully placed nearly circular orbit around the right kind of star, with early nutrient-rich asteroids and comets, and outlying giant planets to protect us from frequent ongoing bombardment by comets. It depends on a solar system placed just so in the galactic habitable zone in a large spiral galaxy formed at the right time. It presupposes the earlier explosions of supernovae to provide us with the iron that courses through our veins and the carbon that is the foundation of life. It also depends on the present rarity of such nearby supernovae. Finally, it depends on an exquisitely fine-tuned set of physical laws, constants, and initial conditions.

But why such a large universe? Why all those other galaxies? Surely our argument breaks down here, right? Doesn't the sheer number and size of galaxies imply that we're simply the lucky winners of a giant cosmic lottery? The devil's advocate could argue that even our universe is just barely habitable. Indeed, some have argued that if life is rare in the universe, then the universe is largely hostile to life. Ignore for a moment the fine-tuning of universal laws and constants that yields a cosmos compatible with life somewhere in it. We do have to account for a selection effect at the local level, since, unlike undetectable alternate universes, we *know* that there are lots of other places within the actual universe. The weak anthropic principle attempts to capture this: If we're living in and observing a just barely habitable universe, we shouldn't be surprised that we will only find ourselves on a habitable planet in a narrow, habitable zone around an unusually life-friendly star in just the right place in a barely habitable galaxy at just the right time in a barely habitable universe that is just the right age to produce terrestrial planets. There are other interesting issues here, but that alone isn't surprising.

But notice how the correlation between life and discovery casts a far sunnier light on a large universe. The very immensity of the universe gives us a wide field for exploration and affords far-flung discoveries that also tell us more about our own place in the cosmos. Other planets, stars, and galaxies allow us to compare and quantify the needs for both life and science. We could not do this if Earth were a lone planet encircling a lone star in an otherwise empty universe. Compared to the starry and sprawling firmament that beckons us and stirs our curiosity, and that finally points us beyond the universe itself, a single isolated planet would be a cosmic cage. There would be far less to discover and inspire us, to draw us out of ourselves, to cause us to appreciate that our existence balances on a razor's edge.

If there were no other galaxy clusters, Edwin Hubble could never have inferred the expansion of the universe from the ruddy hue of distant galaxies. The pull of gravity among the galaxies within our galactic neighborhood, the Local Group, overwhelms the otherwise detectable effects of the cosmic expansion. An observer living in a universe containing only the rich Virgo cluster could not discover Hubble's Law. At the level of the local supercluster, which contains the Local Group and the Virgo cluster, the cosmic expansion is detectable, although the effects of gravity remain. It's only above the level of galactic clusters that the cosmic expansion clearly overcomes the force of gravity and the noise of nearby galaxies. In general, the farther we can sample galaxies, the better we can determine the Hubble constant.

Similarly, if we had existed in the distant future, we would never have detected the cosmic background radiation, the linchpin in deciding between the steady-state and big bang models of the universe. These two pieces of evidence have forced modern people, contrary to their philosophical prejudices, to contemplate the shocking notion that the universe had a beginning.

Some scientists have defined the Copernican principle to capture not just life but science as well. Stuart Clark, for instance, writes that the principle is a "cornerstone in every cosmological model of the Universe: it states that there are no preferred vantage points in the Universe. In other words, from the Earth we see everything that could be seen from any other place within the Universe."[32] Well, cornerstone or not, this is clearly wrong for both our place and our time. If we sat atop a neutron star ten billion years ago, the universe would look quite different. And if we were presently orbiting a red giant in the center of an elliptical galaxy, there would be many things we would not be able to see. (We'd also be dead but let that pass.)

Still, lurking in the background is the impression that since the Earth and its inhabitants are tiny compared to the entire universe, they are insignificant. Even an Old Testament Psalm reflects this thought:

> When I consider your heavens, the work of your fingers,
> the moon and the stars which you have set in place,
> what is man that you are mindful of him,
> the son of man that you care for him? (Psalm 8:3–4, NIV)

Of course, once this impression of insignificance becomes an argument, it doesn't amount to much. Physical size is hardly a reliable indicator of value. Miners sift through tons of rock to discover a single, small diamond. They discard the tons of rock and keep the diamond. Besides, as we discussed in chapter 10, one could just as well argue that we (or the Earth) must be really important since on the size scale leading from quarks to the universe, we're strangely close to the middle.[33] This sort of double-edged reasoning gets us nowhere. Nevertheless, it's good to remind ourselves that we can't answer our deepest questions by comparing the sizes of objects. The answers will hinge on much more subtle factors.

WHAT DOES IT MEAN?

Since many used the Copernican principle as an argument for materialism—the idea that the physical universe is all there is and therefore is unintended—one might suspect that arguments against the Copernican principle would also count against materialism. What's sauce for the goose is sauce for the gander, right? Well, the issue is more complicated, since materialists who reject the Copernican principle have another option: Perhaps, the origin and evolution of life are just astronomically rare accidents. Maybe life is a thin veneer on an isolated speck of dust in an otherwise meaningless and impersonal expanse of space, time, matter, and energy.

Before we consider that prospect, however, and what might count against it, let's consider a popular, and initially more attractive expression of the Copernican principle; namely, the search for extraterrestrial intelligence.

CHAPTER 14

SETI, THE DRAKE EQUATION, AND THE UNRAVELING OF THE COPERNICAN PRINCIPLE

*Nothing in the universe is the only one of its kind . . . there must
be countless worlds and inhabitants thereof.*

Lucretius (98–55 B.C.), *De Rerum Natura*

It was bound to happen. As soon as Copernicus suggested that Earth was a planet, someone was sure to ask whether the other planets, like ours, were inhabited. Of course, people had speculated about other worlds since the dawn of recorded history.[1] But now the speculations had a place to land.

Kepler imagined occupants on the Moon, as did William Herschel. Schiaparelli spoke of Martian "channels," which Lowell took to be Martian construction sites—"canals." As late as the 1950s, some believed that Mars had vegetation.[2]

Discoveries in the latter half of the twentieth century ended the idea of intelligent life in our solar system anywhere but Earth. However, the ET dream made an easy leap from our solar system to planets around other stars. Our neighborhood may be barren, but to quote the opening line of *The Hitchhiker's Guide to the Galaxy*, "Space . . . is big. Really big."[3]

The fascination with ETs first emerged in modern literature at the start of the twentieth century through writers such as H. G. Wells; but it was only with the advent of radio, movies, and television that such interest became a mass force. A realistic 1938 radio adaptation of H. G. Wells's *War of the Worlds* by Orson Welles was revealing. It described an invasion by hostile Martian forces, but inadvertently sparked a nationwide panic since many mistook it for a news broadcast.

Since the 1960s, alien life has become one of the most common themes in entertainment. From Arthur C. Clarke and Stanley Kubrick's 1968 classic *2001: A Space Odyssey* to *Star Trek, Star Wars, Close Encounters of the Third Kind, Alien* and its many sequels and prequels, *E.T., Men in Black, Independence Day, X-Files, Arrival,* and *Avatar,* aliens inhabit the modern mind as much as angels and demons inhabited the medieval one. Technological advances such the Apollo Moon landings from 1969 to 1973 gave the topic the aura of realism. Two-thirds of Americans now believe intelligent life exists on other planets.[4] For years, we too were part of that majority. We're now more skeptical.

The growing allure with alien life matches the growth of reports of UFOs, UAPs (unidentified aerial phenomena), alien artifacts like crop circles, alien encounters, and even alien abductions. In many cases, it's easy to see quasi-religious overtones.[5] Publications such as *UFO* magazine refer to advocates of alien visitation as "believers." In more mainstream circles, belief in extraterrestrial life has become so common that astronomer and science historian Steven Dick described it as the dominant myth of our age, which he calls the new "biophysical cosmology."[6]

PARADOX LOST

None of this, of course, discredits the idea. Do intelligent, extraterrestrial beings exist? That's a legitimate question. The problem is that we have no evidence that they do. In 1950, Nobel Laureate and physicist Enrico Fermi (1901–1954) turned to his colleagues at Los Alamos in New Mexico and asked opaquely, with a curmudgeonly Italian accent, "Where are they?"[7] That is, if aliens exist, why haven't they visited Earth and made themselves known?

His question, and the argument it implies, is called Fermi's Paradox— though it's only a paradox if you assume that ETIs must exist. Let's unpack it. If there were lots of other intelligent civilizations like our own in the

Figure 14.1: Enrico Fermi (1901–1954).

Milky Way, some of them would surely have had a head start on us. At some point, they would run out of room or encounter local hazards or just grow curious and start to migrate. Within a few million years—a mere blink of the eye on galactic timescales—they would have colonized the rest of the galaxy. At least, as some later argued, they would have sent out self-replicating robots.[8] They would target habitable systems and either colonize other habitable planets, terraform nearly habitable planets, or mine asteroids. Yet, our galaxy is upward of twelve billion years old, and there's no trace of such colonization, either now or in the past. The reasonable conclusion: they don't exist.

Notice that Fermi used the logic of the "Copernican principle" to reach a conclusion contrary to the views of most of that principle's partisans. He assumed that if a multitude of advanced civilizations exist, we probably would not be the first ones on the scene. This is an intriguing twist since most ETIs enthusiasts also invoke the Copernican principle.

Granted, Fermi's Paradox is not a knockout argument,[9] since absence of evidence need not be evidence of absence (although it might be). Believers in ETI have offered many responses to Fermi's question. Perhaps interstellar travel is more of a barrier than we suppose. Or maybe smart aliens are not the inveterate colonialists that Fermi assumed. Perhaps the colonization wave passed us by.[10] Or perhaps the little green men are good at hiding, or they're so advanced that they no longer produce detectable radio signatures; or have quarantined us like a nature preserve.[11] Bill Watterson offered his own response in his comic strip *Calvin and Hobbes*. "The surest sign that intelligent life exists elsewhere in the universe," Calvin observes, "is that it has never tried to contact us."

Watterson meant this as a joke, of course, but the serious defenses of ETI are so diverse, and at times so far-fetched, that one might wonder whether belief in ETIs is, for some, unfalsifiable. To their credit, some believers in ETI put their belief in ETI in empirical harm's way. This is the Search for Extraterrestrial Intelligence (SETI), an interesting expression of the Copernican principle.[12]

SEARCHING THE SKIES

Early SETI researchers such as the late radio astronomer Frank Drake sometimes transmitted radio signals into space in the hope that aliens might intercept them. But most SETI folks now spend their time trying to detect intentional or unintentional radio and visible light[13] transmissions from extraterrestrials. Using sophisticated arrays of radio telescopes and pattern-recognition software, they hope to separate intelligent signals (if any ever appear) from the background radio noise that permeates the universe.

As with any research, limited resources[14] have imposed some discipline on SETI. Since they require expensive telescope time, SETI researchers want to aim their radio telescopes at a star system with some chance of harboring detectible, intelligent life. To get some purchase on this question, Frank Drake devised an equation, now called the Drake Equation. He first proposed it at a 1961 meeting as a way to calculate how many advanced civilizations able to communicate with radio signals could be lurking in our galaxy.[15]

To the uninitiated, the Drake Equation can look intimidating; but it requires nothing more than third grade level multiplication and fractions. Although a few symbols have changed over the years,[16] the clearest version is as follows:

$$N = N_\circ \times f_p \times n_e \times f_l \times f_i \times f_c \times f_L$$

N, the product of the equation, is the total number of radios communicating, technological civilizations in the Milky Way at any one time. The answer derives from the number of suitable stars in the galaxy ($N_\circ$), times the fraction of stars with planetary systems (f_p), times the number of habitable planets in each system (n_e—e stands for earthlike), times the fraction of habitable planets on which life emerges (f_l), times the fraction of those planets on which intelligent beings also evolve (f_i), times the fraction of those planets on which the right communications technology arises (f_c), times the fraction of the average planetary lifetime during which it hosts an advanced civilization (f_L). Simple.

Since every f is a number between 0 and 1, and most of them closer to 0 than 1, the product of the equation will be vastly lower than the number of suitable stars in the galaxy ($N_\circ$). In addition, many of the variables are unknown. Nevertheless, walking through the equation helps illuminate what a single environment needs to host technological life. (In Appendix A, we offer our own revised version of the Drake Equation.)

Think of the conditions for producing a habitable planet spread out on a number line with different views represented by points on the line. The point farthest to the left represents the view that there is nothing special about the Earth, its local environment, the Milky Way, the number and types of galaxies, the size and age of the universe, and the laws of physics. While these suffice to make Earth habitable, none are necessary, in this view. That is, there are countless other pathways that have the same outcome—complex, intelligent life. The point farthest to the right represents the view that our existence hinges on every detail, locally and universally. We, and any kind of life, depend not just on the precise mass of the proton and a terrestrial planet but also on the exact number of atoms that make up the planet Mercury and the temperature on a gas giant around a star in the Andromeda Galaxy. So far as we know, no one defends either of these extreme positions. The whole debate takes place to the right of center.

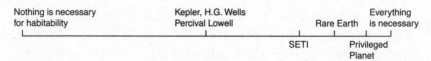

Figure 14.2: On a line representing all views on habitability, the contemporary debate about the frequency of habitable planets in the universe occupies only a tiny segment. It's widely recognized that life requires specific conditions, and that, compared to the total volume of the universe, or even the total number of stars, habitable planets (and civilizations), are uncommon. The debate concerns how many factors are needed to get a habitable planet, and thus, how *uncommon* such planets are.

Compared to the ETI optimism of the last century, in fact, we could cast the current debate as a question of how *rare* habitable planets, and complex life, are in the universe. Carl Sagan sanguinely guessed in the 1960s that there might be one million civilizations in the Milky Way. But even that would mean that only one in a hundred thousand star systems contain ETIs—still a tiny fraction. In any case, what we want to know is just how much is necessary for our existence, and for the existence of complex life-forms similar to us. On the line of habitability, most SETI researchers are well to the left of us, but even they are right of center.

The Drake Equation has two modest qualities: It restricts the search to carbon-based life and planets that can maintain liquid water. As we've seen, these are reasonable assumptions rather than failures of the imagination.

Moreover, it considers only the number of civilizations in the Milky Way. That's good, since even the Andromeda Galaxy, the closest large galaxy to the Milky Way, is more than two million light-years away. The prospects for detecting radio signals from such a distance is nil, not to mention that any signals we did detect would be two million years old. Still, we can calculate the chances of civilizations outside the Milky Way. We just need to add another variable: the fraction of (habitable) galaxies in the universe.

The Drake Equation is now over sixty years old, but until recently only the first term was based on observations beyond our solar system. This has not kept SETI researchers from assuming wildly sanguine numbers for the variables.[17]

Moreover, the equation simplifies the problem, since every factor is the product of another, hidden equation. As SETI supporter Bernard Oliver once said, the Drake Equation, despite its mathematical façade, is "a way of compressing a large amount of ignorance into small space."[18] For example, n_e, the average number of habitable planets per system, depends on many factors, ranging from the metallicity of a host star to planetary size to the architecture of the planetary system. As we discover more about what it takes to build a habitable planet, the values of the equation draw closer and closer to zero.

SETI advocates have drawn inspiration from the Copernican principle, which implies that ETIs should be common. This is obvious in Carl Sagan's famous conjecture, using the Drake Equation, that there are perhaps one million advanced civilizations in the Milky Way.[19] Little more than a guess, it remains entrenched in both the popular and scientific imagination. As we have seen, however, the recent trend of discoveries, and their pace, point in the opposite direction.

Drake, egged along by the Copernican principle and making his calculations before these discoveries, vastly underestimated the barriers for success. However, since his equation has become a standard part of the debate, it's a good place to start.

STARS

In Drake's version of the Drake equation, he plugged in the number of stars in the Milky Way galaxy—which is estimated to be about 200 billion. But as we discussed in chapter 7, that's not a useful number, since the vast majority of stars in our galaxy are not likely candidates for life. To get in the ballpark, we need to include more factors. About 80 percent of the stars in the Milky Way are low-mass red dwarfs, which, as we argued, probably lack habitable zones. About 2 percent are too massive, and therefore short-lived, for technological life. About 4 percent of stars are early G-type, main-sequence stars such as the Sun.

Since we know that our Sun can support life, we might expect the same from other stars the size and type as the Sun. Yet, as we now know, our star seems to contain just the right amount of life-requiring metals, even compared to otherwise similar nearby stars. It contains enough metals to build a habitable planet, but not too much, which might have produced an unstable system with too many massive planets or giant planets that migrate too much. As we discussed in chapter 8, overly massive planets would lumber obtrusively through the system, making a bad neighborhood for rocky planets in stable circular orbits. What's more, at least 50 percent of main-sequence stars are born in binary systems. Most of these are inhospitable for similar reasons.

That might still seem to leave a lot of stars to work with. But remember: Most of those stars aren't in the galactic habitable zone (GHZ), where it's safer from comet impacts and sterilizing radiation, and the metal content is in the right range. Now, these are only the factors related to stars that we've discovered so far. Given the trend, we'll wager there are others. In any case, compared to the rest of the factors in the Drake Equation, this pruning is trivial.

EARTHLIKE PLANETS

Finding extrasolar planets has always been "one of the holy grails of the extraterrestrial life debate."[20] Until recently, however, determining the fraction of stars (f_p) with planets was pure speculation. Astrophysicists like Virginia Trimble used "common sense" and "computer models" to show that "the Milky Way probably still contains at least 10^{10} stars that could have harbored habitable, terrestrial planets for more than five billion years."[21] Translated: she made guesses and plugged them into a computer. This isn't to pick on Trimble. Nobody else was doing any better. But after decades of failure, careful searches in the last few decades have finally begun to pay off. We now know that other sunlike stars do have planets.

These discoveries, however, have been a mixed blessing for SETI enthusiasts. On the one hand, NASA's *Kepler* mission showed that planets around nearby stars are not rare. The average number of planets at least as large as Earth with orbital periods less than 200 days per sunlike star is close to one.[22] This doesn't mean, however, that nearly every star has a planet; just over 20 percent of stars host multiple planets. For this reason, about 50 percent of sunlike stars host one or more planets. And about one in every 100 sunlike stars hosts a hot Jupiter.

In recent years, exoplanet hunters have adopted another, more practical, number. It's called η_{Earth}, pronounced eta-Earth. It's related to n_e in Drake's equation. In short, η_{Earth} is based on the size of a planet and its orbital period about the host star—the two things that photometric transit telescopes like *Kepler* measure directly.[23] Specifically, η_{Earth} is the fraction of planets per sunlike star between 0.5 and 1.5 times the size of Earth and within the habitable zone.[24] There are two ways of setting the inner and outer edges of the habitable zone—the conservative and the optimistic. The optimist sets the inner edge of the zone at the orbit of Venus—at 0.75 AU from the Sun—and the outer edge at the orbit of Mars—1.77. The far more realistic conservative zone puts the inner edge at 0.95 AU and the outer edge at 1.67 AU. These limits are scaled for stars of different masses. η_{Earth} doesn't assume anything else about the planet, including its composition, the shape of its orbit, other planets in the system, or a large moon.

Kepler found only a few planets in these ranges, so estimates are based on extrapolation and are highly uncertain.[25] The best estimates of η_{Earth} range between 0.37 and 0.60 planets per star for the conservative or between 0.58 and 0.88 planets per star for the optimist. Given the uncertainties it could be as low as 0.16 for the conservative case.[26]

Of the over 5,500 confirmed exoplanets known as of this writing, only 24 fall within the optimistic habitable zone boundaries of η_{Earth}.[27] And of these, *none* (zilch, nada) are present around sunlike G stars; the majority orbit M dwarf stars, such as Trappist-1. No doubt this lack is due to the limits of our current ways of detecting exoplanets. In the next few years, we suspect a planet falling within the optimistic η_{Earth} bounds will be found around a G dwarf star.

Chapters 1–8 make clear that n_e, the average number of truly *habitable* planets per system, will be tiny. Be careful not to confuse η_{Earth} with n_e. Theorists who assumed our solar system is typical figured most systems would have about nine planets and a habitable zone. This confidence was premature. Like the other factors, this one is the product of another, larger equation hidden within it. Every new extrasolar system we find confirms how much they can vary. No law of physics or celestial mechanics requires that all smallish terrestrial planets orbit near their star, with all gas giants perched farther out, all in fairly circular orbits. In fact, gas giants can orbit in much smaller, much larger, and highly elliptical orbits. Any such planets would almost surely disturb the orbits of any hapless terrestrial planets unlucky enough to form. No law requires that a well-placed asteroid belt remain after the other planets have formed to deliver just the right amount of water, or that the one terrestrial planet in the habitable zone must have a large moon to stabilize the tilt of its axis. The gravity that controls the orbits of bodies around a star allows enormous freedom in the natures, shapes, and sizes of orbits.

Until recently, most restricted this factor mainly to the average number of right-size planets in a star's habitable zone. We now appreciate how many other details of our solar system play a role in keeping Earth life-friendly. Just since the early 1990s, astronomers have discovered the following:

- Our large Moon helps maintain Earthly life by stabilizing the planet's tilt.
- The climate on a planet in a habitable zone depends on the architecture of the planetary system.

- Gamma ray bursts are powerful events and could sterilize life if such a burst were to occur nearby (within a few thousand light-years).
- Supernovae are more deadly than previously recognized—the estimate of the "lethal distance" has doubled.
- Ravenous giant black holes lurk in the center of nearly all massive galaxies in our cosmic neighborhood.
- Many planetary systems lack asteroid belts.
- Many factors in both space and time on the galactic scale limit habitability in parts of the Milky Way.

How many other such factors are there? Given the trends, we bet there are more to come.

THE ORIGIN OF LIFE

Once we get some habitable planets, we want to know what fraction will be occupied. (Recall that f_l is the fraction of planets with any sort of life.) Despite claims to the contrary, we know very little about the origin of life from inorganic matter and/or organic precursors, or the odds to overcome in the progression of life after that. Even so-called simple life requires a dizzying amount of information. An otherwise habitable planet is necessary but nowhere nearly sufficient for the origin of life. And, as we've noted, there's a chicken-and-egg problem, since simple life helps make a planet hospitable to more complex life. So, in some sense, a planet needs life to get life.

We'll avoid discussing the origin and history of life, since that's a book-length subject in itself. However, we do want to highlight the influence of the Copernican principle on theorists who ponder these questions. On the origin and evolution of life, the Copernican principle takes on a split personality, its explanations refracting into the twin modes of chance and necessity. For some, the principle implies that life is bound to emerge under the right conditions, since otherwise we would be unusual, even special, in the Milky Way. For others, life's a fluke, for if it were all but inevitable, that would heighten the impression that the universe has been fine-tuned to allow for life, which is also a conclusion that sits uncomfortably with the Copernican principle.

The principle's bipolar personality thus exposed; we now see that it's about something more basic than the prevalence of life in the universe.

In the nineteenth century, many thought life on the micro scale was simple. In a letter to Charles Darwin in 1867, for instance, Ernst Haeckel characterized cells as simple "homogeneous globules of protoplasm."[28] We now know that cells are seemingly automated factories possessing nanotechnology beyond the tech industry's wildest dreams. We have what scientists even just a hundred years ago lacked: a sense of the vast chasm between chemistry and biological information.[29] And yet many still pretend that life will emerge if you "just add water." If life on Earth emerged almost as soon as it could have, they surmise, then it would be easy peasy.[30] University of Washington scientists Don Brownlee and Peter Ward penned a book that gave the Copernican principle a black eye (appropriately titled *Rare Earth*), but even they contend, in the same book, that "simple" life may be common in the universe.

Here we can detect a tension between chance and natural law. Brownlee is an astronomer and Ward a paleontologist. In contrast, most biologists, who face life's complexity in their day jobs, see the problem more clearly. They know, or should know, that the chance of getting even the simplest functional proteins through random reactions are absurdly small.[31] In contrast, some origin-of-life researchers, many outside biology, argue that nature may have certain "self-organizing" laws enabling life to emerge anywhere given the right conditions.[32]

Because of this divide, there's not even a biased consensus on the chances of life springing up even on the most life friendly planet we know of—Earth. On one extreme are those who assume that any habitable planet will be inhabited. For them, the fraction of habitable planets with life is close to none. The other extreme could tolerate a universe full of habitable planets, with few or only one inhabited. For these scientists, the fraction is much closer to zero.[33]

INTELLIGENT LIFE

Finally, of those planets with some form of life, we must determine the portion of those on which intelligent life evolves (f_i). Recall our distinction between simple, complex, and technological life. Intelligent life would be some subset of complex life. Judging from the Earthly sample, it's probably a tiny fraction of complex life.

It's here, closer to home, that we reach the core of the Copernican principle. We can sum it up in one statement: *There's nothing special about Homo sapiens.* As with the origin of life, so in the debate over the prevalence of intelligent life we see the tension between chance and necessity, which Charles Darwin combined in his theory. Darwin argued that random variation (chance) combined with natural selection (necessity or law) could explain the appearance of design in biology without a designer. But even the most ardent fans of Darwin's mechanism disagree over which element should have priority.

As a result, the claim that we aren't special can inspire two contrary predictions. On the one hand, it might mean that once life shows up, the evolution of ever more complex and intelligent life-forms is likely if not inevitable. So, we have no reason to think we're unique. On the other hand, it might mean that we're the result of a chain of chance events, catastrophes, and dumb luck, and aren't the crowning achievement of a cosmic drama. If we're unique, it's not because we're designed and doted on, but because we're one-off winners at the cosmic slot machine.

As the Copernican principle trails off to the left and to the right, it's easy to lose the scent. Consider for instance the debate that raged between two paleontologists, the late Stephen J. Gould at Harvard and Simon Conway Morris at Cambridge. According to Gould, Darwin's central message is that evolution is not a rise toward greater perfection, but a meandering march marked by myriad mishaps and mass extinctions. If we took the Earth back to life's starting gate and let the process play out again, life would take radically different forms than it did the first go-around.[34] In short, the Darwinian revolution bears the same message as the Copernican principle: we're not special.[35]

In contrast, Simon Conway Morris argues that natural laws and constraints narrow the role of chance in evolution. He points to the many examples of convergent evolution, such as placental and marsupial mammals, in which similar structural forms appear in creatures that are not closely related.[36] This suggests that strictures for life's evolution have been built into nature, so that the chance of intelligent beings emerging is "probably very high,"[37] even if it takes different pathways. To Conway Morris and others, this may also imply that evolution, and nature itself, has some direction and purpose.

Assuming there are other planets compatible with life, then SETI is much more likely to succeed with Morris's telling than Gould's, but its teleological overtones are an odd fit with the Copernican principle. This leaves SETI advocates in a quandary because most believe that finding ETs would confirm the Copernican principle.

Despite this conflict of interest, its benefits lead most astrobiologists to prefer necessity to chance in the origin and evolution of life.[38] Two exceptions are Brownlee and Ward. In *Rare Earth*, they argue, like Gould, that countless contingent events in Earth's history—many crucial in the history of life—are highly unlikely to recur elsewhere.[39] On Earth, life appeared almost as soon as it could. Intelligent life, however, has only appeared recently. This implies, they think, that even if so-called simple life is common, intelligent life may not be.

In general, then, many astronomers think the Copernican principle implies that life, including intelligent life, is common.[40] Many biologists, in contrast, see life, including intelligent life, as a fluke; yet they too believe their view meshes with the principle. But how can it produce contrary predictions and still be scientifically useful?[41]

ADVANCED CIVILIZATIONS

The last two variables of the Drake Equation, f_c—that fraction of intelligent life that attains radio-communications technology—and f_L—the average lifetime of an advanced civilization—are closely related. (Roughly, f_c is what we refer to as technological life.) We can only guess at both, since we have just one data point to work with: us.

Most SETI researchers assume that intelligent life will often, if not inevitably, develop science and technology. But as we saw in chapter 11, the origin of science and technology depended on all sorts of precursors, and a weirdly long-lasting and stable warm climate. Technology requires dexterity and an aptitude for communicating that, of the millions of known species of life, only humans possess. It also requires access to an oxygen rich atmosphere, dry land, and concentrated ores. Until these factors come together, no civilization can harness radio communication. Even on Earth, this has happened only once. Why assume an inevitable result of life, even intelligent life, everywhere?

A different and ironic wrinkle, known as the doomsday argument, would, if correct, also reduce the chances of chatting with aliens. The argument goes like this. Given the Copernican principle (with the anthropic principle accounting for exceptions), we should assume that we occupy a random rather than privileged sample of space and time. So, we must assume that we don't occupy a special spot along the historical timeline of observers of the human race. Thousands of generations have already lived and died, and the human population has grown exponentially in the last eight generations or so. Moreover, we now have the power to destroy ourselves and most life on the planet. All it would take is a series of geopolitical missteps with, say, Russia, China, Iran, or a terrorist cell with a nuke, to trigger such a cataclysm. Surely, we can't just keep avoiding that fate indefinitely. Therefore, we're likely not in a very early segment of the history of the human race and are more likely near the end.

Since our high-tech civilization has already survived, say, one hundred years, there's a good chance that it will not last much longer than that into the future. Applying such reasoning generally, this would mean that the average civilization advanced enough to use radio communications would survive only a few hundred years. What are the chances that two or more will overlap just when they can communicate with each other?[42]

Some find this reasoning plausible. We don't. We see no good reason to assume that we occupy a random sample of time and space and that the future is finite.[43] It looks to us like a reductio ad absurdum of the Copernican principle, especially since, as we've seen, we have lots of good reasons not to assume these things. The lesson of the doomsday argument is that we should avoid making facile deductions from the Copernican principle. Still, the argument is interesting for its irony, since this pessimistic outgrowth of the Copernican principle implies that SETI, also an inspiration of the Copernican principle, will be unlikely to succeed. Once again, we see the principle turning on itself, like the snake consuming its own tail.

On these two variables of the Drake Equation—that fraction of intelligent life that attains radio-communications, and the lifetime of any intelligent alien race's existence wherein they are advanced—the correlation between life and discovery is a double-edged sword. On the one hand, it gives a ray of hope to SETI researchers. Forget for the sake of argument that the conditions for habitability are extremely rare and that the origin of science depended on many historical contingencies. If intelligent life exists elsewhere in the universe, its local environment is likely as well suited to scientific discovery as ours is. If any ETIs exist, they probably won't be bereft of a large moon, hemmed in by opaque clouds, restricted to deep, murky water, or surrounded by an otherwise empty planetary system bathed in the dull light of thousands of nearby stars. So, they would have similar chances for scientific discovery. Moreover, they would have access to information about long-term climate and nearby objects like asteroids and comets, which would enhance their chances of surviving for long stretches of time. They're also likely to be in the galactic habitable zone, so at least some of them might be nearby.

On the other hand, the correlation between life and discovery strengthens Fermi's argument, since it defuses some of the SETI enthusiasts' responses to it. It would give us reason to suppose that any other civilizations should be as detectable as we are and could detect life on Earth and direct their signals and/or their exploration accordingly. So where are they?

OTHER GALAXIES

Drake limited his equation to the Milky Way galaxy because he was concerned only with ETIs that we have some chance of detecting. However, there are still billions of other galaxies. In the 1990s, astronomers began training the Hubble Space Telescope billions of light-years into space and resolving thousands of distant galaxies never before seen by human eyes. Such a galactic swarm reminds us that the Milky Way is not the only potential habitat in the universe. It motivated probability theorist Amir Aczel to argue that no matter how small the chances, we can be certain that, somewhere else in the universe, there must be intelligent life.[44]

Well, not quite. If the chances are only just barely north of zero, we're back to having to assess the available data and crunching numbers, even with lots of galaxies. We ought to make an evidence-based estimate of how many of the galaxies have a realistic hope of hosting life. Aczel treats galaxies like early SETI researchers treated stars: one's as good as another. As should be evident from chapter 8, this is a mistake. Size, age, type, activity, and metallicity all conspire to winnow the number of galaxies that could harbor even terrestrial planets, never mind all the other conditions needed for life. Our Milky Way galaxy is among the 2 percent most luminous and, hence, most metal-rich galaxies in the local universe, putting it way ahead on the scale of habitability.

We must shake the impression that in the Hubble Deep Fields and recent images from the James Webb telescope, we're seeing pictures of distant galaxies as they now exist. We're really seeing ancient images sent through space billions of years ago during the early stages of cosmic expansion. Many early galaxies were metal poor, which means they didn't have enough materials to produce many habitable systems. In addition, radiation was more intense then; even if a huge galaxy built-up its metals quickly and produced planets, it would have been far more dangerous. Thus, the entire galactic panorama in the Hubble Deep Fields is an ancient portrait of a time and place almost certainly devoid of habitable planets.

We don't know the state of those galaxies now. In the future, when we know more about how galaxies form, we may be able to determine what fraction of these have become like the Milky Way, but right now, this is uncertain. Perhaps a few compare nicely with the Milky Way. Perhaps some now have habitable planets with observers looking back at the early

Milky Way in their own Hubble Deep Field. The truth is, we just don't know. But whether we're unique on a cosmic scale, it's already clear that, in our neighborhood, we might be exceedingly rare. Too much discussion of this issue, however, could distract us from a more basic point. For whether it succeeds, SETI is about something more basic than finding alien races.

SETI AND THE REMAINS OF THE COPERNICAN PRINCIPLE

Complicated and often contradictory, SETI is hard to analyze. It's hard to miss the semireligious overtones of its search for higher, albeit natural, intelligence in the heavens, and its desire for a revelatory message.[45] The 1997 movie *Contact*, based on Carl Sagan's fictional novel about SETI, typifies this spirit. In the movie's climax, the protagonist and SETI researcher, Ellie Arroway (played by Jody Foster), is transported to a distant galaxy to speak directly with an alien representative. She asks him about the purpose of such an interplanetary community, into which humanity has just been inducted. "In all our searching," he tells her, "the only thing that makes the emptiness bearable, is each other."

Later, after this lone encounter, a congressional panel grills Arroway about her experience. Exasperated by their skepticism, she finally describes her discovery as "a vision that tells us how *tiny* and *insignificant* and *rare* and *precious* we are." In one breath, she sums up the paradoxical charm of SETI. On the one hand, its advocates are following the implications of the Copernican principle to its bitter end. On the other, they're embarked upon a quasi-religious quest for those who have lost faith in the traditional object of religious belief. Realizing that there are many other alien civilizations might lead one to regard our own as insignificant. But where does she get *rare* and *precious*? She gets it, perhaps, from the godlike alien being who takes the form of her late, loving father and takes an interest in her. It's not hard to see this fatherlike alien as a proxy for a deceased divine father as well, the father declared dead by nineteenth-century science and philosophy. Every civilization in human history has demanded its god or gods. The SETI enthusiasts are a little different.

At the same time, many SETI researchers have an anti-religious streak.[46] They often claim that finding ETIs would (they usually say *will*) destroy the traditional religious beliefs of most Earthlings.[47] Lucretius in the ancient world also drew this conclusion, as did SETI researchers in Soviet-era Russia.[48] Yet, our insignificance no more follows from a universe abundant with life than our individual significance hinges on a scarcely populated planet. The same is true if intelligent life is extremely rare. A universe teeming with life could just as well be purposefully designed and valued, as could a universe in which life is rare.[49]

Of course, which universe exists—one teeming with life or one where life is rare—is an interesting and worthwhile question. But the interest of both possibilities, and their theological ambiguity, are usually lost on SETI enthusiasts, because the Copernican principle so thoroughly permeates their work.[50]

In fact, as we've noted, various partisans take both possibilities—a universe in which life is rare, or one teeming with living planets—as support for the Copernican principle. Either life is inevitable and hence commonplace, or rare and hence an accident. However, this is a false dilemma since it simply assumes the cosmos has no purpose.

This unstated assumption is apparent in the definition often given to the notion of contingency. Properly speaking, a contingent event is one that happens but didn't have to happen. Philosophers, therefore, contrast contingent and necessary events, the latter being events that have to happen. Most scientists see an event as necessary if it's determined by the laws of physics, but an event can be contingent due to chance or because it's the result of a free choice. Contingency is the arena of both freedom and accident. Materialists collapse all contingencies in the natural world into the category of "chance." Yet, it's not the only option; it's just the only option the materialist will countenance. A good way to foreground the disguised prejudice is to avoid the word *contingency*. Instead, we should split contingency into its two possible forms, and so speak of chance, necessity, and design.

Astronomer and science historian Steven Dick illustrated the point while reflecting on the theological consequences of SETI. He's more theologically astute than many SETI devotees; nevertheless, he concluded that any "monotheistic religions" that wish to "survive extinction" must adjust to new "Cosmo theological principles," by which he means the Copernican principle. The result? In his view, if we are to retain any theological language, we must learn to identify "God" with the cosmos itself.[51]

Unintentionally, Dick has summed up the central aspiration of the Copernican principle: to restrict our gaze to a material universe that, by definition, was not designed. However, what if the truth is otherwise? If science involves thinking hard and open-mindedly about the evidence before us, is it truly scientific to ignore this evidence because it doesn't fit into some philosophical box? And if we choose not to ignore the evidence, then how do we consider it scientifically? Or should I ask, can we?

CHAPTER 15

A UNIVERSE DESIGNED
FOR DISCOVERY

There are more things in heaven and earth, Horatio
Than are dreamt of in your philosophy.
William Shakespeare, *Hamlet*

In the second act of Arthur C. Clarke and Stanley Kubrick's 1968 sci-fi masterpiece *2001: A Space Odyssey*, human beings have made their first small steps into space, and the United States has established a colony on the Moon. During their survey, astronauts unexpectedly uncover a black, domino-like "monolith," buried beneath the lunar surface. When a sunbeam alights upon the mysterious object for the first time in some four million years, it triggers the transmission of a signal toward Jupiter. For obvious reasons, the US government keeps the discovery secret.

No one knows much about the object. But one character announces, without doubt or irony: "This is the first evidence of intelligent life off the Earth." We later learn that the monolith has the geometric proportions of one by four by nine (the squares of the first three natural numbers); but we never see any intelligent life other than human beings. In fact, in the movie we never learn the purpose of the strange object, even after astronauts find a giant replica orbiting Jupiter. Still, there is never any controversy about the origin of these strange structures. No one worries

Figure 15.1: In *2001: A Space Odyssey*, astronauts discover a black rectangular structure on the Moon. Although it's hard to say why, everyone perceives that this object is an artifact of intelligence.

that the lunar monolith might be just a new type of rock. No viewer feels any discomfort that perhaps the characters are jumping to conclusions. Since they haven't seen any aliens, and know nothing of their intentions, why don't the characters suspend judgment about the origin of the object? No one complains that design is unfalsifiable. Everyone can tell that the object bears the hallmarks of intelligent design. No argument, intricate reasoning, calculations, or forensic investigation is needed. Everyone simply *sees* that the object is designed.[1]

Nor do we need strange monoliths planted by standoffish aliens to detect design countless times every day. You're doing it right now, by reading this text. Even if you were looking at a text in a language you didn't understand, you would still know it was text and not, say, a pattern formed from decaying paper. We usually don't need to know the meaning, function, or purpose of an object to know it's designed. We don't even need to know how we do it. We can just do it.

For example, when modern Britons first saw Stonehenge on the Salisbury Plain in Wiltshire, England, they all knew someone had built it. In its discovered state, it was little more than a disheveled circle of large, rough-hewn rocks. (They have since been tidied up a bit.) But most everyone recognized that it was more than the product of wind and erosion. We've since figured out that Stonehenge is aligned with seasonal events like solstices, and that its builders may have used it to predict astronomical

Figure 15.2: Although the purpose and function of Stonehenge has long been a subject of debate, we still recognize that it's designed.

events like eclipses. But even if we didn't know its purpose, we would still "infer design" when we see it.

Archaeologists do this all the time, sifting run-of-the-mill stones from ancient tools, arrowheads, and other artifacts. Such design reasoning plays a key role in several other specialized sciences, such as forensics, fraud detection, cryptography (cracking encoded messages) and, notably, SETI. People are sentenced to life in prison or execution based on a judgment that a death was the result of criminal design rather than mere accident. And everyone assumes that, in principle, SETI researchers could sift out intelligent extraterrestrial radio signals from background radio noise.

In each of these examples, the designer is in nature—either a human being or an ET. But for most of human history, most people also have seen evidence of design in certain natural objects and in nature as a whole, with obvious metaphysical implications. In the West, despite the skeptical arguments of philosophers like David Hume, most still do so, at least in part because of the evidence of the senses. In the English-speaking world, William Paley's famous argument of the watch lying on a heath captured this common sense. As recently as a century ago, his *Natural Theology* was still required reading for undergraduates at Cambridge University. And philosophers have debated, and defended, Thomas Aquinas's arguments for the existence of God for over eight hundred years despite shallow claims of their demise.

Of course, we're told that Charles Darwin made design obsolete in biology, by proposing that natural selection acting on random variations could mimic the work of an intelligent designer. And as we discussed earlier, the dominant view now among scientists and academics is that design is either unscientific or superfluous to natural science.

We have spent a couple hundred pages surveying a class of evidence that contradicts that recent view. With this evidence before us, we can now ask: What's the best explanation for the origin and features of the universe we have described?[2] We have argued at length that the correlation between habitability and discoverability contradicts the Copernican principle and challenges the claim that the findings of science confirm materialism. But it does more than that.

SEPARATING THE CHAFF FROM THE WHEAT

In 1967, a year before the release of *2001: A Space Odyssey*, Cambridge University graduate student and radio astronomer Jocelyn Bell detected an odd extraterrestrial transmission. It consisted of steadily timed pulses. She and her thesis adviser quickly found four sources of such signals. The signals' regularity contrasted with radio signals from then known natural sources. Bell and others wondered if they might be a sign of intelligence. The signal and others like it were dubbed, somewhat tongue-in-cheek, LGMs (for Little Green Men). Research, however, revealed that the signals had a more mundane source. They had come from spinning neutron stars, the remains of supernovae, which they named pulsars. Bell had stumbled across a new kind of star, not the signal of an ETI.[3]

Although SETI has had a few exciting false alarms, so far, it's come up empty. Still, it's easy to imagine what would qualify. In the movie *Contact*, based on Carl Sagan's fictional novel about SETI, researchers receive a repeating signal of beats and pauses, sounding out the sequence of prime numbers from 2 to 101. The viewers and the characters know they have found what they were searching for.

In both Sagan's story and the real-life discovery of Jocelyn Bell, some people intuitively sensed design. This illustrates a couple of things. First, we can't always trust our intuition when it comes to inferring design, especially in natural science, where we try to minimize bias. The signal Bell uncovered turned out to be a pulsar. Kepler was wrong when he thought inhabitants had made the Moon's craters. Percival Lowell was wrong to think Martians had built canals. ET enthusiasts are surely wrong to think that some alien race sculpted the well-known "face on Mars." And it's safe to say that children are wrong to think that clouds really are designed to look like Disney characters.

All the same, we also shouldn't assume that our intuitive judgments are false. The investigators were reasonable to conclude that the radio signal in

Contact was intelligently designed. Ditto the black monolith in *2001*. Or, to take some real-life examples, we're surely right to believe that Egypt's pyramids did not evolve from sand dunes, or the rocks of Stonehenge from unusual seismic events. And you're right to think the black scribbles on this page convey the ideas of authors.

In research, we want to recognize actual design while avoiding false positives; for although we humans are adept and inveterate pattern detectors, we're also pattern imposers. So, how do we separate the chaff from the wheat?

CHANCE, NECESSITY, AND DESIGN

Any event submits to a few basic types of causal explanations: chance, necessity, design, or some combination of the three. These three categories hinge the central disputes in science and philosophy. Within the world, if an event is regular and repetitive, we are prone to attribute it to natural necessity—that is, natural law. Because the repetitive signal Jocelyn Bell received was orderly, she thought it might have come from an intelligent source. But she soon realized that it was the result of a natural object. If a neutron star has certain features, and it's spinning at a certain rate, it will emit a repetitive, pulsing signal, like a lighthouse with a rotating lamp. Bell didn't need to postulate an intelligent agent to explain what she found.

As we noted earlier, if an event is not necessary, then it's contingent. But a contingent event may be the result of either chance or design. If you dump out a box full of Scrabble letters, gravity is responsible for the letters falling to the floor. But gravity doesn't determine that the letters be in any specific order. We would attribute the arrangement of the letters to chance.[4] On the other hand, if you arrange them to spell out a sentence, their order will reflect design. But in both cases, the order of the letters on the floor will be a contingency. So how do we separate chance and design?

A common mistake is to assume that the argument for design is merely a matter of calculating probabilities. The more unlikely an event, many suppose, the more likely it's the result of intelligent design rather than chance. But the fact that an event is merely improbable or that a structure is complex gives little justification for inferring design. After all, improbable events—that is, improbable when viewed in isolation—happen all the time by chance. The world is a big place. Lots of stuff happens. Think how unlikely it is that you should be reading this book just now. What are the chances that the very tree used to make the pages of this book should be your book, and not someone else's? Quite slim. Flip a coin a thousand times, and you've just participated in a rare event. (If you doubt it, try to repeat the same sequence of heads and tails.) Nevertheless, there's no reason to suppose that the particular sequence of heads and tails was contrived.

This is why an argument that inhabited planets are rare does not by itself imply or entail that the universe must be the product of design. Since the universe is old and vast, such planets might just be a rare, chance occurrence. That's why Don Brownlee and Peter Ward can argue in *Rare Earth* that planets with complex life are rare, without concluding that the universe was designed.[5] The rarity of habitable conditions and complex life itself weighs against the idea that the two together flow inevitably from the laws of physics and chemistry. But that rarity alone is an ambiguous signal between chance and design.

Complexity, here a synonym for improbability, is also ambiguous. If you rake up a pile of leaves, the precise arrangement of the leaves in the pile will be complex. If the wind blows them apart, even a clever team of engineers could not reconstruct the pile exactly as it was before. Similarly, if a chimp spends several hours typing on a laptop, he will produce a complex jumble of letters. But in neither case will intelligent design have played a role.

At the same time, complexity, or low probability, seems to be part of the story. It's often necessary, even if not sufficient, for inferring design. One of the reasons the SETI researchers in *Contact* attribute the string of prime numbers to ETIs is that the sequence is complex. But any long, non-repetitive pattern of radio signals would be complex—even if due to chance. So, what led them to reject chance out of hand and infer design?

A SUITABLE PATTERN

Events that lead us to infer design correctly usually have two features: Both low probability (or complexity) and a suitable pattern—what William Dembski calls a "specification."[6] Dembski argues that this joint property— specified complexity—is a hallmark of intelligent agency.

Think of Mount Rushmore. Why do we perceive that the faces of Washington, Jefferson, Roosevelt, and Lincoln were produced by sculptors, but take no notice of the rubble of rocks below them? Neither the faces nor the rock pile is determined by the laws of physics. Both are

Figure 15.3: The faces on Mount Rushmore, in contrast to the more regular configurations of nearby mountains in South Dakota, exhibit both tight specificity and complexity, which together suggest the prior work of a designing intelligence.

complex. But only the faces tightly conform to a pattern that we recognize as meaningful, namely, the likenesses of four American presidents. In fact, even if you didn't know what these presidents looked like, you would still know it was sculpted, and not the product of wind and erosion. Clearly, some intelligent agent (or agents) chose this specific rock shape from the myriad possible shapes.

But the pattern must also be tight. There's a mountain in the New Mexico Rockies called Hermit's Peak. It juts from the ground, as if unrelated to the more gently shaped mountains around it. From a distance, and from just the right angle, it resembles the profile of Abraham Lincoln lying on his back, especially to climbers who have been drinking too much beer in the hot sun. But no one argues that Hermit's Peak was a secret pork barrel project of the Department of Interior. Why? Well, the match between it and Lincoln's face is too loose. The same holds for the famous "Face-on-Mars." It looks like a face from a certain angle and lighting, with low camera resolution. But in a different light and high resolution, it doesn't. It lacks the requisite specificity.

Dembski distinguishes a specification, which in a certain sense is *independent* of the event or structure in question, from a fabrication. It's easy to read a pattern into anything, like we do with clouds or Rorschach blots.

 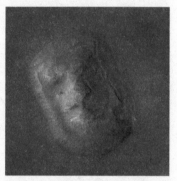

Figure 15.4: The importance of a tight specification. NASA's Viking 1 orbiter took an image of an apparent "face on Mars" in 1976 (left); but the appearance of a face was the result of low resolution and lighting angle. The black dots, one creating a "nostril" on the face, are artifacts of the imaging process. In the much higher resolution image taken by the Mars Global Surveyor in 1998 (right), the structure loses any resemblance to a face.

Figure 15.5: Human beings are not only adept at detecting real patterns, but at imposing patterns subjectively. Rorschach blots exploit this tendency.

When scientists read a pattern into their data, they're cherry-picking. With a large enough body of data, any half-clever scientist can pull out isolated bits that conform to a pattern. But when one looks at the data as a whole, the pattern dissolves.

Dembski gives the example of an archer who draws a small target on a wall, stands back twenty yards, and puts five arrows into the bull's-eye. In this case, we infer that he's skillful rather than just lucky. The bull's-eye is the pattern, and it's independent of the firing of the arrows. Moreover, the target isn't huge. If it were, the pattern match would not be tight enough since it would be too easy for the archer to hit the bull's-eye even if he were an amateur.

But now imagine the archer takes a shot at a large, blank wall and then paints a bull's-eye around the arrow. The pattern will tightly match, but it's just a fabrication since the pattern isn't independent of the event. Therefore, we could not tell from the shot if the archer is skillful.

When we properly infer design, we draw on background knowledge to recognize a significant pattern when we see one. The SETI researchers in *Contact* would not have recognized that they had found an intelligent signal from an ETI if they didn't know the prime number sequence.

In the revised edition of his book *The Design Inference*, cowritten with Winston Ewert, Dembski provides a much simpler way to think of a specification. A specification, unlike some random event, can be described precisely with just a few words. It can be *specified* (and not merely identified, as a proper name can do). The most memorable example comes from the second movie in the original Star Wars trilogy, *The Empire Strikes Back*. At the climax of the movie, Darth Vader tells Luke, "I am your father." Here, the single description "father" packs a lot of punch.

The Mel Brooks parody of Star Wars — *Spaceballs* — provides the telling contrast. In that movie, Dark Helmet (the satirical Darth Vader) tells Lonestar: "I am your father's brother's nephew's cousin's former roommate."[7] Get the joke? Being someone's father is a unique, meaningful relationship. So, the first, short description tells us something significant. The much longer litany of obscure relations does not.

Many designed patterns have other telltale markers. Philosopher Del Ratzsch notes that, historically, arguments for the design of nature have "almost always involved value" to which we attach meaning, and which is habitually associated with mind and intentions.[8]

Such a value is hard to define, but we know it when we see it. Art and music have a value that trash and cacophony lack. A working car has a value that a pile of scrap metal lacks. Being someone's father has a meaning that, say, being the person who went through a random subway turnstile twenty-three days after you did does not. Similarly, a living organism, with its fluid unity, interlocking complexity, end-directedness, and myriad functions, has an intrinsic value that lumps of clay lacks. Learning and discovery have a value that ignorance and loss of knowledge lack. And a universe that can host such life and discovery has an intrinsic value that a barren, opaque one would lack.

So, when we infer design, we're making both a positive and a negative judgment. We judge that chance and blind processes are unlikely to produce something, and we discern certain features, such as complexity and a pattern of value that we can specify with a few words, that we associate with intelligent agents.

A COSMIC DESIGN

We must adapt this explanation a bit since detecting design *within* the natural world differs from detecting the design of the natural world itself. In the world, we normally detect design by contrasting it with what natural laws and chance can do on their own. Designed objects tend to have what Del Ratzsch calls "counterflow." They contrast with the way nature would go if left to operate on its own. A designed event will stand out against the background of nature's normal course. This counterflow may be why the lunar colonists in *2001: A Space Odyssey* saw the monolith as an artifact rather than, say, a type of rock. Unlike the monolith, large rocks don't have perfect, geometric angles. We see geometric shapes in nature at the middling size-scale in quartz and other minerals. And we see spherical structures at the scale of planets and stars. But a large, black, perfectly geometric rectangular box at the human scale that stands out against the more irregular background lunar surface? C'mon. That just screams "design."

How do we see such a contrast when pondering the origin of nature itself, however, since we don't have a background contrast? In fact, as already noted, within nature, we *distinguish* events that arise from natural necessity from ones that are designed. So, if nature and its features are designed, are fine-tuned, how could we tell? What's the relevant contrast against which we discern a meaningful pattern?

In pondering the fine-tuning of the universe, we're distinguishing between logical and natural necessity. Natural necessity is what will happen in the world as long as no one interferes. For instance, if you lift a ball

Figure 15.6: Philosopher and mathematician William Dembski has done seminal work in formalizing important aspects of how we detect the activity of intelligent agents.

off the floor, you haven't violated the law of gravity; you've just interfered with what would otherwise happen. But neither the law of gravity nor the gravitational force constant has the force of logic. Neither *has* to have the value it does. The inverse square law does not have the same status as 2 + 3 = 5 or the claim that "All bachelors are unmarried." Logic doesn't prevent gravity from having a different value. And if no universe had existed, it would have no value. Or to put it differently, there could have been objects that obey a different large-scale attractive force than gravity as it is in our universe. There are other possible worlds in which gravity (or some counterpart to gravity) is different.

When we say the laws of physics are fine-tuned, we're saying that, in contrast to the many other possible universes in its "neighborhood" of possible universes, ours has just the physical laws that make it habitable. We're contrasting the laws in our universe with other, similar (hypothetical) universes with slightly different properties, as well as with the endless sea of chaotic and disorderly universes that might have existed. We're also recognizing a key distinction between universes that host life and ones that lack it. Our habitable universe stands out against the many similar but uninhabitable possible ones. If we were just to pick a universe's properties at random, we would rarely if ever stumble across a habitable one. To use Dembski's terms, our universe and our place and time within it are both complex and specified to make possible that most complex of phenomena — technological civilization. Most hypothetical universes in our universe neighborhood do not.

Sometimes we can calculate the likelihood that an event will occur. Other times, we can't. What are the chances that some universe will exist? Or that this universe would exist? There's no obvious answer unless we restrict the domain of possible universes. So, let's assume that we can't determine the exact improbability of the fine-tuning of fundamental laws and constants, and that we can only compare it with the range of possible universes around it. That is, we can compare it to other possible universes in which the fundamental properties are just slightly different.

This comparison stacks the deck in favor of the no-design hypothesis, since if we were to consider hypothetical universes with settings far different from our own, the odds would be far worse for the chance hypothesis. So, based on this even more focused assessment of our hypothetical-universe neighborhood, what do we find? The range of inhospitable universes still vastly exceeds the range of hospitable ones. And by vastly exceeds, we don't mean ten to one or a hundred to one. We mean by a ratio impossible to imagine and only possible to describe using mathematical exponents representing a number with so many zeros that if zeros were Cheerios, you could feed an army with them.

A common complaint about this argument is that no matter how unlikely our actual universe is, it's no more unlikely than the other possible ones. We can now see why this is not a persuasive rebuttal. When we make valid design inferences, it's not mere rarity that leads us to believe there's something fishy that needs explaining. It's the presence of a telling pattern. We're reminded of the joke about a man, Charles, who comes home early to find his wife naked, in bed, with a flushed face and tousled hair, even though she never naps or sleeps naked. Then he opens the door to his bedroom closet and finds his next-door neighbor, Phil — a notorious philanderer — standing there, naked and in a full sweat. When Charles interrogates him, Phil shrugs and replies, "Don't be so suspicious, Charles. After all, everybody's gotta be someplace."

Get the joke? Of course you do! Why? Because of the meaningful, and highly suspicious, circumstances. The underlying reasoning is sound. We're not comparing the chance of the event in isolation. We're comparing the relative likelihood of competing hypotheses. What's more likely?

(1) Phil just happens to be hanging out naked in Charles's bedroom closet, while Charles's wife—who never naps or sleeps naked—is a few feet away naked in bed.
(2) Phil and Linda are having an affair.

Do we really need to elaborate?

As we mentioned above, design inferences usually involve some tacit judgment of value. When comparing universes, everyone recognizes, unless they're trying to avoid a certain outcome, that a habitable universe with intelligent observers has an intrinsic value that one incompatible with life lacks. Living beings have a value that inanimate objects lack. That value redounds to a habitable universe. Like sweaty Phil naked in the closet, skeptics tacitly betray that they share this judgment when they try to explain away the fact that our universe is fine-tuned for complex life.

THE CORRELATION AS A MEANINGFUL PATTERN

Of course, this argument is quite abstract: We have only one actual universe to observe. Our comparisons must be mediated by theoretical judgments about other possible universes. In contrast, the correlation between habitability and measurability is more concrete: We can confirm it by comparing settings within the universe. We can compare and contrast the conditions needed for life and science and, in theory, calculate the probabilities. We have a large universe in which planets, stars, and galaxies are arranged in all sorts of different ways. Highly habitable and highly science-friendly places stand out from almost all other regions of the universe.

It's because these conditions are so rare that we can see that the correlation between the two forms an interesting pattern. If the universe were homogenous and everywhere compatible with observers and for observing a wide range of phenomena, the most we could say is that the universe is open to scientific discovery and that it is, in some sense, "rationally transparent."[9] This might be suggestive; but we wouldn't have anything tangible to contrast with our immediate surroundings. As it is, we know that the fine-tuned laws and constants of physics and cosmic initial conditions allow a lot of variety when it comes to local settings. It's interesting, and surprising, to find that the rare habitable places are also the ones best suited to diverse types of scientific discovery.

Now, we must guard against reading a pattern of an event or artificially imposing a pattern on it. For instance, there's a matching pattern between the presence of habitable environments and life. We will only find life in habitable settings. Moreover, these settings are exceedingly rare. Nevertheless, one element obviously depends on the other since life can only exist where it's possible. No one should be surprised to find living beings restricted to environments compatible with their existence, to recall the weak anthropic principle. The situation might be the result of design, but this pattern alone provides no evidence for that conclusion.

In contrast, there's no obvious reason to suppose that habitable environments would also be the ones most conducive to diverse types of scientific discovery. Being habitable and being science-friendly are distinct properties. We could compile separate lists of the properties that contribute to life and those that contribute to discovery. We could analyze one without ever referring to the other. There's no logical necessity connecting the two. In fact, few people—even those who study the physical universe full-time—have even bothered to think about the physical prerequisites to science. And yet we can describe this meaningful pattern with just a few words: habitability correlates with discoverability. Or, to state it with a bit less economy: the rare places within our universe where scientists can exist provide those scientists with the best places for observing.

We can't attribute the high degree to which our local environment is congenial to science to a selection effect. While we should expect to find ourselves only in places compatible with our existence, for instance, we have no reason to expect that we must find ourselves in places that allow us to discover the deep truths of the universe at the largest and smallest scales. We have seen that the conditions needed for life provide the best overall conditions for doing science. But doing astronomy, physics, geology, and cosmology is hardly necessary for our existence. If anything, spending countless nights during your prime years fiddling with telescopes surely puts one at a Darwinian disadvantage! Having perfect solar eclipses, or stable polar ice deposits, or tree rings, or being able to view the stars or determine their temperature and composition, or having access to the cosmic background radiation, clearly provided no survival advantage to our ancestors.

If we did not now know otherwise, in fact, we might expect that the more habitable a place is, the less useful it would be for science in general. For instance, the lifeless space between galaxies is better for seeing distant galaxies than is the surface of a planet with an atmosphere near a bright star. We might then suspect that this is true in most or all cases. An Aristotle or a Ptolemy might have conjectured that a setting with the ingredients for life would hinder our knowledge of the universe. But when we take account of the many things that need measuring and observing, we find just the opposite.

It's *surprising* to discover that these conditions correlate in the universe—that observers happen to find themselves in the best places overall for observing. This discovery calls for an explanation beyond an appeal to blind chance or necessity. It's at least a striking coincidence. "A coincidence is always worth noticing," Agatha Christie's detective heroine, Mrs. Marple, wisely observes. "You can always discard it later if it *is* just a coincidence." That is, if it's due to chance. But the correlation between life and discovery is clearly a telling pattern, one worth attending to.

One reason we often prefer an explanation to its competitors is that it resolves our sense of surprise. It offers what philosopher John Leslie calls a "tidy explanation."[10] Design provides just such a tidy explanation here. For instance, imagine a society that assumes that the physical universe is designed so that any observers should find themselves in a place conducive to many kinds of science. In that case, discovering the correlation would be just what they had expected. The presumption that the universe is designed for discovery resolves the surprise by providing a tidy explanation.

This isn't a deductive proof that the cosmos is designed, but it surely confirms the hypothesis. To put it another way, if we assume that the universe is designed at least in part to allow intelligent observers to make discoveries, the correlation between life and discovery is what we would expect. In contrast, if the cosmos exists for no purpose (which we can call chance) and if intelligent observers are simply a purposeless dross in that indifferent cosmos, we would not expect this. It would be an inexplicable fluke. Whatever the probabilities, clearly the correlation is much *more* likely given design than given chance—which here means "no design."

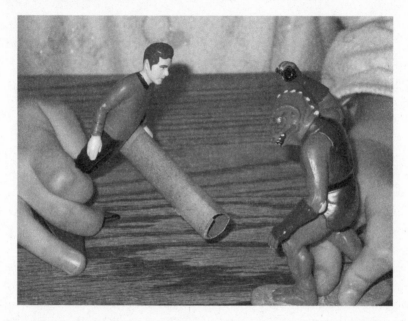

Figure 15.7: In the *Star Trek* episode "Arena," Kirk uses his wits to build a cannon with ingredients supplied by the Metrons. (Gillian Richards reenacts the drama with figurines of Captain Kirk and Gorn Captain.)

A universe designed for discovery resolves our surprise. Shrugging our shoulders and chalking it up to chance, or some blind impersonal process, does not.

The (undesigned) multiverse scenario is also deficient since it does not make the existence of any kind of universe more likely. In fact, it's utterly indiscriminate. No matter what kind of universe existed—a pure hydrogen universe, a black hole universe, an utterly chaotic universe—it would be equally consistent with the hypothesis of a multiverse (of course, no one would be around to form the hypothesis, but that's another matter). Even most of its habitable universes would not contain habitable oases for observers, oases that are also the best overall places for observing.

Contrast that with what we would expect from the design perspective. We all grasp the intrinsic value of a science-friendly universe that can contain complex, living observers like us. And, like winning the lottery twice in a row, to find those observers just where they can best make diverse discoveries is doubly telling. It's the sort of world one could expect an intelligent agent to have some interest in designing.[11] It's a fishy pattern, and we know it.

HOW DID IT HAPPEN?

Okay. But how is the design implemented? We don't know for sure, but nothing in our argument depends on the answer. Any complete answer in our view will require a much richer theory of nature than the one bequeathed to us in the nineteenth century. But as the previous should have made clear, our argument is consistent with a design that's embedded or encoded in the laws, constants, and initial conditions themselves. Although this differs from our usual way of detecting design within the world, there's no reason we can't detect design here.

Imagine if, in 2050, the United States and China are on the brink of war. On Christmas Eve, American and Chinese diplomats meet in Jerusalem to begin their final, fragile round of negotiations. At midnight in Jerusalem, however, while diplomats are reaching for another pot of coffee, some Israeli astronomers notice a striking new pattern of craters on the Moon. Focusing their telescopes, they discover a group of identical craters, which precisely spells out "Glory to God in the highest, and on Earth, peace among those whom he favors!" in Hebrew, English, and Mandarin. The Chinese and American astronomers confirm that the craters were formed by simultaneous impacts from a dense cluster of asteroids moving in normal trajectories through the solar system. Moreover, the asteroids hadn't been perturbed from their normal courses or fired from a giant alien ship hiding in the asteroid belt.

Nevertheless, everyone would still recognize a setup—*based on the effects.* Such highly specified crater patterns are not merely rare but unique. What's more, the Hebrew, English, and Mandarin languages are independent of the impacts, and the crater pattern matches them very closely. We can precisely describe the event in just a few words. Moreover, the timing of the message fits tightly with world events. The event is much more likely if we assume a setup rather than a coincidence.

Similarly, habitable environments are exceedingly rare. The fact that they are also the best overall places for observation forms an independent pattern. So, we have good reason to suspect that this was intended, however that intention was realized.

The correlation not only suggests design but also a specific purpose. Detecting design is often easier than discerning purpose or meaning. For instance, museums often house human artifacts whose purpose has been lost to antiquity. Similarly, the American colonists in *2001: A Space*

Odyssey knew they were dealing with an alien artifact, even though they had no idea what it was, what it did, or who put it there.

Nevertheless, detecting a purpose enhances our confidence that something is designed. In a homicide investigation, a detective may treat a shifty-looking man with blood on his shirt who is walking away from the crime scene as a prime suspect. Finding that the man was married to the victim and had taken out a large life insurance policy on her just before the murder will increase the detective's suspicion.

Notice the pattern the detective discerns is not in an object but in a situation. That's often the case. The first season of *Star Trek* contains an episode called "Arena." The story involves an advanced alien race, the Metrons, who captures Captain Kirk as well as the captain of another alien vessel, of the "Gorn" race. As punishment for invading their space, the Metrons transport Kirk and the Gorn captain to an uninhabited planet to fight to the death. They tell the two combatants that they— the Metrons—will provide the weapons. But, once transported, Kirk and the Gorn captain find nothing but large rocks. The lizard-like Gorn is stronger than Kirk, but not as clever. Before long, Kirk notices a strange abundance of minerals such as diamonds, sulfur, potassium nitrate (saltpeter), and coal.

"This place is a mineralogist's dream!" he exclaims, dismayed that he can find nothing as simple and useful as a large club.

But he soon realizes he has the ingredients for gunpowder, along with the hardest known substance—diamonds—for projectiles. It slowly dawns on Kirk and the viewers that the Metrons have set things up. Rather than supplying him with ready-made weapons, the Metrons have provided the ingredients, which require him to apply his own ingenuity. Once Kirk recognizes this, he begins searching for the rest of the materials he needs. He quickly finds a bamboo stalk that looks a lot like . . . a cannon. With one shot, Kirk wounds the Gorn but decides not to kill him. He thus saves himself while impressing the Metrons with his resourcefulness and capacity for mercy.

Although Kirk's wits play a role in his success, his surroundings clearly were set up for this purpose. In fact, the scenario is a more fitting challenge than a simple cache of weapons would be, because Kirk must use his intellect to forge a weapon himself. Still, no matter how clever Kirk was, he could not have made a cannon without the right ingredients.

The presence of so many disparate elements, all needed to make a cannon, is not just a coincidence. Kirk realizes this in part because, of the countless planets in *Star Trek*, none are so ridiculously well stocked without the need for digging or mining. It becomes clear with a little reflection that the setting is arranged to allow a smart inhabitant to create a cannon. Moreover, it's clear that the Metrons are highly intelligent, and clear precisely because of the intriguing fit between the arena planet and Kirk's background knowledge. A less advanced culture would have put Kirk and the Gorn captain in a fighting rink with ready-made weapons. The Metrons have given Kirk and the Gorn captain a mental as well as physical challenge.

The story is fiction, but Kirk's form of reasoning about his situation is used to good effect all the time in the real world. We often discern a setup—a design—even if we're ignorant of other basic facts.

But what about the situation we've described in this book? Consider one last story. Ken is an avid mountain climber who decides to climb a high, desolate mountain on the island of Hawaii. He doesn't know that the mountain is Mauna Kea. When he reaches the top, he's surprised to

Figure 15.8: When able, astronomers put observatories in the best places for observing. The appearance of such tools in good locations for observing is a meaningful pattern. We discern an intelligent purpose when we discover the large twin telescope domes of the Keck Observatory atop the dormant volcano Mauna Kea in Hawaii, rather than in downtown Honolulu.

see several man-made structures. This includes two large telescopes peering out of the openings in two massive white domes. Although he doesn't recognize the famed Keck telescopes, or know anything about how they are built, he knows that they are telescopes. But, more importantly for our purposes, he realizes *why they're perched atop a high mountain*, and not, say, in downtown Honolulu or the middle of the rain forest. Astronomers put telescopes where they can best view the heavens. Telescopes don't get built just anywhere or grow on mountaintops because of some blind necessity of nature. They're put on mountains for good reason. And appealing to chance in this case is no explanation at all.

Analogously, the rare places with observers are also the best overall places for observing. If an otherwise ignorant climber knows why people put telescopes on high mountains, he should see the same purpose in our natural environment. Like the telescopes atop Mauna Kea, this isn't just an intriguing pattern. It's a meaningful one.

INTIMATIONS OF THE PATTERN

But if so, why have so few noticed this pattern? One reason, perhaps, is that those acquainted with the evidence have been discouraged from considering design or speaking publicly about it. Surely another reason is that most of the evidence is of fairly recent vintage. As a result, we had the good fortune to be the first ones to develop the argument in any detail.

That said, some had caught glimpses of it. In a 1975 book, historian Hans Blumenberg noticed the amazing fit between a habitable atmosphere and a relatively transparent and therefore scientifically useful one.[12] Biologist Michael Denton also made note of this early in his 1998 book *Nature's Destiny*.[13] Denton went further in a later chapter:

> What is so striking is that our cosmos appears to be not just supremely fit for our being and for our biological adaptations, but also for our understanding. Our watery planetary home, with its oxygen-containing environment, the abundance of trees and hence wood and hence fire, is wonderfully fit to assist us in the task of opening nature's door. Moreover, being on the surface of a planet rather than in its interior or in the ocean gives us the privilege to gaze farther into the night to distant galaxies and gain knowledge of the overall structure of the cosmos. Were we positioned in the center of a galaxy, we would never look on the beauty of a spiral galaxy

nor would we have any idea of the structure of our universe. We might never have seen a supernova or understood the mysterious connection between the stars and our own existence.[14]

Denton didn't at the time develop an argument for an astonishingly wide and deep-ranging correlation between life and discovery; but clearly, he had caught the same scent we did.[15]

Similarly, physicist John Barrow has discussed the value of a universe with three spatial dimensions for both life and the high fidelity of information transmission so vital to scientific discovery.[16] And echoing our first chapter, historian of science, Stanley Jaki, argued that the Earth-Moon system not only contributes to Earth's habitability, but to scientific discovery as well.[17]

Philosopher Robin Collins has independently calculated the degree of fine-tuning needed for science, as we noted in chapter 10.[18] He has found that the observed values of several physical constants and initial conditions are within the "optimal discoverability range," which is a subset of the habitability range. Robin's approach differs from ours, but his conclusions are much the same. In fact, he goes a bit beyond us by arguing that the universe is *optimized* for rather than merely friendly to science.[19]

Finally, Michael Mendillo and Richard Hart of Boston University anticipated the discovery described in chapter 1. In 1974, they presented a paper called "Total Solar Eclipses, Extraterrestrial Life, and the Existence of God," which was later excerpted in *Physics Today*.[20] They argued, with comic precision, the following:

Theorem	An exactly total solar eclipse is a unique phenomenon in the solar system.
Lemma	There are observers on Earth to witness the remarkable event of an exactly total solar eclipse.
Conclusion	A planet/moon system will have exactly total solar eclipses only if there is someone there to observe them. As only Earth meets this requirement, there is no extraterrestrial life in the solar system.
Corollary	In a system composed of nine planets and thirty-two moons, for only Earth with its single moon to have exactly total solar eclipses is too remarkable an occurrence to be due entirely to chance. Therefore, there is a God.

Their argument is tongue-in-cheek, of course. But it's funny because there *is* something fishy about finding the best eclipses just where observers can enjoy them. They failed to notice that the same features that produce perfect solar eclipses also play a role in our existence as observers. And they did not imagine, so far as we know, that a similar pattern would turn up, again and again, in other areas. If the correlation stopped with eclipses, we might chalk it up to coincidence. But as we have seen, eclipses are just the beginning.

We live in a universe with laws, constants, and initial conditions finely tuned for complex life. These factors are necessary but not nearly sufficient for such life, however. In tiny pockets of that universe, we find rare oases for complex observers who can study the starry heavens above and ponder the meaning of their existence. In at least one of these places, despite struggle and adversity, some came to believe that the world around them was a rational, orderly *uni*verse. Centuries of study, amplified by technology, have given rise to an unparalleled knowledge of the world around us. Together, those discoveries now give rise to another: The same narrow conditions that have sustained our existence also make possible a stunning array of discoveries about the universe. Observers find themselves in the best places, overall, for observing.

It's thanks to this that our hunger for scientific knowledge can be satisfied. The careful study of nature pays off. With persistence, the world discloses itself to us in ways we do not, and often cannot, anticipate. The thought creeps up quietly but insistently: *The universe, whatever else it is, is designed for discovery*. What better mandate could there be for the natural sciences? The scientific pursuit enjoys cosmic prestige. But that prestige is apparent only to those willing to consider the evidence that the cosmos exists for a purpose.

CHAPTER 16

FIELDING THE BEST OBJECTIONS TO THE PRIVILEGED PLANET

*A fair result can be obtained only by fully stating and balancing
the facts and arguments on both sides of each question.*
Charles Darwin[1]

We have offered evidence and suggested what at the moment is a controversial explanation for that evidence. Even if it weren't controversial, no empirical argument is ever immune to objections. We have drawn on a wide array of scientific disciplines and evidence. The argument is not deductive. We can't just evaluate the premises for contradictions and, having found none, declare victory. To be cogent, our argument should have the preponderance of evidence in its favor. We think it does. Nevertheless, we can anticipate several rejoinders, and our argument would be incomplete if we didn't field them. None of these defeats our argument, but we hope our responses will help us clarify it and further explore its implications.

Objection 1:
It's impossible to falsify your argument.

Because our argument is based on observations, we don't expect perfect agreement between habitability and measurability in every case. In fact, if you isolate the conditions for measurability, it's easy to come up with apparent counterexamples. But this doesn't directly answer our argument. Recall that our claim is not that every condition is *individually* optimized on the Earth's surface for doing science. Our claim is that *our habitable environment is an exceptional compromise among a diverse array of sometimes competing factors, a compromise that achieves a constrained optimization for discoverability. We see this in areas ranging from cosmology and galactic astronomy to geophysics, and that those same conditions also enhance habitability.* This argument accommodates facts that would constitute counterexamples for a less nuanced argument. However, precisely because the argument is empirical and carefully nuanced, a single counterexample probably wouldn't refute it. Thus emerges another problem: How could it be falsified?

Recent work in the philosophy of science has revealed the degree to which high-level theories resist simple refutation; nevertheless, it's a

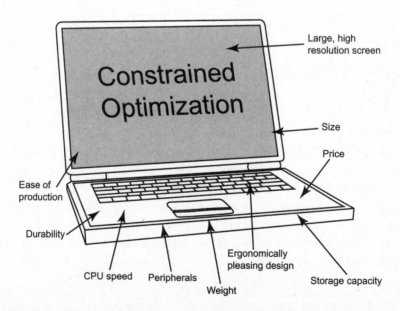

Figure 16.1: A laptop computer, like many well-designed objects, exhibits "constrained optimization." The best-designed laptop computer achieves the best compromise of multiple competing factors.

virtue of scientific proposals to be able to say what evidence would count against it. And that's easy to do here. There are clear discoveries bearing on the needs for life or for doing science that could count against it. More than this, there are two simple and direct ways to falsify our argument as a whole. First: Find a very different environment, which, while quite hostile to life, still offers a superior platform for making as many diverse scientific discoveries as does our local environment. Second: Find a highly habitable place that is a lousy platform for science.

Less devastating but still relevant would be discoveries that contradict parts of our argument. For instance, astronomers could find that complex life does just fine in much wider and more diverse environments than in our account. Imagine, for instance, that NASA finds intelligent life inside a gas giant with an opaque atmosphere, near an X-ray-emitting star in the galactic center, or on a planet without a dark night. That would do serious damage to our thesis.

Or take a less extreme example. We argued in chapter 1 that conditions that produce perfect solar eclipses also contribute to the habitability of a planetary environment. Thus, if intelligent ETs exist, they probably enjoy good to perfect solar eclipses. If instead we find complex, intelligent, indigenous life on a planet without a largish natural satellite, this plank in our argument would collapse.

Our argument presupposes that all complex life in this universe will almost certainly be based on carbon. Find a non-carbon-based complex life-form, and one of our assumptions collapses.

Just as some discoveries would hurt our argument, others would help it. Pretty much any discovery in astrobiology will bear on it one way or the other. If we find that still more ingredients are needed to build a life-friendly planet, for instance, this will strengthen our case. Some evidence comes fast. When we were writing the first edition of this book, several new requirements for habitability came to light—such as the link between a planet's rotation and orbit and the motions in its interior.

Since the first edition, we have discovered a major new instance of our thesis: the solar system appears to be set up to facilitate space exploration.[2] And, as is clear in the updated endnotes for chapters 1–10, research over the past twenty years has continued to build on the original suite of evidence for our thesis. Independently, philosopher Robin Collins has discovered several additional ways the universe is optimized for science.[3]

So, our argument is eminently testable. It can certainly be confirmed or falsified by new evidence.

Other objections involve the structure or nature of our argument itself. Let's take each of these in turn.

Objection 2:
It's inevitable. Whatever environment we find ourselves in, we would find examples conducive to its measurability.

If we were in a different environment, our discoveries would be made based on those conditions. Scientific geniuses would come up with clever ways of testing their hypotheses, no matter what their environment. We must base any argument we make on the conditions we know from our actual setting, not the conditions we don't know. So, at first glance, it might appear that we're under the illusion of a selection effect.

For instance, imagine a loving husband, George, and his wife, Laura, who take the same, eccentric walk every morning. For about a mile they walk along the main road, but they always take a shortcut down a side street, across a wooded trail, and through a park. They have no reason to suppose that anyone else ever takes the same route. One Valentine's Day, George is away on business, so Laura decides to invite her skeptical neighbor to join her on the usual walk. As they walk, the two women find that every hundred feet, a plain, red paper valentine has been placed on a tree, telephone pole, or fence post. Because the course is so unusual, Laura concludes that her husband, knowing the unusual path of their walk, has put up the valentines as a romantic gesture.

Her neighbor, however, an avid reader of the *Skeptical Inquirer*, doesn't buy it. Just a coincidence, she insists. When they arrive back at Laura's house, she tells Laura that they were likely the victims of a selection effect. After all, they didn't inspect the rest of the town. For all they know, the city may have put up valentines everywhere. They only noticed the ones on their walk because, of course, that's where the two of them were. Disheartened, Laura realizes that her neighbor may be right. Perhaps the city is filled with paper valentines, and George has forgotten yet another February 14.

But then it dawns on her that she can easily test her and her neighbor's competing theories by checking the rest of the neighborhood. When she does, she finds she was right all along. The pattern of valentines not only overlaps her route, but the rest of the city's trees, fences, and telephone poles are largely valentine-free.

Notice that as long as Laura couldn't compare the route with the surrounding area, she couldn't determine whether her husband was responsible for the valentines. Only when she had observed the contrast between her route and its surroundings could she be confident that the valentines formed a suitable pattern revealing a very specific design.

We're in a similar situation, since we can compare our environment with others. For the discoveries we have made, we can reflect on the conditions needed for such discoveries, and then compare those with conditions in other settings. For instance, it's obvious that a relatively transparent atmosphere is better for astronomical curiosity and discovery than a murky (translucent) or opaque one. We know that, at least in our solar system, such an atmosphere is rare, and that it correlates with a habitable atmosphere. We can compare the properties of our atmosphere with the properties of other planetary atmospheres we know about. We can also get some sense of the chances that a planet will bear the properties needed for a transparent, habitable atmosphere.

We can compare the features of our planet that allow us to have a dark night with other planets with different configurations. We can compare the properties of our Sun with other known stars. We can compare the shape and size of our orbit with the orbits of other bodies we know. We can also compare it with other possible orbits of planets and moons. We can compare our local galactic environment with the environments in, say, the galactic center or a globular cluster. We can compare the details of our solar system with other systems we discover. We can compare the sedimentation and information storage processes on Earth with those of other planetary bodies. We can compare solar eclipses as seen from Earth's surface to views from other planets and moons.

Using models and theory, we can compare our local circumstances in cosmic time with other cosmic times, both past and future. We can also imagine what sorts of experiments we could perform in different environments and compare their value with what we've been able to discover from Earth.[4] So, we aren't prisoners to our local conditions; we can compare those conditions with other, more common ones.

Notice, moreover, that we can compare our local setting with other similar settings only because we have such a great view. If we had lived in a world with a murky atmosphere, we would not know of other planets or stars or galaxies, and, therefore, we could not compare our home with anything beyond its atmosphere. An observer in such a world would have no way of showing that the correlation exists. As we learn to make more and more such comparisons, and comparisons that would be out of reach and out of view in other environments, this selection-effect objection grows less and less tenable.

Objection 3:
Well, then, it's just a selection effect of a different sort. There are phenomena we cannot observe or measure. The argument is biased toward what can be discovered.

This objection is similar to the previous one and also contains an element of truth. If we are in an environment in which we can't discover something, then we may never know that we can't discover it. We're in the realm of Donald Rumsfeld's unknown unknowns.[5]

Again, however, the argument is overstated.

Contrary to the claims of the anti-realist, who doubts that any of our theories maps onto reality, scientists aren't locked in a box where everything we perceive in the universe is primarily the product of our perception. Many things are hard to discover and measure—especially as the scales of the extremely small and extremely large—and we realize that fact. The math works in quantum theory, for instance, but you don't have to be a hardened skeptic to suspect we might not have the right mental picture of what is really happening at nanoscopic scales.

And there are many less extreme cases. We can't determine the distance and properties of some objects in our sky. Still, we know they exist, since we can detect them either directly or indirectly, and we know that we don't know their distance and many of their properties. We can compare the objects in this category with the objects we can both detect and study and then generalize.

Similarly, we are not so bereft of imagination that we can only ponder those things we directly perceive. Since nature is regular in its operation, we can extrapolate from what we do see to what we don't see. A theorist often predicts the existence of certain objects prior to their discovery, such as planets, dark matter, black holes, and neutrinos. For secure theories, we can imagine what conditions would allow us to detect such objects (and this has happened many times in the past. It's striking how often physicists can devise a way to detect entities that are first predicted by theory.) We can then determine whether our environment allows us to do so and compare it in this regard to other settings in the universe.

Objection 4:
You're cherry-picking. You have used a biased sample to argue for the correlation.

There is always a danger with any general hypothesis like ours. When looking over a large body of data, the theorist may pick out the pieces that form an intriguing pattern and ignore the pieces that don't. As a result, when the data are viewed more completely, the pattern dissolves. This is an especially acute risk for any argument like ours that involves many different scientific disciplines.

This is why we chose important examples from the relevant scientific disciplines. We haven't fixated on obscure experiments or features of our setting that make little difference to science. Some of these are obvious. Just imagine the value of a transparent atmosphere and visible stars for astronomy, or sedimentary processes for geology. Every astrophysicist knows about the role of perfect solar eclipses in the history of stellar physics. No cosmologist would deny the worth of detecting the redshift of distant galaxies or the cosmic background radiation. Without these discoveries, we would grasp little about the history of the universe.

Finally, as we said, other scientists have noticed evidence of the correlation between life and discovery. So, it's unlikely that we're creating it out of thin air.

This is a useful objection, nevertheless. If we're right, the correlation will continue to be confirmed not only in the areas we've considered, but also in areas we have not. We predict that there are still many vital discoveries ahead. Some we can anticipate, some we cannot.

In the first edition of this book, we predicted that an identifiable subset of gamma ray bursts would one day be found to be useful standard candles.[6] Our only reason for predicting this is that, if the correlation is real, then we could expect gamma ray bursts to be prime candidates for helping us measure the universe. Twenty years ago, gamma ray bursts were already known to be at cosmological distances. "Perhaps they will allow tomorrow's astronomers," we wrote, "to probe even greater redshifts than we can with Type Ia supernovae today."

Well, our prediction has since been confirmed. Research over the past several years has shown that gamma ray bursts are indeed standard candles useful for probing the very distant universe.[7] Not only that, but astronomers have also discovered that quasars are also cosmological standard candles and that they are useful to even greater distances than gamma ray bursts.[8] While this latter discovery wasn't one of our predictions, it certainly supports our thesis.

Another such prediction concerns the evidence for early life. As we mentioned in chapter 3, the Earth's geophysical processes have erased much of the early history of life. If our setting is optimal (overall) for doing science, then we have reason for optimism that such information has been preserved somewhere accessible to us. The origin of life is a big question. It would be surprising, assuming the correlation, if it was forever beyond our ken.

In fact, we'll predict that such evidence is available somewhere, if we search diligently enough. It was precisely this expectation that led one of us (Guillermo) to consider the value of lunar exploration for uncovering relatively well-preserved relics of Earthly life from this early period.[9]

While we're at it, we'll also predict that, since carbon and oxygen appear so often among our examples of measurability, they will be central characters in future discoveries as well.

Of course, if we're right about these predictions, this would not prove our position but only further support it. If we're wrong, conversely, it would not destroy our argument; it would put a dent in it. But clearly our argument is fruitful. In contrast, the Copernican and Anthropic Principles in their most unrestrained manifestations seem much less useful. Positing the existence of multiple universes, for instance, offers little in the way of fecund research programs within our universe. It's main use, rather, is to foreclose certain unwelcome metaphysical options.

Objection 5:
Your argument is too speculative. It's based on guesses and a thin empirical base.

Most of our examples are based on well-understood phenomena and solidly grounded in the evidence. Examples include solar eclipses, sedimentation and geophysical processes, the properties of our atmosphere and of the planets in the solar system, the spectra and structure of stars, our place in the Milky Way galaxy, and our place in time. The evidence for some of our other examples, such as the features of exoplanets, is weaker because the evidence is new. But, even in these examples, our arguments are based on solid theory. And the new evidence that has flowed in during the past two decades has consistently supported our argument.

We've noted that there are speculative elements in our discussion of the circumstellar and galactic habitable zones. While we can't yet estimate the precise boundaries of these zones, we strongly suspect that published studies are still missing many relevant factors. We expect that future evidence will narrow estimates of their sizes and strengthen our argument. Again, we're putting our argument in empirical harm's way. It's vulnerable to future discoveries. That's a virtue of a scientific hypothesis.

Objection 6:
Your argument is too subjective. It lacks the quantitative precision needed to make a convincing case.
It's true that we have not tried to quantify the correlation between habitability and measurability. We're the first to make this argument; we can't pretend to offer a mathematically rigorous treatment. In any case, this is not necessary.

For example, we don't need a detailed "habitability index" to see that the Earth's surface is far more habitable than other planets in the solar system. Similarly, we have argued, without benefit of a "measurability index," that the surface of the Earth offers more opportunity for scientific discovery than the other planets.

A skeptic might claim we've arbitrarily weighed the values of just the sorts of observations that are feasible from Earth, while downplaying the opportunity costs of that location. But any such charge will have to deal with our specific examples. We argue, for instance, that our lack of direct access to the giant black hole in our galaxy's nucleus is a small price to pay in exchange for discovering and measuring the microwave background. Is that arbitrary? Clearly not. We give greater weight to the background radiation, because, unlike the galactic nucleus, it: (1) is unique, (2) gives us direct information on the overall properties of the universe, and (3) gives us a glimpse of its origin.

Of course, it doesn't follow that no one could quantify our analysis. In fact, as we noted in chapter 15, Robin Collins has independently discovered a pattern like what we describe in these pages. Although he formulates it differently, Robin does quantify it.[10] In fact, we're hoping an ambitious reader will do so.

Objection 7:
How can you have a correlation with a sample size of one?
It's true that the Earth is, so far, our only example of a habitable planet. But this does not prevent us from finding a correlation between life and discovery. First, our argument doesn't hinge on the quirks of our home

planet and the life we know about. Life anywhere else in the universe, we've argued, will resemble life on Earth, at least biochemically; and a planet very much like ours is required for technological life. Starting with these basics, we have used knowledge from a broad range of disciplines to consider a broad range of environments.

Discovering a correlation between life and discovery, then, is based on our knowledge, not our ignorance. For example, with knowledge of stellar astrophysics and climatology, we can ask if a planet around an M dwarf is habitable and offers better or worse prospects for science than the Earth does. We've used our knowledge of galactic astronomy to ask the same questions of the local and distant universe.

Objection 8:
Since life needs complexity, the correlation is trivial. The greater the complexity, the greater the chance for a correlation between habitability and discoverability.

One of our claims is that even in a universe fine-tuned for life, getting all the right local ingredients for a habitable planet is extremely unlikely and, in that sense, complex. As a result, it might appear that the more complex an environment is, the more habitable and measurable it will be. That is, complex life requires complex conditions, including various stellar, planetary, and geological details. Such a complex environment will by definition be better for doing science, according to this objection. As a result, measurability will attach to habitability in the same way that inhabitants attach to habitability. Perhaps all our argument amounts to is the observation that our environment is complex, and complexity itself entails that the environment is both habitable and measurable.

It's true that complex life and wide-ranging scientific discoveries require a *diverse* environment. But mere complexity in an environment does not correlate with either the needs for life or scientific discovery. For instance, an opaque, chaotic atmosphere like Jupiter's is as complex, if not more complex, than Earth's. But it's much less habitable.

Similarly, a stable planetary system with nine major planets in fairly circular orbits is less complex than an uninhabitable one with a swarm of diverse and irregular planets in various eccentric orbits. A gas giant planet with lots of moons of various sizes and orbits is more complex than the Earth with its single, large Moon.

Having a perfect solar eclipse does not require more complexity than countless other systems. A relatively circular galactic orbit, like our own, is less complex than a wildly elliptical orbit that traverses several different regions of the galaxy and is often perturbed by nearby objects. A single star system like ours is less complex than its closest neighbor, Alpha Centauri, which has three stars. The chaos of two colliding, irregular galaxies is more complex than the orderly spiral of the Milky Way galaxy. We could go on, but you get the point. In many cases, the less complex setting is not only more habitable but also more measurable.

The difference between the conditions on Earth and those of other, more common ones is not simply that Earth's are, in many respects, more complex. Our setting is complex, yes, but it also conforms *to a telling, easily specified pattern*, in which the rare conditions for habitability and discovery correlate.

Objection 9:
There may be separate pathways quite different from ours leading to equally habitable environments.

This argument would have carried more force thirty years ago, but most discoveries since then work against it. Detailed knowledge of other planets in our solar system, theoretical understanding of habitable zones, and the discovery of many examples of fine-tuning in physics and astrophysics all make it less likely that an environment very different from ours would be just as habitable. Recent discoveries about extremophiles notwithstanding, it's starting to look like even the most earthlike locales in the solar system (aside from Earth itself, of course) are too hostile for simple life.

We should not ponder a hardy bug in isolation from its Earthly home but should treat it as part of a system. Even if planets with primitive life turn out to be common, the restrictions on complex and technological life are much, much more severe. Maybe life prospers in some radically different habitat elsewhere. But to affect our argument, those posing this objection will need to offer a realistic example. Otherwise, it's just an assertion.

It's even less likely that life would be based on something other than carbon and water. The evidence from science has continued to reinforce the unique life-essential properties of carbon and water, the latter first used as a design argument by William Whewell over 150 years ago.

This objection is good to keep in mind, though, since finding a large, complex form of life not based on carbon or water would blow a hole in our argument.

Objection 10:
Your argument is bad for science because it encourages skepticism about cosmology.

Cosmologists often claim that they must assume our setting is mundane for theoretical reasons, since it allows them to extrapolate from the part of the universe, we can see to the parts we can't see. If this were correct, then our arguments against the Copernican principle might make us doubt that we can know anything about the cosmos. Consider Martin Rees's question, mentioned in chapter 10: "Why does our universe have the overall uniformity that makes cosmology tractable, while nonetheless allowing the formation of galaxies, clusters and superclusters?"[11] When Rees says cosmology is tractable, he means that it satisfies the following assumptions at some relevant size scale:

(1) Space is homogeneous; the universe looks the same at every location. There is no privileged place.
(2) Space is isotropic; at the largest scales it looks the same in every direction.
(3) The laws of physics are the same everywhere.[12]

As we've seen, the first has required ongoing qualifications as we have learned ever more about the universe. Clearly space isn't homogenous at the size of people, planets, stars, galaxies, or clusters of galaxies. Now, it's assumed that space is homogenous above the level of galactic superclusters, that is, clusters of clusters of galaxies. In fact, recent evidence suggests that such superclusters are arranged in a pattern of voids and filaments, somewhat like large bubbles, throughout the cosmos.[13]

The second assumption in its most unrestrained form is called the perfect cosmological principle. It states that every time in the universe looks just like any other. This claim conflicts with the evidence we've surveyed in this book. How the universe looks depends profoundly on one's location in both space and time. It's no surprise, then, that the perfect cosmological principle inspired the steady-state models and met the same demise. A reformulated, more modest version of this assumption—which doesn't claim that all times look the same—finds support in the smoothness of the microwave background radiation.

The third and final assumption—that the laws of physics apply everywhere in the universe—is perhaps the most important. It's the true legacy of the Copernican Revolution. And it's based solidly on the evidence. For example, spectroscopic observations of distant galaxies over a broad range of wavelengths verify that their atoms have the same properties as those measured in Earth laboratories. Based on such observations, astronomers and cosmologists feel justified in extrapolating the laws of physics from their tiny laboratories to the entire universe.

Does our argument imply that we can no longer extrapolate the universe from the parts we can observe? Does it feed skepticism about our knowledge of the cosmos? Does it require that we restrict ourselves to observational astronomy, and eschew cosmology?

Of course not.

In fact, the success of cosmology depends on a superficially similar, but contrary assumption, namely, that what we see is a good, accurate sample of the universe. Let's dub this the "discovery principle." According to this principle, since our environment benefits so many kinds of science, we can expect to find other ways that our environment benefits the scientific enterprise, features that allow us to gain even more access to the universal laws of physics, as well as the variety of small and large structures in the universe.

Perhaps the most plausible way to put assumptions (1) and (2) above is in the form of a discovery principle: *We should be optimistic that the parts of the universe we can't see are relevantly similar to the parts we can see (from our privileged vantage).* By "relevantly similar" we mean that the sections we can't see don't contain objects or laws fundamentally different from those we can see.

The discovery principle even suggests a properly qualified cosmological principle. Cosmologists have always been fond of the cosmological principle because it allows for "simple" solutions to Einstein's equations (though if you're not well versed in tensor theory you may not find the solutions so simple). This is why mathematical cosmology has been so successful. The universe we see fits the simplest global expression of the equations of general relativity. Could it have been otherwise? Would a less homogeneous universe be just as habitable? We don't yet know for sure.[14] In any case, the cosmological principle (rightly defined) was not born of the Copernican principle and can do just fine without it. And quite apart from demonstrating our mediocrity, it's a prerequisite of a universe designed for discovery by creatures like us, who are restricted (until Elon Musk gets his wish) to a single location.

Something like the discovery principle is crucial if we seek more evidence about the cosmos. Discoveries in astrophysics and cosmology don't come cheap, and future investments in such research would seem prodigal without the reasonable hope that our hypothesis supports.

To those scientists weaned on the Copernican principle, it may seem counterintuitive that to justify conclusions about the universe at large we need a privileged view, not a mediocre one. But the evidence for this is not hard to find. For instance, probably more planetary matter in the cosmos exists in the form of gas giants than in planets like ours. In this respect Jupiter is more mediocre, more typical, than Earth is. But the more mediocre habitat within the opaque depths of a gas giant atmosphere would leave us clueless about the larger universe. We might easily assume that such a heavy atmosphere simply stretched on to eternity.

Similarly, moons around such planets are likely more common than moons around terrestrial planets like our own. And as we argued in chapter 5, these are unlikely to provide the high-quality measuring platform that Earthlings enjoy. Mediocrity is not what science needs. What it needs is the privileged and probably quite rare perch of an earthlike planet. This isn't a mere assumption. It's what the evidence suggests.

Moreover, notice that the discovery principle undermines the reason for postulating unobservable regions with laws and properties different from the observable universe.[15] In other words, cosmology—the study of the cosmos as a whole—is a valid enterprise only if we're right to assume that we have privileged access to a representative sample of the universe. So, cosmologists have a powerful motivation to eschew those many-worlds hypotheses that posit radically different universes from our own.

Of course, it doesn't mean that we'll continue to make major discoveries indefinitely. Perhaps, in some domains, we will reach a knowledge limit. Perhaps, for example, we will be forever stymied from peering behind the curtain of the cosmic microwave background. After all, how could we see all the way back to the time before light decoupled from matter? Well, we couldn't, but we might be able to infer quite a bit about it.

Imagine an observer on the surface of a planet with a thick translucent atmosphere, where one hemisphere continuously faces its host star. The cloud deck on the planet's sunny side would appear uniformly bright, but for all that light, our cloud-befuddled observer could not learn about the distant universe, or even if there was anything beyond the bright clouds. Happily, though, if our alien observer had an advanced degree in physics and the right equipment, he could obtain a spectrum of the light filtering through the clouds and perhaps infer something about its source. Similarly, the background radiation's angular power spectrum tells us something about events just before it formed, even as it imposes a limit on what we can know.

And again, the structure and laws of our universe provide a few more tools that future scientists could exploit to peer even further back. Perhaps one day they will map the neutrino- or gravity-wave background and learn more about what happened less than 300,000 years after the big bang. So, while we surely can't learn every truth about our universe, already we have learned far more than expected, and we will surely learn a great deal more.

Objection 11:
General relativity appears to be a superfluous law of nature, which is not obviously required for habitability. Yet it's an important part of science. Does this not contradict the correlation?
George F. R. Ellis has argued that the major features of the cosmos might be just as well-determined by Newtonian theory alone.[16] In other words, it may be that life does not require a universe where the laws of nature include general relativity. Of course, we don't know if this is correct. True, general relativity seems far removed from our everyday affairs, but it does affect element production, coalescence of neutron stars and black holes, the orbit of Mercury, and the immediate environment around giant black holes in the nuclei of galaxies. Since life is sensitive to the cooling rate of our universe, perhaps a Newtonian universe would fail to meet this requirement.

But let's assume Ellis is right. What does it imply? It implies that at least one known law of nature is not obviously linked to our existence. But, for all we know, perhaps a universe with the same number of space and time dimensions, laws, and/or forces as ours needs general relativity. If we discover a more fundamental law or laws that unite the various forces and reconcile relativity and quantum theory, relativity theory might strike us as an essential outgrowth of the whole, like the light-gathering leaves of a tree. Furthermore, Einstein's theory, much as Newton's theory of gravity before it, may prove critical in helping us move on to that next and deeper level of discovery.

Even ignoring these scenarios, however, it's not clear that Ellis's argument bears much on our argument. We don't argue that everything in the universe is relevant to habitability or to the correlation between habitability and discoverability. There might be facts about the universe that are useful for science that aren't directly related to the needs for life. This doesn't refute the claim that the conditions for both tellingly overlap.

Objection 12:

The correlation of life and discovery isn't mystical or supernatural since it's the result of natural processes.

In a sense, this is true, but it does nothing to weaken our argument. Our argument differs from some design arguments since it does not require or presuppose the direct design of specific objects within the physical world. We have not argued that the correlation between life and discovery transcends natural laws in quite that way. We have focused only on the design of the cosmos as a whole. We claim that the correlation forms a meaningful pattern that gives us very good reason to suppose that the universe is purposefully designed. The issue here is not whether there are natural laws, parameters, and initial conditions, but whether they're evidence of design and purpose. We believe they could be and are evidence of this, in part through the striking correlation between life and discovery.

In the introduction to his *Principia*, Isaac Newton argued that our solar system was designed by God. He concluded: "The most beautiful system of the sun, planets, and comets, could only proceed from the counsel and dominion of an intelligent and powerful Being."[17] It's not clear how Newton thought God implemented this design, but clearly, he saw the solar system itself as a sort of artifact akin to William Paley's watch resting on a heath. He may have thought that God had to directly set things up.[18] For theists, this is always possible. But nothing in our argument requires this. In fact, we have assumed that given the vast chances the universe affords, a solar system like ours *could* come together without direct divine agency.

On the other hand, if in reality the odds are too prohibitive for that, this is not a problem for our argument. Imagine that in the future, we gain detailed knowledge about extrasolar planetary systems and their dynamics. As a result, we can determine that the probability, even in our fine-tuned universe, of getting the non-biological precursors for just one earthlike planet, with its solar system and everything else it needs, is less than 1 chance in 10^{180}.[19] This would mean that, even in a (visible) universe with 10^{11} stars per galaxy and 10^{11} galaxies, totaling 10^{22} available attempts, the chances of getting one such system would still be 1 chance in 10^{158}. This is a staggeringly small probability. Taking habitability as a specification, as it surely is, we would have a strong argument that our solar system is a giant artifact of intelligence, like a watch or a Boeing 747. Although we can presently only guess what the odds are of becoming a habitable planet, perhaps future scientists could nail this down.

The problem with such an argument—that getting even one habitable planet in our universe is statistically impossible—is not that it violates certain philosophical scruples, though it surely does. The problem is that we don't have enough information to evaluate it. We can't base our argument on data that no Earthlings have. Hence, we have assumed that, given the (fine-tuned) laws of physics, the cosmic initial conditions, and the number of stars and planets in the universe, there is some nontrivial chance that at least one habitable planet like the Earth, or at least its non-biological prerequisites, will form. This is, however, just a working assumption. We're now beginning to realize that a single habitable planet requires an array of delicately nested systems ranging from the local to the cosmological level. Perhaps this array is unlikely enough that it renders the "chance" formation of even one such system unimaginably unlikely. Hopefully, one day there will be enough quality data available to tell.

In any case, anyone familiar with the dynamics of planetary systems knows how hard it is to get a system with eight planets like our own to remain "chaotically stable" for several billion years. Given this and the crucial role several of the other planets in our solar system play in promoting Earth's habitability, we suspect true earthlike habitable planets will be much, much rarer than most SETI advocates assume. But we don't yet know exactly how rare such systems are. For one thing, we don't yet know what percentage of stars have earthlike planets, surrounded by a life-nurturing planetary system. Fortunately, we don't need such detailed information to recognize design in this case. Our argument doesn't rest on the notion that Earth is unique or that its environment has been directly fine-tuned. Our argument rests on a particular *pattern*—the correlation between the needs for life and for scientific discovery—of which our environment is an exemplar.

Objection 13:
You haven't really challenged naturalism. You've just challenged the idea that nature doesn't exhibit purpose or design.
Our argument contradicts materialism, but only indirectly challenges naturalism. That's because it's possible to be both a naturalist and to admit design in nature. In fact, in the ancient world, both the Aristotelians and the Stoics did just that. It's possible, for instance, that design is somehow

an inextricable part of an eternal cosmos. The philosopher Thomas Nagel has suggested this in his 2012 book *Mind and Cosmos*.[20]

We can't conclusively rule this out. The problem in our modern setting is that this strategy would require a pantheistic view of nature that most naturalists deny. A cosmos that includes design and purpose as well as chance, matter, and natural law, is quite different from "nature" as most modern materialists understand it.

Moreover, current big bang cosmology discourages the view that the cosmos is eternal, which is necessary if design is coextensive with matter, time, and natural law. A causal agent that transcends the cosmos is a much more sensible explanation for the big bang and the resulting physical universe we know than are purely immanent patterns of design. But either is a better explanation than the currently popular view that the material universe is all there is, was, or ever shall be, and that chance and impersonal necessity exclusively explain its existence.[21]

Objection 14:
You haven't shown that ETs don't exist. In fact, design seems to improve this possibility.

This is true. SETI folks should take note. If you're looking to stay optimistic about finding extraterrestrial life, the materialism bandwagon offers little hope of getting you to the promised land of extraterrestrial contact. ETs seem increasingly improbable if our only resources are chance and necessity. But the odds of their being out there improve dramatically if the design hypothesis is on the table.

By extension, if, on the materialist hypothesis, it's extremely unlikely that life would evolve from nonlife, or that nature would produce lots of habitable planets, then discovering life on far-flung planets would strengthen the argument for design, just as the same person winning a national lottery several times in a row would arouse our suspicions that someone has rigged the system.[22] Improbability would then conspire with a telling pattern. In short, whether the universe is populated with living planets is an open question in our argument, not a reasonable deduction from the Copernican principle, as many seem to think. Our argument does not require that we are the only intelligent life in the universe, nor does it require that we are not. It implies, rather, that if such life exists elsewhere, it will find itself in an environment very much like ours, one friendly to life and discovery.

Objection 15:
God wouldn't do it that way.

One popular retort to cosmic design arguments masquerades as a scientific objection; but it's really a theological claim: "If there were a God, He wouldn't do it that way." This complaint has several problems. Before going into those we should first underscore that our argument points only to a designing agent that transcends matter, space, and time. Those attributes are needed to pull off the big bang and the fine-tuning of the laws and constants of nature, after all. But by itself the argument doesn't entail, or presuppose, the God of the Bible. The argument proceeds from certain signs of design in nature to a design inference.

Moreover, we must distinguish between design on the one hand and perfection or optimality on the other. Something can be designed without being perfect according to this or that criterion. Just because someone can build a better mousetrap doesn't mean that our present mousetraps aren't designed. We're old enough to remember a car from Yugoslavia, which was sold in the US from 1985 to 1992, called the Yugo. It's been called the "worst car in history."[23] Still, everyone could tell the thing was designed.

One version of the "God wouldn't do it that way" complaint often shows up in defenses of the Copernican principle. If the universe were designed for intelligent observers like us, many argue, why is it so large, so old, and so largely uninhabitable? Surely, a wise and benevolent creator would not be so wasteful and so stingy.

But this argument just doesn't bear on the question of design per se. Moreover, it presupposes a superficial, and hardly unimpeachable, view of the very designer the skeptic is most interested in sidelining, namely, the God of the Bible and classical theism. The objection assumes a being limited to finite material resources, since the virtue of efficiency is only relevant to such agents. But theists view God as a supremely powerful Creator. Such a Creator would not be taxed by the most extravagant of universes.

At the same time, it's not wrong to wonder, why is the universe so vast, why there are such a myriad of stars, and worlds innumerable? We have already mentioned one possible reason: the vastness of the universe enhances discoverability, and with it, the scientific enterprise.

And since the objection is theological, it's surely fair to suggest a theological answer. The vastness of the cosmos may hint at the grandeur of what lies behind it. As the psalmist writes:

> When I look at thy heavens, the work of thy fingers,
> the moon and the stars which thou hast established;
> what is man that thou art mindful of him,
> and the son of man that thou dost care for him?

The psalmist, dumbstruck by creation, finally exclaims,

> O Lord, our Lord,
> how majestic is thy name in all the earth! (Psalm 8, RSV)

This, we suspect, is the far more natural response to our vast universe than is the barren shrug of the materialist.

CONCLUSION

READING THE BOOK
OF NATURE

We shall not cease from exploration
And the end of all our exploring
Will be to arrive where we started
And know the place for the first time.

T. S. Eliot[1]

A*pollo* 8 was the first manned mission to orbit the Moon. It was also the mission that gave us the photograph of the Earth "rising" above the Moon's horizon. We've all seen that picture, but few know of the mission's long-term effects on its three crewmen. For Frank Borman and Jim Lovell, seeing the distant Earth from lunar orbit strengthened their conviction that human beings were at home in the universe, that they, and we, existed for a purpose. "The earth from here is a grand oasis in the big vastness of space," Lovell told the world on Christmas Eve, 1968. For Bill Anders, however, who took the *Earthrise* photo, this same view had the opposite effect. The Earth, so tiny and isolated within the cold emptiness of space, suggested a lonely futility. Having abandoned the Catholic faith that he had once held dear, he later described his vague impression that had begun in lunar orbit: "We're like ants on a log."[2]

Three men. Two opposing views of the world.

So, how might the argument and evidence described here inform these two contrary perspectives? Those who already believe in a meaningful world may find our argument satisfying, even expected. The skeptic, though, may accuse us of imposing a pattern on the evidence, much like psychiatric patients impose patterns on Rorschach blots. But such skepticism is a double-edged sword since it may blind one to actual patterns in the world. The reasonable skeptic should keep an open mind.

But even for the open-minded, there's a mental screen that can blind us to real patterns. Since the nineteenth century, we have been discouraged from seeing purpose in nature. To recognize the correlation between life and discovery as a meaningful pattern, rather than a mere curiosity, we must re-awaken atrophied intellectual virtues. Fortunately, scientists have never really neglected them, even if many of them have restricted their application.

Here's what we mean. All of us know far more than we can describe explicitly. As chemist and philosopher Michael Polanyi put it, "We know more than we can tell."[3] How we do this is an open question. But *that* we do it is obvious. Think of how we can recognize a friend or a spouse in a large crowd. We see the specific features of individual faces, but we don't directly attend to them. No one reckons like this: "My wife has brown hair with gray highlights, blue eyes, a rounded nose, high cheek bones, and twenty-six freckles. There's someone with all those features. So that person is probably my wife." Recognition is much more subtle and subterranean. Most of us would have a tough time describing someone's face in such detail that an artist could reproduce it exactly. Nevertheless, we easily distinguish a familiar face from hundreds of other faces.

Likewise, when we read a text, we don't attend to discrete ink marks or pixels or letters. Polanyi said that we attend *from* those letters and words *to* their meaning. We focus on the meaning embedded in the arrangement of ink marks on a page, not on the marks themselves. The text is the vehicle that transmits its meaning. The faculty for discerning that meaning, however, has to be cultivated. We must learn to read.

Like readers, natural scientists foster the skill to read patterns in nature. They learn to *see through* the isolated bits of data to the underlying pattern, as we see through the letters and words on a page to discern their meaning. Before the nineteenth century, for instance, fossil finds were rare and uninformative. People treated them as curiosities and mantelpieces. Today, paleontologists know where to look. Expert fossil hunters often.have an uncanny knack for discerning that a structure is a fossil of a once living thing, rather than a mere rock. To those untrained in fossil hunting, this skill can seem mystical.

Astronomer Edwin Hubble likewise learned to "read" an expanding universe when he photographed redshifted distant galaxies. To do so he had to entertain the possibility that the universe was not in eternal stasis but had changed profoundly over time—that it might even have had a beginning. His discovery led to an explosion of new areas of research and helped secure the legacy of cosmology as a science. When scientists read nature rightly, nature discloses herself in new and surprising ways, like a rich and layered poem to the patient interpreter.

Done well, such readings from the book of nature uncover new lines of research. Herein lies a virtue in recognizing the correlation between life and discovery as a real pattern rather than explaining it away as mere coincidence: We should expect to find it elsewhere and so be spurred to make fresh discoveries. To one who discerns a designed cosmos, this evidence is much like the sublime beauty, the mathematical elegance, of the natural world. It's no longer a troublesome anomaly to be waved away but something fitting and wonderful. To see it as a fluke, in contrast, is both theoretically and aesthetically sterile.

Of course, to see purpose in nature, we may have to do what scientists in so many other respects have learned to do, to cultivate the skill to read the book of nature, to read *through* nature to its meaning. We have argued that the cosmos bears the marks of genius. And all of us are already adept at detecting the work of intelligent agents. Although archaeologists, criminal investigators, cryptographers, and SETI researchers have specialized expertise and well-defined methods for detecting design, everyone who reads a text or understands a language has the same skill. But, just as a grammar book can't replace reading and appreciating great literature, knowing arguments for design in nature is no substitute for developing the skill to see that design.

Such an exercise is, for most of us, a journey. Yet, every journey begins with a single step—in this case, that step involves opening one's mind to an almost forgotten possibility. Thus does our inquiry reduce to a single question: Could this immense, symphonic system of atoms, fields, forces, stars, galaxies, people, and a planet called Earth have sprung not from some inscrutable outworking of blind necessity or an inexplicable accident, but instead from a choice, a purpose? If so, then surely there could be evidence to suggest as much.

We've seen that scientific progress and discovery depend upon nature being more than meaningless matter in motion, and even more than motion that can be generalized with natural laws. Nature has an exquisite structure that preserves vast stores of information about itself and its past. Our life-giving environment provides access to an exceptional and highly sensitive set of devices that encodes information about the natural world. We, in turn, have what we need to create technologies to decode these devices. Technology extends humanity's creativity and vision. As eyeglasses and the light bulb have helped us read written texts, so the microscope and telescope have allowed us to read the book of nature more deeply. In the last few centuries, we have done so with unparalleled speed and success. This, in turn, has allowed us to discover a striking pattern: the myriad conditions that make a place habitable are also the best overall places for discovering the universe in its smallest and largest expressions. This is the central argument, the central wonder of this book.

The central irony is perhaps this: The more we learn how much must go right to get a single, habitable planet, the more we realize that the materialist mindset behind the Copernican principle and SETI actually *reduces* the prospect of finding intelligent life elsewhere. On such a view, that search may slowly wither before the stingy dictates of chance.

In contrast, our argument could satisfy another quite different hope that inspires the search for extraterrestrial intelligence. That same unarticulated hope runs like a red thread through much science fiction. From *2001: A Space Odyssey* and *Star Trek*, to *ET* and Carl Sagan's *Contact*, man is ennobled by cosmic mentors: demigods from an earlier age transformed into natural, alien beings. In *Contact*, Ellie Arroway is even mentored by an alien who appears as her wise and loving father. In these archetypal stories, we see modern man longing for an encounter—groping, half blind, toward some modest form of transcendence.

Arroway's quest involves a signal from space, an encrypted sequence of prime numbers. The story is exciting fiction. In real life, we have found no such radio signal. And yet, as we stand gazing at the heavens beyond our little oasis, not into an abyss, but into a wondrous arena commensurate with our gift for discovery, perhaps we have been staring past a cosmic signal far more meaningful than any mere sequence of numbers, a signal revealing a universe so skillfully crafted for life and discovery that it seems to whisper of an extraterrestrial intelligence immeasurably more vast, more ancient, and more magnificent than anything we've been willing to imagine.

THE REVISED DRAKE EQUATION

"It is the mark of an instructed mind to rest satisfied with the degree of precision which the nature of the subject admits," Aristotle admonished, "and not to seek exactness when only an approximation of the truth is possible."[1] His advice has often gone unheeded. Scientists and intellectuals are often prone to seeking more precision than a subject allows.

The temptation is ever present when trying to determine the odds of cosmic processes building a habitable planet. An optimistic conclusion is all but irresistible. And it's always more interesting to read and write pronouncements of certainty than to hem and haw and hedge with guarded talk of "perhaps" and "possibly" and "it's not wholly out of the question."

But with the ghost of Aristotle on our shoulder, we aim in what follows to resist the temptation, even as we try to update the Drake Equation with data that has flooded in over the past few decades. That data affords us a much better idea than Drake and his contemporaries had of the values of some of the relevant factors.

What follows is not, we should add, a direct implication or prediction of our book's argument. It's more of an afterthought. We include it here because the question of alien life is a profound and fascinating one. And we know from experience that curious readers will ask us for this in any case!

In the year 2000, Donald Brownlee and Peter Ward offered their own revised Drake equation in their book *Rare Earth*.[2] They suggested that the original Drake equation was an abbreviation of myriad unknown and unstated factors, all of which must be satisfied to get a single, radio communicating civilization. The *Rare Earth* version of the equation is as follows:

$$N = N_* \times f_p \times f_{pm} \times n_e \times f_g \times f_l \times f_i \times f_c \times f_L \times f_m \times f_j \times f_{me}$$

This version is longer than the original Drake equation. Compared to the original, the new terms are: f_g, the fraction of stars in the galactic habitable zone; f_{pm}, the fraction of metal-rich planets; f_m, the fraction of planets with a large moon; f_j, the fraction of systems with Jupiter-size planets; and f_{me}, the fraction of planets with a critically low number of mass-extinction events.

But even this longer version of the equation is incomplete.[3] Based on the material in chapters 1–10, we can add several more factors. At the same time, we want to avoid separating factors that aren't really independent. For example, the *Rare Earth* factors f_{me} and f_j are related; some of the mass extinction events are due to comet impacts, which, in turn, depend on the presence of giant planets.

So, instead of simply tacking on a few more factors at the end of the *Rare Earth* equation, we'll propose our own. We have formulated it to reduce such overlap and redundancy. The odds of getting a habitable planet are low as it is. There's no need to exaggerate the challenge. Moreover, since we're not interested in the number of communicating civilizations, just the number of technological civilizations, we won't require a factor for radio communication.

Without further ado, here it is:

$$N_* \times f_{sg} \times f_{ghz} \times f_{cr} \times f_{spir} \times f_{chz} \times n_p \times f_j \times f_{cir} \times f_{oxy} \times f_{mass} \times f_{comp} \times f_{moon}$$
$$\times f_{water} \times f_{tect} \times f_{life} \times f_{imp} \times f_{rad} \times f_{lcomp} \times f_{ltech} \times f_L$$

Here are the terms and their definitions:

N_*:   total number of stars in the Milky Way galaxy

f_{sg}:   fraction of stars that are early G dwarfs and at least a few billion years old

f_{ghz}:   fraction of f_{sg} stars in the galactic habitable zone

f_{cr}:   fraction of remaining stars (from last step) near the corotation circle and with low eccentricity galactic orbits

f_{spir}:   fraction of remaining stars outside spiral arms

f_{chz}:   fraction of remaining stars with at least one terrestrial planet in the circumstellar habitable zone

n_p: average number of terrestrial planets in the CHZs of such systems

f_j: fraction of remaining systems with no more than a few giant planets comparable in mass to Jupiter and in large, nearly circular orbits

f_{cir}:   fraction of remaining systems with terrestrial planets in CHZ with low eccentricities and outside dangerous spin-orbit and giant-planet resonances

f_{oxy}:   fraction of remaining planets near enough inner edge of CHZ to allow high oxygen and low carbon dioxide concentrations in their atmospheres

f_{mass}:   fraction of remaining planets in the right mass range

f_{comp}:   fraction of remaining planets with proper concentration of sulfur in their cores

f_{moon}:   fraction of remaining planets with a large moon and the right planetary rotation period to avoid chaotic variations in its obliquity

f_{water}:   fraction of remaining planets with right amount of water in the crust

f_{tect}:   fraction of remaining planets with steady plate tectonic cycling

f_{life}:   fraction of remaining planets where life appears

f_{imp}:   fraction of remaining planets with critically low number of large impacts

f_{rad}:   fraction of remaining planets exposed to critically low number of transient radiation events

f_{lcomp}:   fraction of remaining planets where "complex" life—that is, multicellular animal life—appears

f_{ltech}:   fraction of remaining planets where technological life appears

f_L: average lifetime of a technological civilization

Even at this level of detail, we've implicitly included some factors within others. For example, we could have listed metallicity as a separate factor, but it's implicit in the definitions of the galactic habitable zone, the fraction of stars with terrestrial planets, the fraction of stars with giant planets, and the fraction of planets with a low number of large impacts. Similarly, a factor dealing with the dynamic effects of a stellar companion is implicit in the factor requiring that terrestrial and giant planets have circular orbits. For instance, most radiation threats are taken care of with the f_{ghz}, the fraction of G dwarf stars of the right age in the galactic habitable zone, and f_{spir}, the fraction of relevant stars outside the spiral arms. But some remain, such as gamma ray bursts, supernovae between spiral arms, and giant flares from the host star. To take care of these cases, we've included the f_{rad}, the fraction of remaining planets exposed to a critically low number of transient radiation events.

Due to the lack of data in some cases, we can't plug in precise numbers for all the factors in our equation.

Okay, that's enough throat-clearing. Let's start with f_j and f_{moon}. Astronomers now know the fraction of sunlike stars with one or more giant planets is near 7 percent.[4] As we noted in chapters 5 and 8, however, most known planets are too massive, orbit too closely to their host stars, or have orbits too eccentric to permit habitable planets to form and maintain nearly circular orbits in their circumstellar habitable zones. Compared to the giant planets being found around other stars, the planets in our solar system have more circular orbits. Not counting the dwarf planet Pluto, the average eccentricity of the planets in the solar system is 0.06. It's true that the average eccentricity of exoplanets goes up as the number of planets goes down. But the multi-planet systems discovered so far are much more compact than our solar system. Finding that giant planets around other sunlike stars can take on a variety of orbits has led us to *reduce* the value of f_j from what we put it at twenty years ago. Thus, f_j is probably no larger than about 0.1 percent.

To estimate f_{moon}, the fraction of terrestrial planets with a moon large enough to stabilize a planet's spin axis, we need to know how the Moon formed. Most astronomers are convinced that a glancing collision between the proto-Earth and a smaller planetary body is the best explanation. As we noted in chapter 1, there's still no consensus on the most likely number and values of impact parameters of the two colliding bodies. To yield a suitable outcome, there are at least five that need to be just right: the timing of the collision, the impact point on the proto-Earth, the direction of spin of the proto-Earth, the plane of the collision relative to the plane of the ecliptic, and the momentum of the impactor. The third value is one half and the fourth is probably about 20 percent, but the other three values are probably small. Therefore, f_{moon} is probably no larger than 0.001 percent.

With these probable upper limits for f_i and f_{moon}, an average of 10 percent for each of the first thirteen factors is surely an overestimate. Most of the remaining six factors are even less well known, but each one is likely to be very small.

But let's be generous to see where it leads. If we assign odds of 10 percent to each of the first thirteen (of a total of twenty) factors, we get a total number of habitable planets in the Milky Way galaxy of 0.01. That is, the chance of getting one habitable planet in our galaxy, given extremely generous terms, is 1 percent. And that's before we get to the origin of life. So, on these calculations, the chances that the Milky Way galaxy contains even one habitable planet is much less than one. The chances of an advanced civilization? Well, that's much, much less than one. This is an interesting result, of course, since we exist.

Does this mean that SETI researchers will come up empty? Probably, and even if the Milky Way, in the face of these odds, was lucky enough to have a few high-tech civilizations. Any search strategy for a rare type of object or event must be very accurate. For example, suppose you're searching for ants with a high IQ. Suppose also that only about one in a billion ants are clever and you have developed a quick, automated search method that's 99.9 percent accurate. That sounds great, but it's not nearly good enough to detect such a tiny fraction of ants. Almost no ant you detect will be a member of the local ant Mensa chapter, since the sample will be swamped by false positives. You'll need other, more time-intensive methods to pick out the ants you want.

If our upper limit estimates for the number of civilizations elsewhere in the galaxy are anywhere near the truth, then SETI false positives will overwhelm genuine detections unless the method is almost perfectly accurate. This means that the present criteria for identifying an ETI signal—a very narrow bandwidth that is repeatedly detected on more than one telescope—may not be enough to eliminate false signals. In fact, there have been false alarms over the past few decades, mostly due to artificial satellites in orbit around the Earth. That's why SETI researchers have adopted more stringent criteria. Maybe it's time to move on to a detection threshold like the one used in the movie *Contact*.

SETI funding probably would be better spent refining theoretical calculations of the chances that extraterrestrial intelligence exists. Astronomers proposing to use large radio telescopes for more mundane research projects, say pulsar timing, must convince a group of their peers that they have a reasonable chance of success.[5] If SETI researchers want to compete for the same public instruments, then they should at least show that their proposals have some chance of success. Of course, much SETI research is now privately funded, and this is a free country.

But what about the other galaxies in the observable universe? Those certainly boost the odds of a habitable planet somewhere. Do they boost the odds enough to make technological life elsewhere in the universe likely? That depends on the galactic properties for habitability and on the values of the remaining "life factors."

We're convinced that, given materialistic assumptions, the odds for both the origin of life (f_{life}) and the appearance of complex life (f_{lcomp}) are so close to zero that they render the rest of the estimate moot. As we've noted, the existence of life elsewhere is far more likely on the design hypothesis than on the materialistic one. We're content, however, to treat the existence of life elsewhere as an open question.

Although the equation, even in its current form, is surely incomplete, it's the best we can do at the moment. There are lots of discoveries still waiting to be made, and many of them will bear on this equation. Still, we're willing to bet that, the more we learn, the smaller the final outcome will become. Any takers?

WHAT ABOUT PANSPERMIA?

In 1971 Francis Crick of DNA fame and Leslie Orgel proposed the theory of directed panspermia at a conference on Communication with Extraterrestrial Intelligence in Soviet Armenia. Two years later they published their theory in the journal *Icarus*.[1] Their idea was that an intelligent spacefaring civilization seeded Earth with life. Implicit in their proposal is the conclusion that the popular origin-of-life theories didn't work. "An honest man, armed with all the knowledge available to us now," Crick wrote, "could only state that in some sense, the origin of life appears at the moment to be almost a miracle, so many are the conditions which would have had to have been satisfied to get it going."[2] Thus, Crick and Orgel sought an external source for the first life on Earth, namely microorganisms sent from inhabited extrasolar planets.

They were not the first to propose such an idea. In 1871 at the British Society for the Advancement of Science, Lord Kelvin introduced panspermia, the idea that the first life on Earth was seeded from microbes arriving from outer space.[3] Svante Arrhenius introduced a different version in 1908.[4] Kelvin's version involved *lithopanspermia*, the transfer of life inside rocks, such as meteoroids. Arrhenius's version involved *radiopanspermia*, the transport of microorganisms, either naked or on small dust grains, by radiation pressure from the sun.[5]

Let's set aside the role of ETs seeding the galaxy and evaluate what's involved in generic panspermia—that is, transferring viable organisms between planetary bodies. There are four broad steps:

(1) Launch from an inhabited planet's surface.
(2) Transit through interplanetary and/or interstellar space.
(3) Arrival at a habitable planet.
(4) Colonization of the habitable planet.

We know the first step can happen naturally, without help from space jockeys, when an asteroid or comet impacts the surface of a planet. Some of the impact ejecta helps escape velocity from the planet and enter space. Analyzing the second requires applying Newtonian gravitational physics to lithopanspermia and stellar radiation pressure forces in radio panspermia. We then calculate the fraction of viable organisms that encounter the target body within a certain timeframe.

The crucial third step—which enthusiasts often forget—involves calculating the fraction of those organisms that survive reentry into the target body's atmosphere and reach its surface. And few of the arriving living organisms will grow and multiply on the target planet because it lacks some key ingredient that was present on the home world.

Prior to the 1990s, most planetary scientists doubted that intact fragments of a planet's crust could be blasted off its surface and hurled into space. That skepticism melted when meteorites from Mars and the Moon were identified. Because Mars is a relatively large planet with an atmosphere, it took a powerful impact by an asteroid or comet to send fragments of its crust on a trajectory that eventually delivered them to Earth. In addition to fist-sized and larger rocks, a big impact also generates a lot of dust-size ejecta—some of it suitable for radio panspermia. These rock-size fragments orbit the Sun, though a small fraction get flung out of the solar system from close encounters with Jupiter or Saturn (lithopanspermia).

Planetary astronomer Jay Melosh estimated that about fifteen Martian meteoroids larger than ten centimeters in diameter are ejected from the solar system each year.[6] Earth and Venus offer far less with their larger surface gravities and thicker atmospheres. Only in the early history of the solar system would much of Earth and Venus have escaped the solar system. But this is little help for the cause of interstellar panspermia since any ejected dust from these planets is unlikely to have contained life at that time.

And even with fragments for a planet like the current Earth not all the fragments will contain viable organisms. Only about 0.2 percent of the ejecta mass comes from the planetary surface, where life resides, but the surface rock is least likely to be melted and strongly shocked.

Radio panspermia only works for tiny dust grains smaller than about 0.2 microns.[7] Most bacteria are larger than this; the smallest of the ultra microbacteria are close to 0.2 microns, as are the smallest plankton.[8] (As it happens, this size limit corresponds to the division between the viruses and other forms of life.) While a few of the smaller species of bacteria and plankton are small enough to be propelled through space by radiation pressure, they would have no shielding from the Sun's sterilizing radiation. And viruses require a host to reproduce. Only when the Sun becomes a luminous red giant star would there be enough light pressure to push shielded and much larger bacteria out of the solar system. But by then Earth would be lifeless.

Another snag is step 3. Once they reached an exoplanet, the grains and their delicate cargo will get burned to a crisp entering the atmosphere.

We can safely conclude, then, that radiopanspermia is a hopeless prospect for getting life from one system to another.

Lithopanspermia, which involves bigger rocks, can transport larger organisms, but far more slowly than with radiopanspermia. The typical transfer time from Earth to Mars is millions of years, though the first transfer could occur after about 100,000 years.[9] The greatest threats to life in space are solar electromagnetic radiation (UV, X-ray) and solar and galactic cosmic ray particle radiation. Microorganisms inside boulder-size impact ejecta fragments are protected from these hazards if they are larger than about 11 meters.[10] However, the natural radiation from uranium (including U-238, U-235 isotopes), thorium and potassium 40 (K-40) in the boulder pose other threats to organisms deep in its interior.

As mentioned above, almost all the fragments launched from the planets were ejected within the first few hundred million years of the solar system's history, when large impacts were common. (Today, the terrestrial planets lose very little material.) But the early solar system was a more dangerous place for panspermia. The radiation dose rate for organisms in crustal rocks would have been about 2.5 times the present rate. The young Sun, too, was more hazardous, producing more frequent powerful flares.

If we trust the simulations, anywhere from 5 to 20 percent of the impact ejecta from Earth escape the solar system; for Mars the fraction is more like 15 to 20 percent.[11] This happens almost entirely thanks to Jupiter. And it takes anywhere from about 700,000 to 4 million years for the first Earth ejecta fragment to leave the solar system; the *typical* timescales for loss from Earth is 10 to 50 million years.[12]

With a typical ejection speed of five kilometers per second, a meteoroid ejected from the solar system will travel about five light-years in 300,000 years. This is optimistic, since it assumes a straight-line trajectory from the solar system to the target system. A few lucky pieces may pull this off, but most would dawdle and meander like drunken sailors who've lost their way home.

Planetary systems like ours, with giant planets in large orbits, are about 100,000 times more likely to capture such a payload than stars lacking such planets. But the capture by a binary star system is about 100,000 times more likely than it is for a single star with an orbiting Jupiter.[13] For every planet orbiting a single star with a Jupiter that captured life rocks, there are 100,000 planets in stellar binary systems that captured them. So, yes, Jupiter does make it more likely that our solar system and systems like it capture interstellar vagabonds than the typical planetary system, but not as likely as planets in binary systems. You could say this is a step where the solar system has an advantage for interstellar panspermia.

Once captured, a rock from our solar system will dance around the planets in its new home system, eventually colliding with one of them. If there are terrestrial planets in the target system, they will receive a tiny fraction of the captured rocks. And of that tiny bit, only a tiny fraction will survive entry to the planet's surface; most will enter the planet's atmosphere so fast that they will vaporize before reaching the surface.

How many organisms will survive depends on the survival time raised to the fourth power.[14] Careful experiments with bacterial DNA show that it degrades on timescales of hundreds of thousands of years.[15] Researchers have managed to revive frozen nematodes from permafrost dated between 30,000 to 40,000 years old.[16] And prokaryotes have been cultured from fluid inclusions in halite crystals from core samples in dry lake beds several tens of thousands of years old.[17] This may seem like a long time, but it's tiny compared to the travel time between planetary systems.

What if viable organisms do reach the surface of another planet? The researchers in the study above concluded that "most ancient prokaryotes in halite are dead or viable but not culturable, or that our culturing conditions were simply not suitable."[18] This gives us a glimpse of the obstacles to achieving step 4—colonization of another planet. Even in a controlled laboratory under the guidance of intelligent scientists, cultivating ancient microorganisms is just barely possible.

Given these obstacles, the whole idea of panspermia seems far-fetched. Indeed, it seems fanciful even without any detailed analysis. For instance, imagine that we take some bacteria, grind them up, break up their DNA into smaller fragments, and sprinkle the resulting bits onto a sterilized petri dish with agar growth medium. We can add some liquid and even stir it up to aid the reactions. If we're ambitious, we could try this hundreds or thousands of times. What will happen? Nothing of interest. And if nothing results after many trials, then why should we expect dead bacteria to seed a distant planet around another planetary system?

There's little question that plenty of viable organisms from Earth landed safely on the surfaces of Mars and Venus—our two closest neighbors and the two most earthlike planets known other than Earth. So why didn't life take on these worlds?

We get it. The origin of life is a hard problem, and, if one assumes materialism, finding a solution can lead to desperate measures. If researchers want to explain the origin of life on Earth, however, and their assumptions lead to far-fetched conjectures and dead ends, then perhaps they should rethink their assumptions.

As mentioned above, the challenges are so great that when Crick and Orgel took up the idea, they opted for a souped-up version of panspermia— *directed panspermia*. Directed panspermia could, in principle, overcome the problem of getting microbes to successfully hitchhike across trillions of miles of cosmic ray infested interstellar space inside a rock, finding Earth via blind luck, surviving a hellish atmospheric entry, and then springing to life out of cold storage. In this revised scenario, high-tech space aliens engineer the whole affair, taking the blind luck, and the absurdly inadequate mode of travel, out of the equation.

Note: Crick and Orgel invoke intelligent design to explain the appearance of the first life on Earth, albeit with space aliens rather than a transcendent or divine agent filling the role of the designer. Crick and Orgel opted for this no doubt because they realized that undirected panspermia just wouldn't cut it.

Of course, their proposal does nothing to address the problems bedeviling the origin-of-life question in general. They just moved the problem to another—unknown—planet. After all, even if we have space aliens tooling about the universe, we still need the first microbial life to spring up some-where to get the ball rolling. Orgel and Crick's scenario simply moves offstage the problem of getting from mere chemistry to biology without the role of intelligence. It doesn't solve the problem.

Then, too, there are all the other challenges discussed in Appendix A for getting a planet with a technological civilization. So, Crick and Orgel's solution leaves much to be desired. It's primary virtues, we suggest, are revealing the problem with materialist origin-of-life stories and point-ing up the fact that even some eminent scientists who want panspermia to work regard undirected panspermia as thoroughly implausible.[19]

Figure B.1: Early in the history of the solar system, when large impacts were still common, the terrestrial planets exchanged material. It is possible that Earth contaminated Mars with its microbes during these violent episodes.

Acknowledgments

Many people deserve credit for their help with this book—so many, in fact, that we fear we may leave some unmentioned. Several people were kind enough to review parts or all of our manuscript. Others contributed needed information, crucial distinctions, and insightful arguments.

We are especially grateful to Jonathan Witt. Jonathan provided countless editorial suggestions to an early version of our manuscript in its first edition and once again in this edition. He was bold enough to encourage us to cut some unneeded text.

Andrew Sperling provided indispensable help in creating many of our line drawings. Without his help, the text would have been much more sparsely, and much less impressively, illustrated. We would also like to thank our agent Sam Fleishman for seeing the potential in this project when it was little more than an idea with a few examples.

Ben Wiker and Mike Keas gave us excellent advice on the manuscript, especially on our historical sections. Bruce Nichols provided keen editorial advice in the very early stages of our book. Bill Dembski helped us through the tricky thickets involving probability theory.

332 The Privileged Planet

Thanks also to Kerry Magruder, Dennis Danielson, Michael Crowe, and Ted Davis for reading our historical chapter—and for saving us from making several mistakes. We also benefited from suggestions by Nancy Pearcey, Allan Sandage, Kyler Kuehn, Peter Hodgson, Josh Gilder, Phil Skell, Daniel Bakken, David Snoke, Scott Minnich, and Ginny Richards.

Thanks also to our friend Bob Cihak for his unflagging support over many years.

Philosophers Robin Collins and Del Ratzsch offered some very helpful technical advice. Robin and Del were also participants in the symposium held at Notre Dame, "The Mathematics of the Fine-Tuning Argument," in April 2003, along with Neil Manson, Tim and Lydia McGrew, Nick Bostrom, Roger White, Alex Pruss, and Brian Pitts. We were delighted to be included in this meeting and benefited greatly from it.

Thanks to John Armstrong, Don Brownlee, Tom Quinn, and Forrest Mims for providing scientific data and illustrations.

Thanks to Ellie Richards, Rachel Schroder, Mollie Black, Elizabeth Oliver, Emma Roberts, and Amelia Kuntzman at the Heritage Foundation for their careful copyediting of parts of the manuscript for the new edition. Thanks also to Jennai Gonzalez for helping to prepare illustrations for the new edition. Thanks to our first editor at Regnery, Miriam (Moore) Bridges, and to Tom Spence, for agreeing to let us do a new edition of a book. Thanks also go to our editor at Skyhorse Publishing, Sarah Janssen.

We would like to thank the John Templeton Foundation, which provided financial support to Guillermo early on for research on this book within the context of its "Cosmology & Fine-Tuning Research Program" (Grant ID #938-COS187).

Finally, our thanks to the Discovery Institute, and especially Bruce Chapman, Steve Meyer, and John West, for their many years of friendship and unflagging support while we have worked on this project and the new, revised edition.

Any errors remaining are, alas, our own.

FIGURE CREDITS

Many of the line drawings were created with the valuable help of Andrew Sperling at Discovery Institute. Of these, most are original, but some were drawn, with certain changes, after other published line drawings. For many of these, complete bibliographic information is included in the endnotes of the relevant chapters.

Figures 1.1, 1.3, and 1.7 are drawn after Littmann et al., p. 8, p. 131, and p. 88, respectively. Figure 2.4 is after Figure 1 of Zachos et al., "Trends Rhythms, and Aberrations in Global Climate 65 Ma to Present," Science 292 (2001): 686–693. Figure 3.1 is after Figure 17.10 of Tarbuck and Lutgens (1999). Figure 3.2 is after Figure 19.28 of Tarbuck and Lutgens (1999). Figure 3.5 is after Van der Voo (1990). Figure 3.6 is after P. Cloud, *Oasis in Space: Earth History from the Beginning* (New York: Norton), 1988. Figure 8.2 is after Vallée (2002). Figure 8.8 is after Figure 39 of Timmes et al. (1995) and Figure 1 of Samland (1998). Figure 9.3 is after Figure 7 of Perlmutter et al. (1999). Figure 9.5 is after Figure 17-13, Neil F. Comins and William J. Kaufmann III, *Discovering the Universe*, sixth edition (New York: W.H. Freeman and Company), 2003. Figure 10.2 is after Figure 7.15 of Rolfs and Rodney (1988). Figure 10.3 is after Figure 4 of Tegmark (1998). Plate 16 is after Figure 9-15 of Michael A. Seeds, *Stars and Galaxies*, 2nd edition (Pacific Grove, CA: Brooks/Cole, 2001).

Figures 5.5, 12.2, and Plates 1, 16, are courtesy NASA. Figure 8.10 is courtesy NASA/STScI/AURA. Figure 5.3 and Plates 2 and 11 are courtesy NASA/JPL/Cal Tech. Figure 15.4 is courtesy NASA, Viking Project and Malin Space Science Systems. Figure 15.8 is courtesy Keck Observatory. Plate 6 is courtesy Boston University/R. Myneni. Plate 13 is courtesy R. Evans, J. Trauger, H. Hammel, the HST Comet Science Team and NASA. Plate 19 is courtesy Robert Williams, Hubble Deep Field Team, and NASA. Plate 20 is courtesy NASA/WMAP Science Team.

Figures 1.5, 6.1 (Einstein portrait), 9.2, and 9.8 are courtesy US Naval Observatory Library. Figure 1.6 is courtesy Lick Observatory. Figure 2.2 is courtesy Jeremy Young and the Natural History Museum in London. Plate 7 is courtesy the National Geophysical Data Center's Marine Geology & Geophysics Division/NOAA. Figure 3.3 is from A. D. Raff and R. G. Mason, "Magnetic Survey off the West Coast of North America, 40°N Latitude to 52°N Latitude," *Geological Society of America Bulletin* 72 (1961): 1267–1270, courtesy the Geological Society of America.

Figures 4.2 and 8.6 are courtesy Donald Brownlee. Figures 6.2, 8.1 (left), 8.4, 8.11, 15.3, 15.7, and Plates 3 (top), 9, 10, 14, and 15, are by Guillermo Gonzalez. Figure 8.3 is courtesy Ronald Drimmel. Figure 9.1 is courtesy of the Observatories of the Carnegie Institution of Washington. Figure 14.1 is from University of Chicago, courtesy AIP Emilio Segrè Visual Archives.

Figure 15.6 is courtesy William Dembski. Plate 3 (bottom) is courtesy Mike Reynolds. The original photo used in Plate 4 is courtesy Jagdev Singh of the Indian Institute of Astrophysics, Bangalore, India. Plate 12 is courtesy Robin Canup and Bill Ward.

Figure 2.1 is based on data described in Petit et al. (1999). Figure 2.3 is based on ice core data from Vinther et al. (2009) and on marine data from Rosenthal et al. (2013). Figure 2.6 is based on data from the 1.9-mile-deep Greenland summit ice core obtained as part of the Greenland Ice Sheet Project 2 (GISP2), and data from the Quaternary Isotope Laboratory at the University of Washington. In Figure 5.2, data for Earth are courtesy Thomas Quinn, and data for Mars are courtesy John Armstrong. In Figure 7.4, the comet data are from JPL's DASTCOM database as of June 21, 2003. Asteroid data are from the Minor Planet Center and include all objects in their database larger than about half a kilometer, as of June

21, 2003. Both sets of data are restricted to asteroids and comets with known orbital elements. Figure 8.7 is from Figure 2 of G. Gonzalez, "The Metallicity Dependence of Giant Planet Incidence," *Monthly Notices of the Royal Astronomical Society* 443 (2014): 393–97.

The historical images in Figures 6.1, 6.3, 11.1, 11.2, 11.4–11.14, and 13.4 are courtesy University of Oklahoma, History of Science Collections, and some of the information in the captions for these figures is courtesy Kerry Magruder, History of Science Collections, University of Oklahoma.

NOTES

FOREWORD TO THE 2024 EDITION

1   John Gribbin, *Alone in the Universe: Why our Planet is Unique* (New York: Wiley, 2011), 143.
2   Guillermo Gonzalez, "Mutual Eclipses in the Solar System," *Astronomy & Geophysics* 50, no. 2 (2009): 2.17–2.19.
3   Dave Waltham, "Anthropic Selection for the Moon's Mass," *Astrobiology* 4 (2004): 460–68.
4   Dave Waltham, "Testing Anthropic Selection: A Climate Change Example," *Astrobiology* 11 (2011): 105–14. Waltham also discussed his ideas in his popular science book, *Lucky Planet: Why Earth Is Exceptional—and What That Means for Life in the Universe* (New York: Basic Books, 2014).
5   Robert M. Hazen *The Story of Earth: The First 4.5 Billion Years, from Stardust to Living Planet* (New York: Penguin Books, 2012).
6   Michael J. Denton *Fire-Maker: How Humans Were Designed to Harness Fire and Transform our Planet* (Seattle: Discovery Institute Press, 2016).
7   Michael J. Denton, *The Wonder of Water: Water's Profound Fitness for Life on Earth and Mankind* (Seattle: Discovery Institute Press, 2017).
8   Michael J. Denton, *Children of the Light: Astonishing Properties of Sunlight that Make us Possible* (Seattle: Discovery Institute Press, 2018
9   See https://exoplanetarchive.ipac.caltech.edu/ for up-to-date statistics on the known exoplanets.
10 Recent review papers on planetary habitability: Sarah R. N. McIntyre, Charles H. Lineweaver, Michael J. Ireland "Planetary Magnetism as a Parameter in Exoplanet Habitability," *Monthly Notices of the Royal Astronomical Society* 485 (2019): 3999–4012; Manasvi Lingam and Abraham Loeb. "Physical constraints on the likelihood of life on exoplanets," *International Journal of Astrobiology* 17, no. 2 (2018): 116–26; Aditya Chopra and Charles Lineweaver,

"The Case for a Gaian Bottleneck: The Biology of Habitability." *Astrobiology* 16 (2016): 7–22; Guillermo Gonzalez, "Setting the Stage for Habitable Planets," *Life* 4 (2014): 35–65; M. Güdel, R. Dvorak, N. Erkaev et al., "Astrophysical Conditions for Planetary Habitability," in *Protostars and Planets VI*, Henrik Beuther, Ralf S. Klessen, Cornelis P. Dullemond, and Thomas Henning (eds.) (Tucson: University of Arizona Press, 2014), 883–906. Preprint at: https://arxiv .org/abs/1407.8174.

11 Their paper appeared in multiple places: Lawrence M. Krauss and Robert J. Scherrer "The Return of a Static Universe and the End of Cosmology." *General Relativity and Gravitation* 39 (2007): 1545–50; *International Journal of Modern Physics* D 17 (2008): 685–90. Version for a popular readership: Lawrence M. Krauss and Robert J. Scherrer, "The End of Cosmology?: An Accelerating Universe Wipes out Traces of Its Own Origins," *Scientific American* (March 2008): 46–53.

12 Tony Rothman and George F. R. Ellis, "The Epoch of Observational Cosmology," *The Observatory* 107 (1987): 24–29.

13 Guillermo Gonzalez, "Eschatology of Habitable Zones," in *The Story of the Cosmos: How the Heavens Declare the Glory of God*, Paul M. Gould and Daniel Ray, eds. (Eugene, OR: Harvest House Publishers, 2019); Guillermo Gonzalez, "The Solar System: Favored for Space Travel," *Bio-Complexity* 1 (2020): 1–8.

INTRODUCTION

1   In I. Good, ed., *The Scientist Speculates* (New York: Basic Books, 1962), 15.

2   Although unmanned missions had orbited the Moon and photographed the side facing away from Earth (often called inaccurately the dark side of the Moon), Apollo 8 was the first manned mission to break loose of Earth's gravity and orbit the Moon. For the story of Apollo 8, see Robert Zimmerman's fast-paced and moving account in *Genesis: The Story of Apollo 8: The First Manned Flight to Another World* (New York: Dell, 1998).

3   Zimmerman, 234.

4   Carl Sagan, *Pale Blue Dot* (New York: Ballantine Books, 1994), 7.

5   To avoid tedium, we will often refer to this claim as "the correlation," or use rough synonyms.

6   Carl Sagan, *Cosmos* (New York: Ballantine Books, 1993), 4.

7   See discussion in Steven J. Dick, *Life on Other Worlds: The 20th Century Extraterrestrial Life Debate* (Cambridge: Cambridge University Press, 1998), 209–220.

8   Amir Aczel, *Probability 1: Why There Must Be Intelligent Life in the Universe* (New York: Harcourt Brace, 1998).

9   Stuart Ross Taylor, *Destiny or Chance: Our Solar System and its Place in the Cosmos* (Cambridge: Cambridge University Press, 1998).

10 Peter Ward and Donald Brownlee, *Rare Earth: Why Complex Life is Uncommon in the Universe* (New York: Copernicus, 2000). Gonzalez, Ward, and Brownlee have pursued technical research confirming this hypothesis. See Gonzalez, Brownlee, and Ward, "The Galactic Habitable Zone: Galactic Chemical Evolution," *Icarus* 152 (2001): 185–200.

11 Stuart Ross Taylor in *Destiny or Chance* puts it with refreshing bluntness:

> The message of this book is clear and unequivocal: so many chance events have happened in the development of the solar system that any original purpose, if it existed, has been lost. Superimposed on these chance events from the physical world are those of biological evolution, which has managed to produce one highly intelligent species out of tens of billion attempts over the past four billion years (204).

12 Henry Petroski, *Invention by Design* (Cambridge: Harvard University Press, 1996), 30.

13 We're only considering observational sciences like comparative planetary geology, solar physics, stellar, galactic, and cosmological astronomy, rather than strictly laboratory or experimental sciences, which are much less sensitive to location. To minimize the risk of cherry-picking only evidence that fits our theory, and ignoring contrary evidence, we've restricted ourselves to examples especially important in their disciplines. A similar argument

may be possible in some other observational or historical sciences, such as archaeology and paleontology. Perhaps some budding archaeologists and paleontologists will expand our argument into their disciplines.

Chapter 1: Wonderful Eclipses

1   Martin Amis, *London Fields* (New York: Vintage Books, 1991), chap. 22.
2   In this book, we will capitalize "Sun" and "Moon" when referring to our own, to avoid confusing them with other suns and moons.
3   Since the first edition of this book was published, both of us witnessed (in separate places) the solar eclipse of August 21, 2017, and one on April 8, 2024, in Waxahachie, Texas.
4   The results of my experiment were published in "Ground-level humidity, pressure and temperature measurements during the October 24, 1995, total solar eclipse," *Kodaikanal Observatory Bulletin* 13 (1997): 151–54. The temperature dropped 25 degrees Fahrenheit during the middle of the eclipse, relative to the previous day.
5   These included pregnant women staying indoors during the eclipse, ceasing from ordinary bodily activities during the eclipse, and bathing and changing into clean clothes after the eclipse.
6   S. Brunier and J. P. Luminet, *Glorious Eclipses: Their Past, Present, and Future* (Cambridge: Cambridge University Press, 2000), 6.
7   Brunier and Luminet, *Glorious Eclipses*, 17.
8   J. Laskar, et al. "Stabilization of the Earth's obliquity by the Moon," *Nature* 361 (1993), 615–17; M. Saillenfest, J. Laskar, G. Boué, "Secular Spin-axis Dynamics of Exoplanets," *Astronomy & Astrophysics* 623 (2019), 623: A4; O. Neron de Surgy and J. Laskar, "On the Long Term Evolution of the Spin of the Earth," *Astronomy & Astrophysics* 318 (1997): 975–89. The Moon stabilizes Earth's obliquity by exerting a torque on Earth that reduces its precession period by about a factor of three, taking it far from a dangerous resonance. For a planet of a given size, certain combinations of rotation period, precession period, and orbital period produce resonances that cause the obliquity to vary chaotically over a wide range. Earth will approach such a resonance in about 1.5 billion years as its rotation period continues to slow due to the action of the tides. Thus, to have a stable obliquity, several factors relating to a planet's physical and orbital properties must be met simultaneously.

If the Earth were moved to a star less than 23 percent the mass of our Sun, then the tidal torque produced by the star alone would be equal to the tidal torques produced by our Sun and Moon. One could argue that tides and stabilization of the planet's rotation axis would therefore not require a large moon. However, planets around stars in this mass range are less likely to be habitable to complex life for a variety of reasons. See D. Waltham, "Star Masses and Star-Planet Distances for Earth-like Habitability," *Astrobiology* 17, no. 1 (2017): 61–77.

A rotation period of twelve hours would also produce a stable obliquity, with or without the Moon. So, it would seem that a stabilizing moon is not needed if a planet has a rapid spin. The problem with this solution is that Earth's originally high spin was likely caused by the collision event that formed the Moon (see below). The Moon has been gradually robbing from Earth the high spin it gave it, but, as a trade, it has maintained its stable obliquity.
9   D. M. Williams and D. Pollard, "Earth-Moon interactions: Implications for Terrestrial climate and life," in *Origin of the Earth and Moon*, R. M. Canup and K. Righter, eds. (Tucson: University of Arizona Press, 2000), 513–525.
10   The prediction that wind and tides are the main drivers of global ocean circulation was made by W. H. Munk, and C. Wunsch, "Abyssal Recipes II: Energetics of Tidal and Wind Mixing," *Deep-Sea Research* 45 (1998): 1977–2010. Their prediction was confirmed by Topex/ Poseidon satellite altimeter data: G. D. Egbert, and R. D. Ray, "Significant Dissipation of Tidal Energy in the Deep Ocean Inferred from Satellite Altimeter Data," *Nature* 405 (2000): 775–78.

11   One of the oceanographers who suggested the deepwater tidal theory notes, "It appears that the tides are, surprisingly, an intricate part of the story of climate change, as is the history of the lunar orbit." C. Wunsch, "Moon, Tides, and Climate," *Nature* 405 (2000): 744.

12   Calculations indicate that the Moon may have formed from a glancing impact by an object two to three times the mass of Mars when Earth was only about half-formed; A. G. W. Cameron, "Higher-Resolution Simulations of the Giant Impact," in *Origin of the Earth and Moon*, eds., R. M. Canup and K. Righter (Tucson: University Of Arizona Press, 2000): 133–44. Other simulations indicate that the impact more likely occurred near the end of Earth's formation and with an impactor the mass of Mars; R. M. Canup, and E. Asphaug, "Origin of the Moon in a Giant Impact Near the End of the Earth's Formation," *Nature* 412 (2001): 708–12. This more recent set of simulations makes the formation of the Moon a more probable event than Cameron's simulations imply, because smaller impactors are more common than bigger ones. But caution is appropriate here, since this is a rapidly developing area of research.

13   Recent measurements of the isotopes of tungsten in Earth and primitive meteorites help to establish when Earth's iron core formed and confirm the Moon's contribution to that event. This is possible because radioactive hafnium presents early on (with a half-life of nine million years, after which it decays to tungsten) and tungsten have different affinities for iron. Thus, when the core formed, some tungsten went down with the iron while hafnium remained behind in the mantle and crust. The data indicate that the core formed about thirty million years after the beginning of the solar system, about the same time the Moon is thought to have formed. See R. Fitzgerald, "Isotope Ratio Measurements Firm Up Knowledge of Earth's Formation," *Physics Today* 56, no. 1 (January 2003): 16–18, and references cited therein.

14   For an entertaining and informative, though sometimes speculative, account of the many ways the Moon is important for life on Earth, see N. F. Comins, *What If the Moon Didn't Exist?: Voyages to Earths That Might Have Been* (New York: HarperCollins, 1993). See also C. R. Benn, "The Moon and the Origin of Life," *Earth, Moon, and Planets* 85-86 (2001): 61–66 and D. Waltham, "Is Earth Special?" *Earth-Science Reviews* 192 (2019): 445-70.

15   This is because the Earth would have transferred some of its angular momentum to the Moon as the Moon receded away from the Earth, and to the Earth's orbit too via the solar tides.

16   Waltham, "Is Earth Special?," 450–51. Note, this is larger than Waltham's original estimate of a maximum mass increase of 10 percent for the Moon. This is still small and corresponds to a maximum angular size only 2 arc minutes larger than the Moon's angular size at its present distance.

17   C. R. Benn has also suggested a connection between the occurrence of total solar eclipses and life on Earth. He notes that the strength of the tides from the Moon or the Sun depends on its angular size raised to the third power. Thus, the tides induced by the Moon and Sun are comparable in strength as a result of their similar angular sizes. Benn argues that the beating this produces in the tides may have been instrumental in the origin of life, since the long period of the beating would permit slow chemical reactions on the surface. However, this argument breaks down when one considers that the Moon was much closer to Earth when life first appeared on Earth. It is not clear why life would benefit from the similarity in solar and lunar tidal strengths at the present.

18   M. L. Lantink et al., "Milankovitch cycles in the banded iron formations constrain the Earth-Moon system 2.46 billion years ago," *Publications of the National Academy of Science* 119(40) (2022): e2117146119. T. Eulenfeld and C. Heubeck, "Constraints on Moon's orbit 3.2 billion years ago from tidal bundle data," *Journal of Geophysical Research: Planets* 128 (2023): e2022JE007466. We'll explain just how astronomers know this in chapter 2.

19   In this eclipse, the Moon's apparent size was only about 45 arc seconds greater than the Sun's photosphere. This works out to about 2.5 percent of their mean angular size.

20 Neither is quite a geometric sphere, but the Sun comes closer than just about any natural object we know of. The oblateness or flattening of the lunar profile in Earth's sky is a mere 0.06 percent. See, S. K. Runcorn, and S. Hofmann, "The Shape of the Moon," in *The Moon*, S. K. Runcorn, H. C. Urey, eds. (Dordrecht, Holland: D. Reidel, 1972), 22. The lunar axis pointing toward Earth, however, is larger than the other two axes by about five kilometers. Had the Moon not yet achieved a rotationally synchronized configuration, the longer axis would result in a less round lunar profile. (A planetary body in a rotationally synchronized orbit keeps the same face aimed toward the body it's orbiting; for the Moon, this state probably set in within about a million years of its formation.)

21 Although the Moon may have started with a relatively round profile, it probably attained its present shape about 3.8 billion years ago. At that time large impactors were still frequent, and they injected sufficient energy into the Moon to reshape it according to the stronger tidal forces from Earth at that time. See V. N. Zharkov, "On the History of the Lunar Orbit," *Solar System Research* 34 (2000): 1–11.

22 Incidentally, only one moon in the solar system, Titan, has a substantial atmosphere. The pioneering space artist Chesley Bonestell rendered on his canvas a view of a total lunar eclipse as it would appear from the Moon. See C. Bonestell, *Rocket to the Moon* (Chicago: Children's Press, 1961).

23 Note that Figure 1.4 does not include the many small moons discovered around the outer planets in recent years. As of Summer 2023, the total number of known moons in the Solar System is 219, excluding Pluto's moons. All these new moons belong to the irregular class with highly eccentric and inclined orbits, are very small, and do not produce total solar eclipses. For the latest listing of the new moons, see https://solarsystem.nasa.gov/moons/overview/ .

24 G. Gonzalez, "Mutual Eclipses in the Solar System," *Astronomy & Geophysics* 50:2 (2009): 2.17–2.19.

25 D. Veras and E. Breedt, "Eclipse, Transit and Occultation Geometry of Planetary Systems at Exo-Syzygy," *Monthly Notices of the Royal Astronomical Society* 468 (2017): 2672–83. The most famous multiple exo-planetary system is Trappist-1, with 7 planets orbiting a red dwarf. As seen from one of the planets in the Trappist-1 system, the other planets would appear comparable in size to our moon in our skies. None of the solar eclipses in the Trappist-1 system, however, are total. https://physicstoday.scitation.org/do/10.1063/PT.6.1.20170817a/full/.

26 Veras and Breedt, "Eclipse, Transit and occultation geometry of planetary systems at exo-syzygy," 2676.

27 In the 1930s Bernard Lyot of France invented the coronagraph and employed it to study the Sun's bright inner corona without the benefit of a total solar eclipse. This was not an easy feat. It required an exceptional site. Lyot observed from Pic du Midi Observatory in France) and carefully designed and built optics. Even today, the best coronagraphs on high mountaintops cannot reveal as much detail in the corona as a total solar eclipse.

28 While eclipse observations made it possible to identify hydrogen as a major constituent of prominences, it was not until the 1920s that Cecilia Payne-Geposchkin first demonstrated quantitatively that hydrogen is the main constituent of the Sun and other stars.

29 Cited by J. B. Zirker, *Total Eclipses of the Sun* (New York: Van Nostrand Reinhold Company, 1995), 18.

30 To understand what is going on here, imagine the Sun's chromosphere as a very thin shell enveloping the entire surface of the Sun. To produce an absorption line spectrum, the continuous spectrum from the underlying photosphere must pass through this shell. But, because its glow is so feeble compared to the photosphere, we cannot observe the chromosphere directly in white light. It is only during a total eclipse that the bright photosphere is completely blocked and the cross section of the faint, thin shell is seen against a dark background.

31 From experiments in his lab, Gustav Kirkhoff developed three laws of spectroscopic analysis, which together account for the three basic types of spectra seen in nature: continuous, emission, and absorption.

32 One could generate an artificial rainbow during an eclipse with a misty spray of water in front of a dark cloth and thus conduct this experiment without the benefit of a natural rainbow.

33 See D. Kennefick, *No Shadow of a Doubt: The 1919 Eclipse that Confirmed Einstein's Theory of Relativity* (Princeton: Princeton University Press, 2019). The story of the expeditions to observe the 1919 eclipse involves much more than technical aspects and makes for a very interesting read. Kennefick argues that the real hero of the expedition was not Eddington but rather Astronomer Royal Frank Dyson. In a couple of ways, the 1919 eclipse was nearly optimal for the planned observations. First, the Sun was near the Hyades star cluster, providing several bright stars that could be photographed during totality. Second, the duration of totality was the longest in 503 years!

34 Zirker, *Total Eclipses of the Sun*, 175–79.

35 F. R. Stephenson, L. V. Morrison, and C. Y. Hohenkerk, "Measurements of the Earth's rotation: 720 BC to AD 2015," *Proceedings of the Royal Society* A472 (2016): 20160404; G. Gonzalez, "New Constraints on ΔT prior to the second century AD," *Monthly Notices of the Royal Astronomical Society* 482 (2019): 1452–55. The second reference describes ancient observations used to determine Earth's past rotation that don't involve solar or lunar eclipses, but they still involve the Moon and are fewer than the eclipse observations.

36 Once the theoretical tidal component is subtracted from the observed rate of spin-down of the last century, it is clear that Earth's rotation is slowing at an irregular rate. Because of this, we cannot extrapolate our present knowledge of Earth's rotation more than a few centuries into the past. Even if our estimate of the length of the day is off by just a tiny amount, this leads to large errors over long periods of time, since clock errors accumulate. The insertion of a leap second in our atomic clocks about once every 1.5 years shows the rate that a steady clock errs in reflecting Earth's rate of rotation.

37 For an illustration of this see M. Littmann, K. Willcox, and F. Espenak, *Totality: Eclipses of the Sun* (Oxford: Oxford University Press, 1999), 130–31.

38 This would become important if the Moon's apparent size were at least twice its present value, since the various twentieth-century starlight deflection experiments generally did not employ stars within about two solar radii of the Sun's limb.

39 For a review on what can and cannot be done from space with respect to study of the Sun's corona, see J. M. Pasachoff, "Solar-Eclipse Science: Still Going Strong," *Sky & Telescope* 101, no. 2 (2001): 40–47.

40 This is due to tides induced by the Moon. J. O. Dickey et al., "Lunar Laser Ranging: A Continuing Legacy of the Apollo Program," *Science* 265 (1994): 482–90.

41 See F. de. S. Mello and A. C. S. Friaca, "The End of Life on Earth is Not the End of the World: Converging to an Estimate of the Life Span of the Biosphere?," *International Journal of Astrobiology* 19 (2020): 25–42. The carbon dioxide level in the atmosphere is expected continue to decline as the sun brightens. It will eventually decline below the level that plants can continue to function, depending on the species. After about 200 million years the level of carbon dioxide will be too low for C3 plants to function. About 85 percent of plant species are C3. This includes trees. C4 plants (about 0.4% of plant species) are expected to endure for another ~850 million years. Both the biological productivity and diversity will steadily decline over the next few hundred million years. The authors of this study assume that Earth is not strongly perturbed in their modelling. However, it is very likely Earth will be hit by asteroids and comets and have major volcanic eruptions in this time period. These perturbations could be enough to push Earth's biosphere over the edge in its anemic state. For these reasons, I estimate the future lifetime of the biosphere to be somewhat less than 500 million years.

42 My (GG) discovery that viewing perfect solar eclipses correlates with habitability was a rather surprising conclusion, and its publication ["Wonderful Eclipses," *Astronomy & Geophysics* 40, no. 3 (1999): 3.18–3.20] generated some interest—including articles in several newspapers ["Eclipse shows signs of Life," *The Daily Telegraph* (June 23, 1999): 16; "Right Distance for a 'Perfect' Total Eclipse," *The Irish Times* (July 12, 1999): 12; "In the Shadow of Brilliance," *Chicago Sun-Times* (June 27, 1999): 32; "Leben wir unter einem einzigartigen

Stern?," *Spectrum* (July 17, 1999): 8], a BBC radio interview, which aired on August 11, 1999 (the day of the European total solar eclipse), and a mention in "The Year in Weird Science" section of the January 2000 *Discover* magazine. This last item is the most revealing one—the moniker "weird" implies that the result of my study contradicted the expectations of *Discover*'s editorial staff.

CHAPTER 2: AT HOME ON A DATA RECORDER

1   New American Standard Bible.
2   D. A. Hodgson, N. M. Johnston, A. P. Caulkett, and V. J. Jones, "Paleolimnology of Antarctic Fur Seal *Arctocephalus gazella* Populations and Implications for Antarctic Management," *Biological Conservation* 83, no. 2 (1998): 145–54; Dominic A. Hodgson and Nadine M. Johnston, "Inferring seal populations from lake sediments," *Nature* 387 (May 1, 1997): 30–31.
3   For an overview of the diverse array of natural records of the ancient climate, see the special issue of *Science* on paleoclimatology (April 27, 2001).
4   Researchers Kurt Cuffey and Edward Brook write on ice-core paleoclimatology, "The reward for this effort is an astonishing expansion of our knowledge of past environments, remarkable both for its implications and its level of detail." ["Ice Sheets and the Ice-core Record of Climate Change," in *Earth System Science: From Biogeochemical Cycles to Global Change*, M. C. Jacobson, R. J. Charlson, H. Rodhe, and G. H. Orians, eds. (San Diego: Academic Press, 2000), 466.] This is the opening quote in a section of their article titled "Ice Sheets as Paleoclimate Archives."
5   Specifically, scientists can relate the amount of deuterium (an isotope of hydrogen) in ice to the local temperature when it was deposited, and the isotope ratio in oxygen in trapped bubbles to global ice volume and, more generally, to the hydrological cycle—the global evaporation, transport, and precipitation of water. For a non-technical, personal account of fieldwork and laboratory measurements of Greenland ice cores, see R. B. Alley, *The Two-Mile Time Machine* (Princeton: Princeton University Press, 2000); R. B. Alley and M. L. Bender, "Greenland Ice Cores: Frozen in Time," *Scientific American* (February 1998): 80–85; and R. J. Delmas, "Environmental Information from Ice Cores," *Reviews of Geophysics* 30, no. 1 (1992): 1–21.
6   R. B. Alley and M. L. Bender, "Greenland Ice Cores: Frozen in Time," *Scientific American* (February 1998): 13.
7   J. R. Petit et al., "Climate and Atmospheric History of the Past 420,000 Years from the Vostok Ice Core, Antarctica," *Nature* 399 (1999): 429–36. EPICA community members, "Eight Glacial Cycles from an Antarctic Ice Core," *Nature* 429, no. 6992 (2004): 623–28.
8   See C. M. Zdanowicz, G. A. Zielinski, and M. S. Germani, "Mount Mazama Eruption: Calendrical Age Verified and Atmospheric Impact Assessed," *Geology* 27:3 (1999): 621–24.
9   Alley and Bender, "Greenland Ice Cores," 160–161; M. Christl, C. Strobl, and A. Mangini, "Beryllium-10 in deep-sea sediments: a tracer for the Earth's magnetic field intensity during the last 200,000 years," *Quaternary Science Reviews* 22 (2003): 725–39.
10   For other examples, see M. Ram and M. R. Stolz, "Possible solar influences on the dust profile of the GISP2 ice core from Central Greenland," *Geophysical Research Letters* 28, no. 8 (1999): 1043–46; J. Donarummo Jr., M. Ram, and M. R. Stolz, "Sun/dust Correlations and Volcanic Interference," *Geophysical Research Letters* 29 (2002): 75-1-75-4. Strong evidence that the 1,300-year cycle seen in several proxies is caused by the Sun is given by G. Bond et al., "Persistent Solar Influence on North Atlantic Climate During the Holocene," *Science* 294 (2001): 2130–36. They write:

> Earth's climate system is highly sensitive to extremely weak perturbations in the Sun's energy output, not just on the decadal scales that have been investigated previously, but also on the centennial to millennial time scales documented here.

The Sun's signature has also been detected in other types of proxies discussed later in this chapter. For an example of proxies obtained from stalagmites in caves, see U. Neff et al., "Strong Coherence between Solar Variability and the Monsoons in Oman between 9 and 6 kyr ago," *Nature* 411 (2001): 290–93. For additional discussions of evidences for Sun-Earth climate links, see F. M. Chambers, M. I. Ogle, and J. J. Blackford, "Palaeoenvironmental Evidence for Solar Forcing of Holocene Climate: Linkages to Solar Science," *Progress in Physical Geography* 23 (1999): 181–204; B. van Geel et al., "The Role of Solar Forcing upon Climate Change," *Quaternary Science Reviews* 18 (1999): 331–38; E. Haltia-Hovi, T. Saarinen, and M. Kukkonen, "A 2000-year record of solar forcing on varved lake sediment in eastern Finland," *Quaternary Science Reviews* 26 (2007): 678–89; S. Engels and B. van Geel, "The Effects of Changing Solar Activity on Climate: Contributions from Palaeoclimatilogical Studies," *Journal of Space Weather and Space Climate* 2 (2012): A09; B. Zolitschka et al., "Varves in Lake Sediments—A Review," *Quaternary Science Reviews* 117 (2015): 1–41.

11   The abundance variations of beryllium 10 and chlorine 36 in ice and carbon 14 in tree rings have also been used to reconstruct the sunspot cycle from pre-telescope times. See I. G. Usoskin et al., "Solar Cyclic Activity over the Last Millennium Reconstructed from Annual ^{14}C Data," *Astronomy & Astrophysics* 649 (2021): A141; N. Brehm et al., "Eleven-year Solar Cycles over the last Millenium Revealed by Radiocarbon in Tree Rings," *Nature Geoscience* 14 (2021): 10–15. An especially interesting research program uses all three isotopes to reveal strong solar particle events in the distant past. See C. I. Paleari et al., "Cosmogenic Radionuclides Reveal an Extreme Solar Particle Storm Near a Solar Minimum 9125 Years BP," *Nature Communications* 13 (2022): A214. This event corresponds to 7176 BC; other strong solar particle events have been found to occur in 774/5 AD, 993/4 AD, and 660 BC. This unique knowledge of past extreme solar events allows us to assess their risk to our advanced but vulnerable modern civilization, a risk that we otherwise would be ignorant of.

12   A. Wallner et al., "Recent Near-Earth Supernovae Probed by Global Deposition of Interstellar Radioactive ^{60}Fe," *Nature* 532 (2016): 69–72; P. Ludwig et al., "Time-resolved 2-million-year-old Supernova Activity Discovered in Earth's Microfossil Record," *Publications of the National Academy of Sciences* 113 (2016): 9232–37; G. Korschinek et al., "Supernova-Produced ^{53}Mn on Earth," *Physical Review Letters* 125 (2020): 031101.

13   A. L. Melott, F. Marinho, and L. Paulucci, "Hypothesis: Muon Radiation Dose and Marine Megafaunal Extinction at the End-Pliocene Supernovae," *Astrobiology* 19 (2019): 825–30; M. J. Orgeira et al., "Statistical Analysis of the Connection Between Geomagnetic Field Reversal, a Supernova, and Climate Change During the Plio-Pleistocene Transition," *International Journal of Earth Sciences* 111 (2022): 1357–72.

14   See R. C. L. Wilson, S. A. Drury, and J. L. Chapman, *The Great Ice Age* (Routledge: London, 2000), 163–86.

15   Fossilized terrestrial mollusk shells, too, help us reconstruct past climate. See O. Moine et al., "Paleoclimatic Reconstruction Using Mutual Climatic Range on Terrestrial Mollusks," *Quaternary Research* 57 (2002): 162–72.

16   Wilson, Drury, and Chapman, *The Great Ice Age*, 71–77. Another method makes use of the magnesium/calcium ratio in planktonic forams and its chemical dependence on temperature. Magnesium atoms often replace calcium atoms in the formation of their shells, and the degree to which this occurs depends on temperature. When researchers combine this information with the oxygen isotope ratios from the same shells, they can derive both temperature and global ice volume. See D. W. Lea, T. A. Mashiotta, and H. J. Spero, "Controls on Magnesium and Strontium Uptake in Plantonic Foraminifera Determined by Live Culturing," *Geochimica et Cosmochimica Acta* 63 (1999): 2369–79. Another promising paleothermometer makes use of calcium isotope ratios. See T. F. Nagler, A. Eisenhauer, A. Muller, C. Hemleben, and J. Kramers, "The ^{44}Ca-Temperature Calibration on Fossil and Cultured *Globigerinoides sacculifer*: New Tool for Reconstruction of Past Sea Surface Temperatures," *Geochemistry, Geophysics, Geosystems* 1 (2000): 2000GC000091.

17   B. M. Vinther et al., "Holocene Thinning of the Greenland Ice Sheet," *Nature* 461 (2009): 385-388; Y. Rosenthal, B. K. Linsley, and D. W. Oppo, "Pacific Ocean Heat Content During the Past 10,000 Years," *Science* 342 (2013): 617–21.

18   Biogenic silica produced by diatoms serve as a temperature proxy, with cold periods producing more diatomaceous ooze. See A. A. Prokopenko et al., "Orbital Forcing of Continental Climate During the Pleistocene: A Complete Astronomically Tuned Climatic Record from Lake Baikal, SE Siberia," *Quaternary Science Reviews* 25, no. 23–24 (2006): 3431–57.

19   Among the quantities measured in marine sediments are carbon and oxygen isotope ratios in foram skeletons, detritus from melting icebergs, and extraterrestrial helium 3. Marine sediments do suffer from a slight problem—mixing by worms, or bioturbation. Bioturbation decreases the time resolution of marine cores, but some locations are much less affected than most. Ice cores, for obvious reasons, don't suffer from this problem.

20   J. O. Dickey et al., "Lunar Laser Ranging: A Continuing Legacy of the Apollo Program," *Science* 265 (1994): 482–90.

21   J. C. G. Walker and K. J. Zahnle, "Lunar Nodal Tide and Distance to the Moon during the Precambrian," *Nature* 320 (1986): 600–602; M. L. Lantink et al., "Milankovitch cycles in the banded iron formations constrain the Earth-Moon system 2.46 billion years ago," *Publications of the National Academy of Science* 119(40) (2022): e2117146119. T. Eulenfeld and C. Heubeck, "Constraints on Moon's orbit 3.2 billion years ago from tidal bundle data," *Journal of Geophysical Research: Planets* 128 (2023): e2022JE007466.

22   For more on the method, see Walker and Zahnle, "Lunar Nodal Tide," 600.

23   Researchers also assume that the size of Earth's orbit has not changed significantly over the last few billion years. This assumption receives support from long-term numerical simulations of the orbits of the planets in the Solar System.

24   For an example of this type of study, see C. P. Sonett and M. A. Chan, "Neoproterozoic Earth-Moon Dynamics: Rework of the 900 Ma Big Cottonwood Canyon Tidal Laminae," *Geophysical Research Letters* 25 (1998): 539–542. More general reviews of paleoastronomy as applied to tidal rhythmites is given by G. E. Williams, "Geological Constraints on the Precambrian History of Earth's Rotation and the Moon's Orbit," *Reviews of Geophysics* 38, no. 1 (2000): 37–59 and R. Mazumder and M. Arima, "Tidal Rhythmites and their Implications," *Earth-Science Reviews* 69 (2005): 79–95. By reading modern tidalites, astronomers can confirm the lunar orbital period to better than one percent.

25   Continued study of tree rings may reveal other ways they record their local environment. The records from old tree trunks, such as those preserved in bogs and lake bottoms, can be combined to form a long, complete record extending beyond living trees. This is possible when there is overlap among several tree trunks. The irregular pattern of ring thickness greatly facilitates such reconstructions. Had rings and ring patterns been uniform, a given set of tree rings would look like any other, making reconstruction uncertain. The University of Arizona maintains an excellent website on dendrochronology at: http://www.ltrr.arizona.edu/.

26   For example, Forrest Mimms III has told us of his discovery of what he believes is a proxy in tree rings for the UV level reaching the ground. The mechanism involves the transport of tannin from the leaves to the trunk and branches in response to changing UV levels.

27   G. J. Retallack, "A 300-Million-Year Record of Atmospheric Carbon Dioxide from Fossil Plant Cuticles," *Nature* 411 (2001): 287–90. See also: W. M. Kurschner, "Leaf Sensors for CO_2 in Deep Time," *Nature* 411 (2001): 247–48; M. Rundgren and O. Bennike, "Century-Scale Changes of Atmospheric CO_2 During the Last Interglacial," *Geology* 30, no. 2 (2002): 187–89; G. J. Retallack and G. D. Conde, "Deep time Perspective on Rising Atmospheric CO_2," *Global and Planetary Change* 189 (2020): 103177. More accurately, these studies reconstruct the ancient variations in the carbon dioxide partial pressure, which is just the pressure contributed by a given constituent of the atmosphere. The "stomatal index," the proportion of leaf cells that are stomata, is insensitive to other environmental changes, such as sunlight and relative humidity.

28 V. J. Polyak et al., "Wetter and Cooler Late Holocene Climate in the Southwestern United States from Mites Preserved in Stalagmites," *Geology* 29, no. 7 (2001): 643–46.

29 Wilson, Drury, and Chapman, *The Great Ice Age*, 45. The following study estimates that the Great Barrier Reef in Australia began growing about 600,000 years ago: International Consortium for Great Barrier Reef Drilling, "New Constraints on the Origin of the Australian Great Barrier Reef: Results from an International Project to Deep Coring," *Geology* 29 (2001): 483–86.

30 The study of remnant magnetic variations in sediments and rocks is called *magnetostratigraphy*. For more detailed discussion on this topic, see W. Lowrie, *Fundamentals of Geophysics* (Cambridge, UK: Cambridge University Press, 1997), 297–302.

31 For more background, see W. S. Broecker and G. H. Denton, "What Drives Glacial Cycles?" *Scientific American* (January 1990): 49–56; R. C. L. Wilson, S. A. Drury, and J. L. Chapman, *The Great Ice Age* (London: Routledge, 2000), 61–65; A. Berger and M. F. Loutre, "Astronomical Forcing through Geological Time," in *Orbital Forcing and Cyclic Sequences: Special Publication of the International Association of Sedimentologists*, P. L. De Boer, D. G. Smith, eds. (Oxford, UK: Blackwell Scientific Publications, 1994), 15–24.

32 P. L. De Boer and D. G. Smith review Milankovitch cycles and their effects on Earth processes in "Orbital Forcing and Cyclic Sequences," in *Orbital Forcing and Cyclic Sequences: Special Publication of the International Association of Sedimentologists*, P. L. De Boer, D. G. Smith, eds. (Oxford: Blackwell Scientific Publications, 1994), 1–14. A more recent review is L. A. Hinnov, "Chapter One—Cyclostatigraphy and Astrochronology in 2018," *Stratigraphy & Timescales* 3 (2018): 1–80. For a recent detailed study of Milankovitch cycles in lake sediment cores, see H. Huang et al., "Astronomical Forcing of Middle Permian Terrestrial Climate Recorded in a Large Paleolake in Northwestern China," *Palaeogeography, Palaeoclimatology, Palaeoecology* 550 (2020): 109735.

33 For example, they can compare the amount of dust content or oxygen isotope ratios of ice and marine sediment cores.

34 Biologists now know that, strictly speaking, there is no "simple" life, since even the simplest single-celled organisms are highly complex and more technologically sophisticated than the most advanced human technology. But "simple life" remains in common usage as a category among biologists. We can distinguish between simple and complex life in several ways. We can distinguish single-celled and multicellular (or metazoan) organisms, eukaryotes and prokaryotes (with and without a nucleus, respectively), aerobic and anaerobic organisms (oxygen-breathing or not, respectively), or microscopic and macroscopic organisms. Simple life dominated Earth for most of its history.

35 Some scientists have offered "alternative chemistry" proposals for life forms. See Gerald Feinberg and Robert Shapiro, *Life Beyond Earth: The Intelligent Earthling's Guide to Life in the Universe* (New York: William Morrow Co., 1980).

36 See N. R. Pace, "The Universal Nature of Biochemistry," *Publications of the National Academy of Sciences* 98 (2001): 805–8.

37 Biochemist A. E. Needham comments on the metastability of carbon reactions: "As in so many other respects, carbon seems to have the best of both worlds, in fact, combining stability with lability, momentum with inertia. Most organic compounds are metastable . . . that is to say they are not in complete equilibrium with their environmental conditions and are easily induced to react further." *The Uniqueness of Biological Materials* (Oxford, UK: Pergamon Press, 1965), 30.

38 Michael Denton, *The Miracle of the Cell* (Seattle, WA: Discovery Institute, 2020), 21–38.

39 In contrast, silicon forms a mineral, silica, when oxidized. This results from silicon's tendency to form single bonds with the oxygen atoms, rather than double bonds like carbon does. The double bonds that carbon forms with each oxygen atom do not leave the oxygen atoms free to form bonds with other carbon atoms nearby. However, this is what happens with the single-bonded oxygen atoms attached to silicon: they bond with other silicon atoms, forming a stable crystalline matrix of SiO_2 that is wholly unsuitable for the hustle and bustle that is cellular life.

40 Biologist Michael Denton argues that the weak chemical bonds that allow large organic molecules to form three-dimensional shapes are also an essential requirement for life. *Nature's Destiny: How the Laws of Biology Reveal Purpose in the Universe* (New York: The Free Press, 1998), 114.

41 It also doesn't work to try to create a carbon equivalent by combining several kinds of atoms. For example, silicones—alternating chains of silicon and oxygen atoms—offer perhaps the best alternative to carbon-based, or organic, chemistry. However, they fail on several grounds. First, there is no good solvent for silicone chemistry. Second, silicone chemistry becomes more biologically useful if organics are available as side chains to attach to the silicone base; but this makes the silicones superfluous. In other words, if the environment had carbon in the first place, building the life form from silicone would be like building golf club heads out of clay when you had titanium on hand. Finally, we do not see silicone chemistry in nature, only in the laboratory, but we do see long-chained carbon molecules in many places, including Earth, meteorites, and the matter between the stars, called the interstellar medium.

42 Denton, *Nature's Destiny*, 115–116.

43 The water molecule, it is also interesting to note, is composed of the two most abundant reactive elements in the universe.

44 *The Fitness of the Environment: An Inquiry into the Biological Significance of the Properties of Matter* (New York: Macmillan, 1913). Henderson's work was a quantitative extension of the nineteenth-century work by William Whewell: *Astronomy and General Physics, Considered with Reference to Natural Theology* (London, 1833), ninth edition published in 1864.

45 See Denton's *Nature's Destiny* as well as *The Wonder of Water: Water's Profound Fitness for Life on Earth and Mankind* (Seattle: Discovery Institute Press, 2017); J. D. Barrow and F. J. Tipler, *The Anthropic Cosmological Principle* (Oxford, UK: Oxford University Press, 1986), 524–41; *The Uniqueness of Biological Materials*. Unlike the "Face on Mars," which disappeared after images with greater resolution were obtained with newer orbiters, the apparently high fitness of water for chemical life has only become more impressive as our "resolution" in chemistry has increased since the nineteenth century.

46 The element bismuth also has this property. Water ice actually has higher heat conductivity than liquid water. However, ice forming on water still acts to slow heat loss from the water below. First, considerable heat is given off as ice forms at the base of the ice sheet, slowing the formation of new ice. Second, the ice does not convect, so heat must flow through it only by conduction; as its thickness grows, the insulation increases. Third, snow forming on the surface of the ice provides extra insulation, since snow is a better insulator than ice.

47 We find water in all three states at Earth's surface, and the mean surface temperature is near the triple point of water—a unique combination of pressure and temperature where all three states can coexist. Not only does this provide a diverse set of surfaces (the relevance of which will be covered in chapter 4), but it also best exploits water's anomalous properties for regulating the temperature.

48 Needham, *The Uniqueness of Biological Materials*, 13.

49 K. Regenauer-Lieb, D. A. Yuen, and J. Branlund, "The Initiation of Subduction: Criticality by Addition of Water?" *Science* 294 (2001): 578–80; G. Lu et al., "Reviewing Subduction Initiation and the Origin of Plate Tectonics: What Do We Learn from Present Day Earth?," *Earth and Planetary Physics* 5 (2021): 123–40.

50 Denton also describes water's role in regulating body temperature in animals. Its value derives from its viscosity, thermal capacity, heat conductivity, and heat of vaporization. As we noted above, some of these physical properties are also key to regulating the climate. In addition, he describes several properties of water (some of which we described earlier) that work together to preserve it in a liquid state. See *The Wonder of Water: Water's Profound Fitness for Life on Earth and Mankind* (Seattle, WA: Discovery Institute, 2017).

51 F. H. Stillinger, "Water Revisited," *Science* 209 (1980): 451.

52 J. S. Lewis, *Worlds Without End: The Exploration of Planets Known and Unknown* (Reading, UK: Helix Books, 1998), 199. Henderson was also struck by the fitness of carbon and water, in *The Fitness of the Environment*, 248.

53 There is as much carbon dioxide in a liter of water as there is in a liter of air. Without the high solubility of carbon dioxide in water, creatures such as us could not rid our cells of this product of oxidative metabolism. While doing so, the dissolved carbon dioxide forms a weak acid in the blood, which helps buffer and regulate acidity in organisms (Denton, 132–137). This synergy between carbon and water is also important on the planetary scale. The carbon dioxide dissolved in rainwater forms a weak acid important in the chemical weathering of exposed rocks.

54 The point is explored in Denton's *Nature's Destiny* and in less detail in Henderson's *The Fitness of the Environment*.

55 R. E. Davies and R. H. Koch, "All the Observed Universe Has Contributed to Life," *Philosophical Transactions of the Royal Society of London* B 334 (1991): 391–403; V. Trimble, "Origin of the Biologically Important Elements," *Origins of Life and Evolution of the Biosphere* 27 (1997): 3–21.

56 Molybdenum is key to the operation of two enzymes, nitrogenase and nitrate reductase, which are involved in nitrogen fixation. Nitrogen in gaseous form is not useful to life, and so its conversion to a form that life can use is essential. The fact that no other transition metal is used by life for nitrogen fixation, even though molybdenum is rare in some parts of the land, implies that there is no substitute for it. Francis H. C. Crick and Leslie E. Orgel cited the scarcity of molybdenum in Earth's crust as possible evidence that Earth was "seeded" by an extraterrestrial civilization (called directed panspermia); "Directed Panspermia," *Icarus* 19 (1973): 341–46, or what a skeptic might call the Little-Green-Man-in-the-Gap theory.

57 Note also that many chemical elements are found in Earth's oceans only because they have been washed away from the continents. On a planetary body with oceans but without continents, many of the life-essential elements may not be available in sufficient concentration.

58 B. M. Jakosky and E. L. Shock, "The Biological Potential of Mars, the Early Earth, and Europa," *Journal of Geophysical Research* 103 (1998): 19359–364.

59 Denton notes that chemical reactions with fluorine are too energetic for the stability of organic reactions, and the product of hydrogen and fluorine is a very reactive acid. *Nature's Destiny*, 121.

60 Henderson, *The Fitness of the Environment*, 247–48.

61 "Extremophiles" are microorganisms that can live in extreme environmental conditions, far from the average in temperature, pressure, moisture, salinity, and acidity. See Michael Gross, *Life on the Edge: Amazing Creatures Thriving in Extreme Environments* (Cambridge, MA: Perseus, 2001); L. J. Rothschild and R. L. Mancinelli, "Life in Extreme Environments," *Nature* 409 (2001): 1092–1101.

62 For more information on the dependence of subsurface life on surface life, see R. A. Kerr, "Deep Life in the Slow, Slow Lane," *Science* 296 (2002): 1056–1058.

63 That's between 294 and 325 degrees Kelvin (21 to 52 degrees Centigrade), with optimum growth at 309 degrees Kelvin (36 degrees Centigrade). Méndez develops an equation of state of life for prokaryotes, in which he relates the growth rate of prokaryotes to the temperature, pressure, and water concentration of their local environment. See "Planetary Habitable Zones: The Spatial Distribution of Life on Planetary Bodies," paper presented at the 32nd Lunar and Planetary Science Conference, March 12–16, 2001.

64 The diversity of vascular plants correlates with the productivity of an ecosystem. See S. M. Schneider and J. M. Rey-Benayas, "Global Patterns of Plant Diversity," *Evolution and Ecology* 8 (1994): 331–47. More recent studies confirm that biodiversity generally correlates with productivity; see R. B. Waide et al., "The Relationship Between Productivity and Species Richness," *Annual Review of Ecology and Systematics* 30 (2000): 257–300. Productivity, in turn, depends on such factors as temperature and nutrient availability; see A. P. Allen, J. H. Brown, and J. F. Gillooly, "Global Biodiversity, Biochemical Kinetics, and the Energetic-equivalence Rule," *Science* 297 (2002): 1545–48. Complex life is even less tolerant of

warmer temperatures. Denton notes that the solubility of oxygen in water drops rapidly as temperature increases, and the metabolic demand for oxygen doubles with every 18-degree rise in temperature; these factors alone limit large complex life to temperatures below 115 degrees Fahrenheit. See Michael Denton, *Nature's Destiny*, 124.

65   Research on thermophiles indicates that these organisms resist high temperature by incorporating certain key amino acids in their protein structure. These alterations (relative to their mesophilic cousins) make the amino acid sequences much more restrictive; they also make the proteins more rigid at lower temperatures. See C. Vielle and G. J. Zeikus, "Hyperthermophilic Enzymes: Sources, Uses, and Molecular Mechanisms for Thermostability," *Microbiology and Molecular Biology Reviews* 65 (2001): 1–43.

66   James Lovelock concurs in *The Ages of Gaia* (New York: W. W. Norton & Company, 1988), 108.

67   Another example of ancient records that may enhance future survival is the field of paleotempestology—the study of ancient severe storms from their effects on marine and lake sediments and coastal areas. While still a young field of study, the hope is that enough historical hurricanes and severe storms will be found in the sediment record to discern correlations with the global climate. From this, it may be possible to prepare long-term forecasts of hurricane threats (not specific ones, mind you). Sample studies are J. P. Donnelly et al., "Sedimentary Evidence of Intense Hurricane Strikes from New Jersey," *Geology* 29, no. 7 (2001): 615–18; A. J. Noren et al., "Millennial-scale Storminess Variability in the Northeastern United States during the Holocene Epoch," *Nature* 419 (2002): 821–24; J. C. Bregy et al., "2500-year Paleotempestological Record of Intense Storms for the Northern Gulf of Mexico, United States," *Marine Geology* 396 (2018): 26-42; T. S. Winkler et al., "Revising Evidence of Hurricane Strikes on Abaco Island (The Bahamas) over the last 700 years," *Scientific Reports* 10 (2020): 1–17.

68   The paleoclimate records of the past half-million years show that the times of greatest climate stability are the relatively short-lived "peak interglacials," like the one we are presently living in (though ours is already longer-lived). See J. P. Helmke et al., "Sediment-color Record from the Northeast Atlantic Reveals Patterns of Millennial-scale Climate Variability during the Past 500,000 Years," *Quaternary Research* 57 (2002): 49–57, and J. F. McManus et al., "A 0.5 Million-year Record of Millennial-scale Climate Variability in the North Atlantic," *Science* 283 (1999): 971–74. Using otoliths (bony structures in fish for acoustics and balance) as temperature proxies, the following study determined climate conditions about 6,000 years ago: C. F. T. Andrus, "Otolith ^{18}O Record of Mid-Holocene Sea Surface Temperatures in Peru," *Science* 295 (2002): 1508–11. They showed that sea surface temperatures were three to four degrees centigrade warmer and El Niño events less severe at that time.

69   In their study "Future Climatic Changes: Are We Entering an Exceptionally Long Interglacial?" *Climatic Change* 46 (2000): 61–90, they stated, "This insolation variation . . . is really exceptional and has very few analogues in the past."

70   If the Milankovitch astronomical cycles are the main trigger of the ice ages, this could help explain the highly anomalous climate stability of the Holocene from the past 12,000 years. But there's more going on, because Berger and Loutre predict moderate variations up to 60,000 years into our future. Moreover, the observed range of the Milankovitch variations since the start of the Holocene has been surprisingly small. Their prediction seems well founded, but there seems to be more going on than they realize. Their explanation for the stability of the Holocene is the low amplitude of the Milankovitch cycles from 5,000 years ago until 60,000 years into the future, compared to the past three million years or so. This seems to be an odd coincidence. But how can you use the predicted changes in the future as part of the explanation for the anomalous stability of the Holocene so far?

71   S. Talento and A. Ganopolski, "Evolution of the Climate in the Next Million Years: A Reduced Complexity Model for Glacial Cycles and Impact of Anthropogenic CO_2 Emissions," *Earth System Dynamics Discussions* (2021): 1–31.

72   Carbon dioxide only needed to be about 40 ppm lower. A. Ganopolski, R. Winkelmann, and H. J. Schellnhuber, "Critical Insolation-CO_2 Relation for Diagnosing Past and Future Glacial Inception," *Nature* 529 (2016): 200–203.

73   W. F. Ruddiman et al., "The Early Anthopogenic Hypothesis: A Review," *Quaternary Science Reviews* 240 (2020): 106386. The "Early Anthropocene Hypothesis" has gained support from archeological evidence for widespread deforestation starting about 7,000 years ago, and wet-rice farming and livestock tending starting about 5,000 years ago. These activities would have increased the warming from increased carbon dioxide and methane emissions, respectively. The timing of the increase in these greenhouse gases measured in the ice cores agrees with the timing of the rise in human agricultural activities. It is especially revealing that the most similar (in terms of Milankovitch orbital parameters) interglacial (MIS 19) to our present one (MIS 1) in the ice core record (from 777,000 years ago) revealed declining carbon dioxide at the equivalent time to our present interglacial when it started rising from human activities. The time period in MIS 19 that corresponds to our present (since the start of the interglacial) was already experiencing the early stages of descent into an ice age cold period. We can only figure this out because of the diverse and informative set of paleoarchives that are available to us.

74   See https://www.nasa.gov/feature/goddard/2016/carbon-dioxide-fertilization-greening-earth; S. Piao et al., "Characteristics, Drivers and Feedbacks of Global Greening." *Nature Reviews Earth & Environment* 1, no. 1 (2020): 14–27. The global greening has also resulted in a global cooling effect, partly offsetting the recent warming. See Y. Li et al., "Biophysical Impacts of Earth Greening can Substantially Mitigate Regional Land Surface Temperature Warming," *Nature Communications* 14 (2023): 121.

75   Piao, "Characteristics, Drivers."

76   We must interpret this data properly if we are to avoid costly and even deadly mistakes, For instance, this evidence suggests that hindering worldwide economic growth for a slight reduction in the global temperature won't make Earth more habitable for life in general. In fact, it would probably make it much less hospitable for civilization. We agree with Peter Huber in *Hard Green* (New York: Basic Books, 1999): a healthy economy leads to a healthy environment.

In addition, there is a huge volume of published material demonstrating the benefits to the biosphere and to civilization of moderately elevated carbon dioxide levels. See C. D. Idso, "Earth's Rising Atmospheric CO2 Concentration: Impacts on the Biosphere," *Energy & Environment* 12 (2001): 287–310; I. M. Goklany, "Reduction in Global Habitat Loss from Fossil-Fuel-Dependent Increases in Cropland Productivity," *Conservation Biology* 35 (2021): 766–74; Indur M. Goklany, *Impacts of Climate Change: Perception and Reality,* report 46 (London, UK: The Global Warming Policy Foundation, 2021), https://www.thegwpf.org/content/uploads/2021/02/Goklany-EmpiricalTrends.pdf. The increased plant growth resulting from higher carbon dioxide levels and modestly warmer temperatures will greatly aid in food production in the coming century as population grows. The overall climate would also be more hospitable. For example, the growing season in cold latitudes would be lengthened; the warming is expected to occur mostly in the winter and nights; evaporation and precipitation would increase, likely resulting in more rainfall throughout the world and more snowfall in Arctic regions.

The moderate increase in global temperature, too, should benefit civilization. The effects of the climate fluctuations of the past millennium support the notion that warm periods are preferable for civilization. The Medieval Warm Period, also called the *Little Climate Optimum*, peaked near 1200 AD, reaching perhaps half a degree centigrade warmer than the present. Afterwards, the Little Ice Age saw its coldest temperatures in the seventeenth century. There's much historical anecdotal evidence that the Little Ice Age was much harsher on European peoples than the Medieval Warm Period. For a historical treatment, see H. Lamb, *Climate, Change, and the Modern World*, 2nd Ed. (London: Routledge, 1995). There is also increasing evidence from paleoclimate archives that the Medieval Warm Period and Little Ice

Age were global phenomena; see http://co2science.org/data/mwp/mwpp.php. If the many predicted catastrophes resulting from warmer temperatures (such as the shutdown of the Atlantic circulation or the sudden release of methane from the ocean floor) did not occur during the Medieval Warm Period (or during the even warmer "Holocene maximum" about 7,000 years ago), then they are not likely to occur in the coming centuries either.

77 Apocalyptic environmentalist literature often claims that we're conducting a dangerous experiment on the climate. The quote, "Through his worldwide industrial civilization, Man is unwittingly conducting a vast geophysical experiment," was an early expression of this view. It originated in R. Revelle, W. Broecker, H. Craig, C. D. Keeling, and J. Smagorinsky, "Restoring the Quality of Our Environment," in *Report of the Environmental Pollution Panel, President's Science Advisory Committee* (Washington, DC: The White House, 1965), 126.

CHAPTER 3: PEERING DOWN

1 Brownlee and Ward, *Rare Earth* (New York: Copernicus, 2000), 220.

2 We could learn something of Earth's deep interior without earthquakes. Seismic waves generated by underground nuclear tests have been used to measure the depth of the inner core, for instance. Of course, there are extreme limits on the number and locations of underground nuclear detonations. As a result, they can't compete with natural earthquakes as probes of Earth's interior.

3 A seismograph can record the arrival times and amplitudes of the two types of waves that traverse Earth's interior: compressional (or longitudinal) waves and shear (or transverse) waves. Compressional waves are basically sound waves. Shear waves cannot propagate through a liquid medium. Thus, the relative timings of compressional and shear waves and their presence or absence at a given seismograph station tell a geologist much about density variations.

4 The proverb about lightning never striking the same place twice is, of course, false.

5 P. G. Richards, "Earth's Inner Core—Discoveries and Conjectures," *Astronomy & Geophysics* 41, no. 1 (2000): 20–24; J. Zhang et al., "Inner Core Differential Motion Confirmed by Earthquake Waveform Doublets," *Science* 309 (2005): 1357–60.

6 Y. Yang and X. Song, "Multidecadal Variation of the Earth's Inner-Core Rotation," *Nature Geoscience* 16 (2023): 182–87.

7 See Lowrie, *Fundamentals of Geophysics*, 295–306.

8 A firsthand account of the research leading to the discovery of the seafloor remnant magnetic field is R. Mason, "Stripes on the Sea Floor," *Plate Tectonics: An Insider's History of the Modern Theory of the Earth*, N. Oreskes, ed. (Boulder, CO: Westview Press, 2001), 31–45.

9 David Sandwell, "Plate Tectonics: A Martian View," *Plate Tectonics: An Insider's History of the Modern Theory of the Earth*, N. Oreskes, ed. (Westview Press: Boulder, 2001), 342.

10 Ibid., 343.

11 Ibid., 343.

12 An introduction to this subject as applied to the Atlantic is J. G. Sclater and C. Tapscott, "The History of the Atlantic," *Scientific American* (June 1979): 156–74.

13 As late as 1960, geology textbooks were still confident of the geosyncline theory. For example, Clark and Stearn write in their textbook, "The geosynclinal theory is one of the great unifying principles in geology. In many ways its role in geology is similar to that of the theory of evolution, which serves to integrate the many branches of the biological sciences. . . . Just as the doctrine of evolution is universally accepted among biologists, so also the geosynclinal origin of the major mountain systems is an established principle in geology." T. H. Clark and C. W. Stearn, *The Geological Evolution of North America* (New York: Ronald Press, 1960), 43.

14 The sunspot cycles' effects on the geomagnetic field are quantified by the "Ap" and "aa" indices. A description of these indices can be found in P. N. Mayaud, "The aa Indices; a

100-Year Series Characterizing the Magnetic Activity," *Journal of Geophysical Research* 77 (1972): 6870–74.

15   The values of the aa and Ap indices at sunspot minimum correlate very well with the strength of the ensuing sunspot maximum, making them useful predictors of a given sunspot cycle. G. Gonzalez and K. Schatten, "Using Geomagnetic Indices to Forecast the Next Sunspot Maximum," *Solar Physics* 114 (1987): 189–92.

16   The technique of combining data from widely separated radio telescopes pointed at the same astronomical target is called Very Long Baseline Interferometry (VLBI). For a review, see O. J. Sovers, J. L. Fanselow, and C. S. Jacobs, "Astrometry and Geodesy with Radio Interferometry: Experiments, Models, and Results," *Reviews of Modern Physics* 70 (1998): 1393–1454.

17   We say "ironically" because earthquakes can cause death and destruction, which might seem to count against the correlation. But the ability to detect and measure earthquakes allows advanced civilizations the freedom to prevent such deaths. Many large population centers are located near plate boundaries, where earthquakes are most frequent, but it doesn't have to be that way. We now know where strong earthquakes are likely to happen with a fairly high degree of confidence. They don't just happen haphazardly, though we still cannot predict their timing. Also, better building codes have greatly improved earthquake survivability of buildings in threatened regions, for those people who choose to remain there. Advanced economies could virtually eliminate deaths from earthquakes if they chose to do so.

18   See Z. Mian, "Understanding Why Earth is a Planet with Plate Tectonics," *Quarterly Journal of the Royal Astronomical Society* 34 (1993): 441–448; K. Regenauer-Lieb, D. A. Yuen, and J. Branlund, "The Initiation of Subduction: Criticality by Addition of Water?," *Science* 294 (2001): 578–80; R. Boyle, "Why Earth's Cracked Crust May be Essential for Life," *Quanta Magazine* (2018), https://www.quantamagazine.org/why-earths-cracked-crust-may-be -essential-for-life-20180607/.

19   The carbon cycle is too complex for us to cover in detail here. For a general review of the carbon cycle and its interaction with the climate, see L. R. Kump, J. F. Kasting, and R. G. Crane, *The Earth System*, 4th Ed. (Dubuque, IA: Kendall Hunt Publishing, 2022); J. F. Kasting, "The Goldilocks Planet: How Silicate Weathering Maintains Earth 'Just Right'," *Elements* 15 (2019): 235–40.

20   J. C. G. Walker, P. Hays, and J. F. Kasting, "A Negative Feedback Mechanism for the Long-term Stabilization of Earth's Surface Temperature," *Journal of Geophysical Research* 86 (1981): 9776–82.

21   P. Chowdhury, et al., "Magmatic Thickening of Crust in Non-plate Tectonic Settings Initiated the Subaerial Rise of Earth's First Continents 3.3 to 3.2 Billion Years Ago," *Publications of the National Academy of Sciences* 118, no. 46 (2021): e2105746118.

22   Biological productivity in Earth's oceans depends on certain key nutrients, especially phosphorous and iron. Most of the needed nutrients flow into the oceans from the rivers. But even this may not be enough for all regions of the oceans. A simple oceanographic experiment conducted in 1995 caught many by surprise. (See K. H. Coale, et al., "A Massive Phytoplankton Bloom Induced by an Ecosystem-scale Iron Fertilization Experiment in the Equatorial Pacific Ocean," *Nature* 383 (1996): 495–501.) Small amounts of water-soluble iron were spread in a small patch of the eastern Pacific Ocean known to have a low biological productivity. Quickly, the populations of phytoplankton and zooplankton skyrocketed. Apparently, iron was the key limiting nutrient in these waters. Due to global wind patterns, some regions of the oceans receive very little dust from the continents, and the eastern equatorial Pacific is one of them. The iron in the dust is in a form that the biota can use. A planet without continents would lack both rivers and dust as sources of iron. A very small amount can also be delivered via meteoritic dust.

23   G. Gonzalez, D. Brownlee, and P. D. Ward, "The Galactic Habitable Zone: Galactic Chemical Evolution," *Icarus* 152 (2001): 185–200; Y. Alibert, "On the Radius of Habitable Planets," *Astronomy & Astrophysics* 561 (2014): A41.

24 We aren't yet certain when or from where Earth received most of its water, but there are several reasons a larger Earth-twin would probably start out with deeper oceans. First, it would collect a larger fraction of the water from asteroid and comet impacts, because a smaller fraction of the ejecta would be lost to space. Second, the greater gravitational focusing factor would cause more asteroids and comets to impact than a simple geometrical calculation would indicate. Third, as noted in the text, a larger terrestrial planet will have less surface relief.

25 See S. Jin and C. Mordasini, "Compositional Imprints in Density-Distance-Time: A Rocky Composition for Close-in Low-mass Exoplanets from the Location of the Valley of Evaporation," The Astrophysical Journal 853 (2018): 163.

26 Even with tectonic plates churning with more vigor, rocks on a planet with gravity greater than Earth's could not support mountains as high as those we enjoy. For a discussion on the physics of mountain heights and other interesting physics problems, see V. F. Weisskopf, "Of Atoms, Mountains, and Stars: A Study in Qualitative Physics," Science 187 (1975): 605–612.

27 In some waterworlds a natural desalinization pump would deplete salts too efficiently. See A. Levi and D. Sasselov, "A New Desalinization Pump Helps Define the pH of Ocean Worlds," The Astrophysical Journal 857 (2018): 65. The outcome is an acidic ocean with a pH in the range 2 to 4. On Earth all fish die below an acidity of 4.2.

28 Here, the relevant scaling relation is the fact that the mass of a sphere of uncompressed matter increases with diameter raised to the third power.

29 Smaller terrestrial planets follow the uncompressed scaling relations fairly closely but begin to deviate significantly before they get as big as Earth. Earth's mean density is 5.52 grams per cubic centimeter. Mars, about half the size of Earth, has a mean density of 3.93 grams per cubic centimeter.

30 J. S. Lewis, Worlds Without End: the Exploration of Planets Known and Unknown (Reading, UK: Helix Books, 1998), 64–66. If we had ignored compression, we would expect such a planet to be only about eight times more massive with twice the surface gravity.

31 A larger terrestrial planet will have a larger impact cross-section for asteroids and comets. The impact cross-section increases with increasing planet mass for two reasons. First, the physical size of the planet is greater. Second, gravitational focusing is greater, irrespective of its size.

32 Systems that can form giant Earth siblings (super Earths) will also likely form other smaller terrestrial planets (S. N. Raymond and A. Morbidelli, "Planet Formation: Key Mechanisms and Global Models," Demographics of Exoplanetary Systems: Lecture Notes of the 3rd Advanced School on Exoplanetary Science (2022): 3–82.). Therefore, the problems noted above for a very large terrestrial planet in a given system would still not prevent it from having a terrestrial planet in the required mass range. However, there are other problems associated with a system that forms massive terrestrial planets. More on this in chapter 5.

33 Ground-based optical telescopes achieve such high resolution by partially compensating for distortions of images of astronomical objects caused by Earth's atmosphere. These kinds of corrections work best for high altitude observatories. See the incredible ground-based image of Ganymede in Richard Dekany, et al., "PALM-3000: exoplanet adaptive optics for the 5 m Hale telescope," The Astrophysical Journal 776, no. 2 (2013): 130.

34 Fire is not a typical exothermic reaction. It releases great heat in a slow, nonexplosive manner. This makes it very useful for numerous technological applications. See Michael Denton, Nature's Destiny (New York: The Free Press, 1998), 122–23; and Michael Denton, Fire-Maker: How Humans were Designed to Harness Fire and Transform our Planet (Seattle, WA: Discovery Institute Press, 2016).

35 Certainly, some building materials can be shaped without metal tools. For example, wood can be shaped with stone axes, but many more construction options are available with bronze- and iron-age tools. Fire is also required to run a steam engine and prepare ceramics and durable bricks. The steam engine and the dynamo generator were arguably the most important prerequisites for the transition from basic to high technology. Both require the ability to shape iron.

36 In retrospect, it seems surprising that human beings learned how to purify metals from ores. About seven thousand years ago, people in the Middle East learned how to smelt copper from colorful azurite and malachite ore. Early potters probably discovered copper smelting by accident when they experimented with powdered malachite and azurite for use as pigments in their carbon-fueled and oxygen-starved kilns. This turned out to be just the environment needed for removing oxygen atoms from the copper oxides in the ore to produce pure copper metal. Soon afterward, the early smelters improved their skills at eliminating impurities by adding a "flux"—another mineral that chemically combined with impurities and formed easily removed slag. Hematite, an iron oxide, was the most popular flux used in the Middle East for copper smelting. The very same conditions that allowed humans to purify the copper ore allowed them to produce iron metal from the hematite flux. This may have been how iron smelting was discovered. See Stephen Sass, *The Substance of Civilization: Material and Human History from the Stone Age to the Age of Silicon* (New York: Arcane Publishing, 1998).

37 Both types of environments, waterworlds and gas giants, have been proposed as possible abodes for life. Neither is likely to support life. On a true waterworld, it would be very difficult to get nutrients to the surface where they are needed; also, see J. Krissansen-Totton, et al., "Waterworlds Probably do not Experience Magmatic Outgassing," *The Astrophysical Journal* 913 (2021): 107. On Earth, nutrients make their way from the continents into the oceans via rivers and deltas. If the ocean is not deeper than Earth's, then some methanogens might survive in mid-ocean ridges in a waterworld, but large, complex life is unlikely. Mastery of fire and the invention and use of electronic circuits also would be considerably more difficult. As for life in the atmosphere of a gas giant, no known organism on Earth has its complete life cycle in the atmosphere, making it less likely that such life exists in the atmospheres of gas giants. The basic problem is that solids in an atmosphere will fall downward and cease to be available to life. Additionally, while fire might not be impossible in such a setting, it would be difficult to maintain.

38 The modern industrial revolution was actually preceded by the industrial revolution of the European Middle Ages, which took place largely in monasteries. Monks harnessed the power of rivers with waterwheels attached to various types of labor-saving machinery. They also greatly increased the use of iron in agriculture and everyday life. Water power requires mountains to catch the rain and provide the gradient needed to make the water flow rapidly. Windmills began to appear in the twelfth century as another practical source of power. Thus, most of the precursors of the machinery of the modern industrial revolution were already in place in the Middle Ages; the biggest change is the source of power. For more information on the technological advances of the Middle Ages, see J. Gimpel, *The Medieval Machine: The Industrial Revolution of the Middle Ages* (New York: Holt, Rinehart & Winston, 1976).

39 "Thorium," World Nuclear Assocation, https://world-nuclear.org/information-library/current-and-future-generation/thorium.aspx. [last accessed April 29, 2024]

40 George Brimhall, "The Genesis of Ores," *Scientific American* (May 1991): 84.

41 C. M. R. Fowler, *The Solid Earth: An Introduction to Global Geophysics* (Cambridge, UK: Cambridge University Press, 1990), 225.

42 See P. Karuda, *The Origin of Chemical Elements and the Oklo Phenomenon* (New York: Springer-Verlag, 1982); G. A. Cowan, "A Natural Fission Reactor," *Scientific American* (July 1976): 36–47.

43 Methane gas could be used as a fuel, but nonbiological methane is not likely to coexist with oxygen in planetary atmospheres. Some bodies in the solar system do have methane, but not with oxygen. Methane may have been abundant on Earth for the planet's first couple billion years, but it likely became scarce when the atmosphere became oxygen-rich.

44 Denton, *Nature's Destiny*, 117. See also Amedeo Balbi and Adam Frank, "The Oxygen Bottleneck for Technospheres," *Nature Astronomy* (2023): https://doi.org/10.1038/s41550-023-02112-8. Amedeo and Frank argue that an oxygen partial pressure of at least 18 percent is required for high technology.

45   Robert M. Hazen, *The Story of Earth: The First 4.5 Billion Years, from Stardust to Living Planet* (New York: Penguin Books, 2012).

46   J. Thompson, "Life Helps Make Almost Half of All Minerals on Earth," *Nautilus* (July 8, 2022): https://nautil.us/life-helps-make-almost-half-of-all-minerals-on-earth-238511/.

CHAPTER 4: PEERING UP

1   Well into writing this book, we found that twentieth-century German philosopher Hans Blumenberg came very close to reflecting on the correlation between habitability and measurability in this introductory sentence to his monumental volume *The Genesis of the Copernican Revolution*, translated by Robert M. Wallace (Cambridge, MA: MIT Press, 1987), 3, originally published as *Die Genesis der kopernikanischen Welt* in 1975. Regrettably for the progress of natural philosophy and happily for our personal ambitions, the sentence serves only to introduce his case for the importance of historical contingencies in scientific progress, and he does not develop the idea in detail.

2   Isaac Asimov, *Nightfall and Other Stories* (Greenwich, Conn.: Fawcett Publications, Inc, 1969), 11–43.

3   See Denton, *Nature's Destiny*, 50–61.

4   *Encyclopaedia Britannica*, fifteenth ed. 18 (1994): 203.

5   For a more detailed discussion on this point, see Denton, *Nature's Destiny*, 47–70, and references cited therein. For a technical discussion, see W. H. Freeman and A. P. Lightman, "Dependence of Microphysical Phenomena on the Values of the Fundamental Constants," *Philosophical Transactions of the Royal Society of London A* 310 (1983): 323–336; J. D. Barrow and F. J. Tipler, *The Anthropic Cosmological Principle* (Oxford, UK: Oxford University Press, 1986), 338. They derive the typical surface temperature of a star (and thus the typical photon energy) and the typical energy of biological reactions from first principles and confirm that the two are surprisingly close. They consider it a genuine coincidence.

6   G. Greenstein, *The Symbiotic Universe* (New York: William Morrow, 1988), 96–97.

7   Here's a riddle. How can astronomers produce images of astrophysical sources like gamma rays, which are stopped high in the atmosphere? This sounds impossible. But thanks to a phenomenon called Cerenkov emission, simple light detectors on the ground can locate and determine the energy of individual gamma rays. When a high-energy gamma ray enters the top of the atmosphere, it hits an atom and triggers a nuclear reaction cascade. The first interaction produces a few particles, and these, in turn, interact with other atoms in the atmosphere, and so on. Eventually, all the original energy is used up and the cascade stops. Along the cascade, any charged particles traveling fast enough (faster than the speed of the light in the atmosphere) will produce optical Cerenkov radiation, which can be detected on the ground. Modern Cerenkov detectors can produce images of the gamma ray sources on the sky. For a sample of this type of experiment, see A. Kawach, et al., "The Optical Reflector System for the CANGAROO—II Imaging Atmospheric Cherenkov Telescope," *Astroparticle Physics* 14 (2001): 261–69. See also the VERITAS web site: http://veritas.sao.arizona.edu /. So, from the ground we can see the universe in three separate slices of the spectrum—the gamma ray, optical, and radio. Combined, these three make up a sizeable swath of the total range. Still, the optical region is by far the most important.

8   How far we can see into the atmosphere of a star depends on its opacity, that is, to its resistance to the transmission of light. The opacity of the gas in a star increases quickly at the temperature that molecules and neutral atoms can form—in other words, where chemistry starts to become important. We can see into the Sun's atmosphere just where molecules can start forming. (This has implications for the appearance of the solar spectrum, which we will discuss in chapter 7.) Molecules dominate the visible surfaces of significantly cooler stars, while few if any molecules are present in hotter ones. Most of the photons from hot stars would be absorbed in Earth's atmosphere and break apart molecules, including those essential for life. Most of the photons from cool stars would be absorbed by molecules

and excite low energy vibrations in them but would not be useful for stimulating chemical reactions. Thus, the appearance of the solar spectrum is intimately linked to life on Earth and the transparency of Earth's atmosphere in the optical part of the electromagnetic spectrum. Also, the surface temperature of a star turns out to be very sensitive to the constants of physics. (We'll discuss this at length in chapter 10.).

9   Typically, Earth's atmosphere limits angular resolution to about one second of arc from the surface. Any sources of radiation with angular sizes smaller than this, such as stars, will appear as unresolved points. Some high-quality mountaintop sites often achieve about one-third to one-half second of arc resolution.

10 Getting a dark sky also requires that we not have ever-present, overly bright moons polluting the sky with their light (although a less polluting bright moon is valuable, as we will discuss in chapter 6). Finally, it requires that the sky not be so filled with nearby stars that we are blind to the distant universe (discussed in chapter 8).

11 For informative accounts of the folklore, art, and science of rainbows, see R. L. Lee Jr, and A. B. Fraser, *The Rainbow Bridge: Rainbows in Art, Myth, and Science* (University Park: The Pennsylvania State University Press, 2001); C. B. Boyer, *The Rainbow: From Myth to Mathematics* (New York: Thomas Yoseloff, 1959). A more technical treatment is K. M. Carlson, et al., "Global Rainbow Distribution under Current and Future Climates," *Global Environmental Change* 77 (2022): 102604.

12 M. D. King, S. Platnick, W. P. Menzel, et al., "Spatial and Temporal Distribution of Clouds Observed by MODIS Onboard the Terra and Aqua Satellites," in *IEEE Transactions on Geoscience and Remote Sensing*, 51, no. 7, pp. 3826-3852, July 2013, doi: 10.1109/TGRS.2012.2227333. The authors also find that the cloud fraction over land is 55 percent, and over the oceans, 72 percent. We don't yet know how cloudiness varies over long periods of time or how it varies with composition of an atmosphere. However, it does vary by a few percent over decades.

13 See J. Lovelock, *The Ages of Gaia: A Biography of Our Living Earth* (New York: W. W. Norton & Company, 1988), and J. Lovelock, *Gaia: The Practical Science of Planetary Medicine* (Oxford, UK: Oxford University Press, 2000).

14 While we find much to laud in the Gaia hypothesis, we would not call the Earth system a "superorganism" or consider people to be infectious "disease agents," as Lovelock and especially some of his followers have done. We also disagree with Lovelock's characterization that the present is in a "fever" state, while the ice ages are "healthy"; the evidence we presented in chapter 2 suggests otherwise.

15 A. J. Watson and J. E. Lovelock, "Biological Homeostasis of the Global Environment: The Parable of Daisyworld," *Tellus* 35B (1983): 284–89. A more recent study is by T. M. Lenton and J. E. Lovelock, "Daisyworld Revisited: Quantifying Biological Effects on Planetary Self-regulation," *Tellus* 53B (2001): 288–305.

16 Such flexibility is most effective while both species of daisies exist. This is analogous to partly cloudy skies providing better climate stability than all clouds or no clouds. Daisyworld's stability requires that the daisies grow best over a specific temperature range; this is reasonable, since real daisies can't grow in arctic or desert climates. Any other organisms on the planet that require similar temperatures will benefit from this regulation.

17 R. A. Berner, "Paleozoic Atmospheric CO_2: Importance of Solar Radiation and Plant Evolution," *Science* 26 (1993): 68–70; T. W. Dahl and S. K. M. Arens, "The Impacts of Land Plant Evolution on Earth's Climate and Oxygenation State—An Interdisciplinary Review," *Chemical Geology* 547 (2020): 119665.

18 Lovelock was among the first to connect CCNs produced by life to the climate. R. J. Charlson, J. E. Lovelock, M. O. Andrea, and S. G. Warren, "Oceanic Phytoplankton, Atmospheric Sulfur, Cloud Albedo, and Climate," *Nature* 326 (1987): 655–661; R. J. Charlson, et al., "Reshaping the Theory of Cloud Formation," *Science* 293 (2001): 2025-2026.

19 Recent studies of dimethyl sulfide (DMS) and its influence on the carbon cycle confirm its role in the climate system but indicate it is a more complex topic than originally envisioned. See S.

Wang, *et al.*, "Influence of Dimethyl Sulfide on the Carbon Cycle and Biological Production," *Biogeochemistry* 138 (2018): 49–68.

20 We can't yet say how cloudy Earth was during the last glacial period. Determining how clear the atmosphere was in the distant past is tricky since there are no known proxies for cloudiness. But there are clues. One could try to relate the content of aerosols in polar ice known to be important in CCN to cloudiness. But this would probably be somewhat ambiguous, since the deposition of aerosols in the ice would also depend on the wind speeds over the oceans surrounding the polar regions. Another approach would employ computer simulations of ancient climates based on the global conditions recorded in the ice cores. Such a study would be beyond our current understanding of all the processes involved in cloud formation.

21 A. A. Pavlov, L. L. Brown, and J. F. Kasting, "UV Shielding of NH_3 and O_2 by Organic Hazes in the Archean Atmosphere," *Journal of Geophysical Research* 106 (2001): 23, 267–23, 288; G. Arney, Shawn D. Domagal-Goldman, et al., "The Pale Orange Dot: The Spectrum and Habitability of Hazy Archean Earth," *Astrobiology* (2016): 873–99. According to these studies, the organic haze played a role similar to that of ozone in the present atmosphere in shielding the ground from dangerous ultraviolet radiation.

22 See https://cneos.jpl.nasa.gov/stats/totals.html. [last accessed April 29, 2024]

23 O. B. Toon, et al., "Environmental Perturbations caused by the Impacts of Asteroids and Comets," *Reviews of Geophysics* 35 (1997): 41–78; C. R. Chapman, "Impact Lethality and Risks in Today's World: Lessons for Interpreting Earth History," in Proceedings of the Conference on Catastrophic Events and Mass Extinctions: Impacts and Beyond, Vienna Snowbird IV Conference, *Geological Society of America Special Paper* (2001).

24 See S. W. Paek, et al., "Optimization and Decision-Making Framework for Multi-Staged Asteroid Deflection Campaigns Under Epistemic Uncertainties," *Acta Astronautica* 167 (2020): 23-41.

25 The Earth, the Solar System, and even its place within the Milky Way all seem set up for space exploration. See G. Gonzalez, "The Solar System: Favored for Space Travel," *BIO-Complexity* (2020): 1–8.

26 "Yukon Meteorite Bonanza," *Sky & Telescope* 99 (June 2000): 22. This was a fortunate landing, since it's much easier to see dark meteorites against the white backdrop of ice and snow. Deserts and the ice fields of Antarctica have yielded many good quality specimens. Meteorite hunters prefer ice fields, in particular, since they can easily pick out the dark space rocks against the blue-white background. What's more, ice flows against mountain ranges concentrate meteorites on the surface, while the cold temperatures and isolation from the atmosphere help to minimize their erosion. Here again, the fact that our atmosphere contains water vapor and clouds benefits scientific discovery. For a first-hand account of meteorite collecting in Antarctica, see B. Livermore, "Meteorites on Ice," *Astronomy* 27 (July 1999): 54–58. See also the readable personal account of Guy Consolmagno, *Brother Astronomer: Adventures of a Vatican Scientist* (New York: McGraw-Hill, 2000).

27 The Yarkovsky effect also transfers asteroidal material to the inner planets. This effect results from the recoil force on a spinning asteroid from the emission of thermal radiation. It can significantly change the mean orbital distance of an asteroid on timescales of about one million years. As the Sun continues to brighten, this timescale will become shorter, resulting in an increased risk of asteroid impacts. See W. F. Bottke et al., "The Yarkovsky and YORP Effects: Implications for Asteroid Dynamics," *Annual Review of Earth and Planetary Sciences* 34 (2006): 157–191.

28 T. Hiroi, C. Pieters, and M. Zolensky, "The Tagish Lake Meteorite: A Possible Sample from a D-Type Asteroid," *Science* 293 (2001): 2234–2236.

29 As of November 9, 2023, there are 51 meteorites with orbits determined from optical observations of their meteor trails. See https://www.meteoriteorbits.info/.

30 For an extensive review of the subject of pre-solar grains, see E. Zinner, "Stellar Nucleosynthesis and the Isotopic Composition of Presolar Grains from Primitive Meteorites," *Annual Review of Earth and Planetary Sciences* 26 (1998): 147–188; L. R. Nittler, "Presolar

Stardust in Meteorites: Recent Advances and Scientific Frontiers," *Earth and Planetary Science Letters* 209 (2003): 259–73.

31   In particular, they can measure the important carbon 12 to 13 isotope ratio both in the grains and in cool star atmospheres. (Isotope ratios of oxygen and magnesium can also be measured in the spectra of some stars.) Another method compares isotopic ratios with theoretical "stellar nucleosynthesis," which is the production of isotopes via nuclear reactions at the high temperatures and pressures found deep inside stars.

32   J. E. Chambers and G. W. Wetherill, "Planets in the Asteroid Belt," *Meteoritics & Planetary Science* 36 (2001): 381–99.

33   Carbon is only a trace element in the bulk Earth. Had it not been delivered to Earth's surface near the end of its formation by asteroids and comets, there may not have been enough carbon near the surface for life to flourish. Water was also brought to Earth's surface by asteroids. Some asteroids contain large amounts of water chemically bonded in the minerals (called hydration). Once hydrated, the minerals in asteroids can better retain their water than if the water had not been present in hydrated form. The hydration of the minerals with water requires special conditions. One possibility is that the short-lived radioactive isotopes present in the very early history of the solar nebula heated the asteroid building blocks enough for hydration to occur. Another possibility is that shock waves in the gas provided enough heating and water vapor pressure to hydrate the minerals in the dust. Jupiter may have been the source of the shock wave. See M. D. Suttle, et al., "The Aqueous Alteration of CM Chondrites, a Review," *Geochimica et Cosmochimica Acta* 299 (2021): 219–56.

34   The significance of Jupiter for the habitability of a terrestrial planet is complex and multifaceted. See N. Georgakarakos, S. Eggl, and I. Dobbs-Dixon, "Giant Planet: Good Neighbors for Habitable Worlds?" *The Astrophysical Journal* 856 (2018): 155.

35   This impressed Donald Brownlee, a planetary astronomer and meteoriticist at the University of Washington:

> Common planetary systems may not have Jupiters or asteroid belts and hence these systems will be also highly deficient in meteoriticists. Without our benevolent asteroid belt, meteoritics might be a science based only on Nakhla, Shergotty, Chassigny, and perhaps Layayette and Zagmi [likely Martian meteorites] and even a few lunar meteorites if Earthlings would have ever convinced themselves such exceedingly rare rocks could actually fall from heaven. Unfortunately the most common interplanetary rocks in any solar system are probably those from comets. Comet rocks do not appear to survive atmospheric entry to become meteorites, so an earth in an asteroid-free solar system would be a pretty boring place for people with names like Anders, Wasserburg, Walker, Arnold, Wood, Wasson or Urey [famous meteoriticists].

D. Brownlee, "Mysteries of the Asteroid Belt," *Meteoritics & Planetary Science* 36 (2001): 328–29.

CHAPTER 5: THE PALE BLUE DOT IN RELIEF

1   As translated and published by G. E. Hunt and P. Moore in *The Planet Venus* (London: Faber & Faber, 1982), 74–76.

2   Even though the northern Martian polar cap does not preserve annual layers nearly as well as Earth's polar regions, high-resolution orbiter images of large cracks in the ice have revealed layering. The layering pattern resembles simulations of long-term obliquity and eccentricity changes of Mars over the past million years; see S. Byrne, "The Polar Deposits of Mars," *Annual Review of Earth and Planetary Sciences* 37 (2009): 535–560.

3   Laskar et al. "Orbital Forcing of the Martian Polar Layered Deposits," *Nature* 419 (2002): 375; J. Laskar, and P. Robutel, "The Chaotic Obliquity of the Planets," *Nature* 361 (1993): 608–12.

4   The possible Martian climatic responses to large tilt (obliquity) variations are discussed in B. M. Jakosky, B. G. Henderson, and M. T. Mellon, "Chaotic Obliquity and the Nature of the Martian Climate," *Journal of Geophysical Research* 100 (1995): 1579–84; T. Nakamura and

E. Tajika, "Evolution of the Climate System of Mars: Effects of Obliquity Change," *Lunar and Planetary Science* XXXIII (2002): 1057. The second study concludes that Mars cannot maintain residual ice caps when its tilt goes above forty-five degrees. Mars does have two very small moons, Deimos and Phobos. Close-up images from the Viking orbiters in the late 1970s revealed shapes and surface details very similar to those of asteroids visited by more recent probes. For this reason, planetary astronomers believe they are captured asteroids. Phobos, the larger and closer of the two moons, orbits Mars in a little under eight hours. Since it is well inside the synchronization distance (where its orbital period equals the planet's rotation period), Phobos is slowly spiraling inward from tidal interactions with Mars. It should impact Mars in about 40 to 50 million years. Thus, not only can Mars's moons not produce total solar eclipses, but they cannot stabilize its obliquity. Indeed, one of them will commit a final insult to it.

5   A. Mittelholz et al., "Timing of the Martian Dynamo: New Constraints for a Core Field 4.5 and 3.7 Ga ago," *Science Advances* 6.18 (2020): eaba0513.

6   David Sandwell doubts that Martian remnant fields are evidence of ancient tectonic spreading because "one cannot be this lucky twice." See his chapter "Plate Tectonics: A Martian View," in *Plate Tectonics: An Insider's History of the Modern Theory of the Earth*, N. Oreskes, ed. (Boulder, CO: Westview Press, 2001), 343. By "lucky," he means it's highly improbable that just the right set of circumstances necessary for plate tectonics (discussed in chapter 3) would also occur on the very next planet out from Earth. The following study presents evidence that the Martian banded magnetic anomalies are due to a process like the one that produced the North American Cordillera: A. G. Fairen, J. Ruiz, and F. Anguita, "An Origin for the Linear Magnetic Anomalies on Mars through Accretion of Terranes: Implications for Dynamo Timing," *Icarus* 160 (2002): 220–23.

7   Mars has a greater surface area to volume ratio. A smaller planet has more surface area compared to the volume of its interior and, hence, cools faster.

8   S. M. McLennan, "Crustal Heat Production and the Thermal Evolution of Mars," *Geophysical Research Letters* 28 (2001): 4019–22.

9   The hydrological cycle has been the most important factor in Earth's very large mineral diversity. See R. M. Hazen and S. M. Morrison, "On the Paragenetic Modes of Minerals: A Mineral Evolution Perspective," *American Mineralogist* 107:7 (2022): 1262–87. "Of the 5659 minerals examined, 4583 (81.0%) either incorporate essential OH- or H_2O, accounting for 3150 of 5659 mineral species (55.7%), or they are anhydrous phases that were formed through the action of aqueous fluids, for example, by dehydration, evaporation, or precipitation from a hydrothermal solution (1433 species)," the authors write. "No other physical or chemical factor comes close to this dominant role of water in creating mineral diversity. . . . The sharp contrast between Earth's large complement of minerals and the relative mineralogical parsimony of the Moon and Mercury, as well as the modest diversity found on Mars, stems from the differing influences of water."

10   If its core is stratified, then it lacks convection, which is also required for generating a magnetic field. A leading theory posits that Earth's early giant moon-forming collision homogenized its core, allowing it to be convective. If Venus lacked a similarly large collision, then that would also account for its lack of a magnetic field. For an up-to-date review of the past and present habitability of Venus, see F. Westall, et al., "The Habitability of Venus," *Space Science Reviews* 219, no. 2 (2023): 17. The recent claimed discovery of phosphine (PH3) in the atmosphere of Venus (which remains controversial as of this writing) has stimulated discussion of possible life in its atmosphere. However, the proposed biological mechanisms to produce phosphine in the atmosphere of Venus have been ruled out because they would produce abundances of other compounds incompatible with the observed composition of the atmosphere. See S. Jordan et al., "Proposed Energy-Metabolisms Cannot Explain the Atmospheric Chemistry of Venus," *Nature Communications* 13 (2022): 3274. What's more, there is no form of Earthly life that spends its complete life cycle in the atmosphere, and this suggests that this may be an unworkable mode of life anywhere.

11 S. A. Jacobson et al. "Formation, Stratification, and Mixing of the Cores of Earth and Venus." *Earth and Planetary Science Letters* 474 (2017): 375–86.

12 D. C. Catling and K. J. Zahnle, "The Planetary Air Leak," *Scientific American* 300.5 (May 2009): 36–43.

13 There is evidence for some past tectonic activity on Mars, but it's controversial. See M. H. Acuna et al., "Global Distribution of Crustal Magnetization Discovered by the Mars Global Surveyor MAG/ER Experiment," *Science* 284 (1999): 790–93. Even if past tectonic activity on Mars, it does not appear to have been the full-blown type we see on Earth.

14 C. C. Reese and V. S. Solomatov, "Non-Newtonian Stagnant Lid Convection and Magmatic Resurfacing on Venus," *Icarus* 139 (1999): 67–80.

15 See M. Turbet et al. "Day–Night Cloud Asymmetry Prevents Early Oceans on Venus but Not on Earth," *Nature* 598.7880 (2021): 276–80; J. F. Kasting and C. E. Harman, "Venus Might Never Have Been Habitable," *Nature* 598 (2021): 259–60. The authors argue that water never condensed on the surface of Venus, so it never formed oceans.

16 This means that the main fluid cycle period relates to one of the motions of the solid body of a planet by some small integer multiple.

17 J. Touma and J. Wisdom, "Nonlinear Core-Mantle Coupling," *Astronomical Journal* 122 (2001): 1030–50.

18 The precession of the perihelion of Mercury was an unexplained problem for Newtonian gravity discovered in the nineteenth century. Observations showed that the orientation of the orbit of Mercury was changing faster than could be explained by classical physics, although there were proposals that claimed to explain it within this paradigm. Einstein argued that he could explain it with his theory of General Relativity. But this was counted as a "postdiction" rather than a prediction since the Mercury problem was known prior to Einstein's theory.

19 Observations also suggest the presence of an ocean in Ganymede, the third Galilean Moon of Jupiter, but it is covered by a thicker ice layer than Europa's ocean. Additionally, Saturn's moons Enceladus and Titan show evidence of subsurface oceans. See J. C. Castilo-Rogez and K. Kalousova, "Ocean Worlds in our Solar System," *Elements* 18 (2022): 161–66.

20 See C. F. Chyba and C. B. Phillips, "Europa as an Abode of Life," *Origins of Life and Evolution of the Biosphere* 32 (2002): 47–68; M. Lingam and A. Loeb, "Subsurface Exolife," *International Journal of Exobiology* 18.2 (2019): 112–41.

21 The following study reports on biological activity of bacteria at pressures up to 16,000 times the surface pressure on Earth: A. Sharma et al., "Microbial Activity at Gigapascal Pressures," *Science* 295 (2002): 1514–16. The biological activity at high pressure was observed to be about one percent of that at ambient conditions.

22 It's not even clear if Europa can have seafloor volcanism. See M. T. Bland and C. M. Elder, "Silicate Volcanism on Europa's Seafloor and Implications for Habitability," *Geophysical Research Letters* 49, no. 5 (2022): e2021GL096939.

23 Chyba and Phillips, "Europa as an Abode of Life," 56.

24 J. Lovelock, *The Ages of Gaia: A Biography of Our Living Earth* (New York: W. W. Norton, 1988), 106. Most life on Earth is limited to salt concentrations to less than 0.8 molar; the salinity of seawater reaches about 0.68 molar. Molar concentration is the number of moles of a solute per liter of water.

25 W. B. McKinnon and E. L. Shock, "Ocean Karma: What Goes Around Comes Around on Europa (Or Does It?)," *Lunar and Planetary Science* 32 (2001): abstr. 2181.

26 Even with all these limitations on life in Europa, some still speculate that a niche may be available a few meters below the surface in the ice cracks, where the pressure is low.

27 While not nearly as dynamic as a giant planet's atmosphere, Earth surface is also in constant (slow) motion. For this reason, little remains of Earth's Hadean period between 4.5 and 3.8 billion years ago. (One exception are tiny zircon crystals, sometimes described as "time capsules" of Earth's geologic history. See Cypress Hansen, "Keeping Time with Zircons," *Knowable Magazine*, September 4, 2021, https://knowablemagazine.org/article/physical -world/2021/keeping-time-zircons.) At the same time, the churning that erased this early

history is not only vital for life, but it gives geologists a great way to map Earth's interior. So, it's hard to fault Earth for mucking up our Hadean period data.

28   Four of the six seismometers left on the Moon by Apollo astronauts recorded over 12,000 seismic events between 1969 and 1977. Only eighty-one of these were good enough to reveal the Moon's internal structure. Of these, thirty-four were deep quakes and fourteen were shallow ones; the remainder resulted from artificial (slamming probes into the Moon) and natural impacts. The biggest lunar quakes are very weak by our standards, less than about three on the Richter scale. Without water, the lunar rocks allow the seismic waves to propagate with relatively little attenuation, but its very heterogeneous and fractured crust scatters them, making it more difficult to interpret the data. A. Khan, K. Mosegaard, and K. L. Rasmussen, "A New Seismic Velocity Model for the Moon from a Monte Carlo Inversion of the Apollo Lunar Seismic Data," *Geophysical Research Letters* 27 (2000): 1591–94.

29   S. C. Stahler et al., "Seismic Detection of the Martian Core," *Science* 373 (2021): 443–48; W. Sun and H. Tkalčić, "Repetitive Marsquakes in Martian Upper Mantle," *Nature Communications* 13 (2022): 1695.

30   See D. M. Williams and D. Pollard, "Earth-like Worlds on Eccentric Orbits: Excursions Beyond the Habitable Zone," *International Journal of Astrobiology* 1 (2002): 61–69; J. Jernigan et al., "Superhabitability of High-obliquity and High-eccentricity Planets," *The Astrophysical Journal* 944 (2023): 205. The "superhabitability" that the second study refers to concerns increased productivity of simple marine organisms in their models of planets with high obliquities and/ or eccentricities. There are at least two problems for complex life in these types of worlds. First, these planets experience large seasonal variations in ocean chemistry (especially oxygen levels). Ocean deoxygenation is associated with some of the largest mass extinctions in Earth's history. Second, planets with high obliquity suffer more rapid water loss; W. Kang, "Wetter Stratospheres on High-obliquity Planets," *The Astrophysical Journal Letters* 877, no. 1 (2019): L6.

31   It was only in the late 1980s that astronomers discovered that our solar system is chaotic, though not catastrophically so. J. Laskar, "A Numerical Experiment on the Chaotic Behaviour of the Solar System," *Nature* 338 (1989): 237–38. One need only spend a little time with a gravity simulation to see how quickly multiple planet systems grow chaotic and catastrophic. For example, see the website: https://testtubegames.com/gravity.html. (last accessed April 29, 2024)

32   J. Laskar, "Large-scale Chaos in the Solar System," *Astronomy & Astrophysics* 287 (1994): L9–L12.

33   This is based on simulations. Lecar et al., "Chaos in the Solar System," 47.

34   The following study considers the stability of the primary variations in Earth's orbital parameters and how that relates to their use in paleoclimatology and in the "geologic orrery": P. E. Olsen et al., "Mapping Solar System Chaos with the Geological Orrery," *Proceedings of the National Academy of Sciences* 116, no. 22 (2019): 10664–73.

35   This is the limit of a purely mathematical orbital simulation. J. Laskar et al., "La2010: A New Orbital Solution for the Long-term Motion of the Earth," *Astronomy & Astrophysics* 532 (2011): A89.

36   For an application of astrochronology beyond the roughly 50-million-year Solar System chaos limit, see G. Charbonnier et al., "Astrochronology of the Aptian Stage and Evidence for the Chaotic Orbital Motion of Mercury," *Earth and Planetary Science Letters* 610 (2023): 118104.

37   For the latest news on extrasolar planet research, see the NASA Exoplanet Archive website: https://exoplanetarchive.ipac.caltech.edu/.

38   The first planet found around another star was 51 Pegasi, with a four-day orbital period. M. Mayor and D. Queloz published their discovery in "A Jupiter Mass Companion to a Solar-Type Star," *Nature* 378 (1995): 355–39.

39   Astronomers have found most of these using the photometric transit method. They have also found many with the Doppler method, which makes use of the wobble of a star as a planet

orbits around it. Small changes in the Doppler shift of a star's light can reveal an otherwise invisible planet.

40 These are: interaction between a planet and density waves set up in the protoplanetary disk by the planet, gravitational scattering of rocky bodies in the vicinity of a planet, or mutual gravitational interactions among two or more planets, which leave one in a smaller orbit. Such mutual gravitational interactions tend to leave at least one planet in a highly eccentric orbit.

41 There is another consequence of giant-planet migration. As a giant planet migrates inwards, it not only tends to scatter away any intervening terrestrial planets, but it also clears away planetesimals. Without plenty of planetesimals, Earth-size terrestrial planets cannot form. See P. J. Armitage, "A Reduced Efficiency of Terrestrial Planet Formation Following Giant Planet Migration," *Astrophysical Journal Letters* 582 (2003): L47–L50. Another study showed that very water-rich planets tend to form in the wake of a giant planet undergoing extreme migration: S. N. Raymond, A. M. Mandell, and S. Sigurdsson, "Exotic Earths: Forming Habitable Worlds with Giant Planet Migration," *Science* 313 (2006): 1413–16. However, such water-rich worlds are not likely to be habitable.

42 The details of the formation of terrestrial planets turn out to be very sensitive to the configuration of the giant planets. See A. Morbidelli et al., "Building Terrestrial Planets," *Annual Review of Earth and Planetary Sciences* 40 (2012): 251–75.

43 Statistical analysis of multiple planet systems has revealed that the greater the number of planets in a system, the smaller the average eccentricity of its planets. When the trend is extrapolated to the solar system, it is found to be a close fit. See N. Bach-Møller and U. G. Jørgensen, "Orbital Eccentricity–Multiplicity Correlation for Planetary Systems and Comparison to the Solar System," *Monthly Notices of the Royal Astronomical Society* 500, no. 1 (2021): 1313–22. However, the multiple-planet systems discovered as of this writing are otherwise very different from the solar system. About 60 percent have super-Earths, and these systems are much more compact (smaller than about 0.5 AU). They also tend to be close to dynamical instability. See K. Volk and R. Malhotra, "Dynamical Instabilities in Systems of Multiple Short-Period Planets Are Likely Driven by Secular Chaos: A Case Study of Kepler-102," *The Astronomical Journal* 160, no. 3 (2020): 98.

44 For example, see D. M. Williams, J. F. Kasting, and R. A. Wade, "Habitable Moons around Extrasolar Giant Planets," *Nature* 385 (1997): 234–236; D. Vera, R. Heller, and E. L. Turner, "The Effect of Multiple Heat Sources on Exomoon Habitable Zones," *Astronomy & Astrophysics* 601 (2017): id.A91.

45 Of course, not every planetary system will have the same comet flux as ours, due to different formative histories and giant planet configurations. A system that forms an Earth-size terrestrial planet, however, is almost certainly going to be accompanied by plenty of comets and asteroids and possibly one or more gas giants.

46 P. M. Schenk et al., "Cometary Nuclei and Tidal Disruption: The Geologic Record of Crater Chains on Callisto and Ganymede," *Icarus* 121 (1996): 249–74.

47 An Earth-size moon need not generate an intrinsic magnetic field in its core. The Galilean Moon, Europa, has a field thought to be induced in water with dissolved salts in its interior as the moon passes through Jupiter's magnetic field lines as it orbits. They form conducting electrolytes that generate magnetic fields as the moon crosses the giant planet's magnetic field lines. Ganymede's magnetic field is thought to be intrinsic, making it the only moon in the Solar System with such a field. Ganymede's magnetic field strength is only about a tenth of Earth's, and Europa's only about a fortieth. Thus, while such induced fields offer moderate protection against charged particle radiation, this does not seem to be as effective as a magnetic field generated in an iron core.

48 If Jupiter were orbiting closer to the Sun, say at the Earth-Sun distance, then its radiation belts would be even more intense, further undermining the habitability of any of its moons.

49 Vera, Heller, and Turner, "The Effect of Multiple Heat Sources on Exomoon Habitable Zones." To be effective as a source of internal heat, other planet-sized moons must also be present to

maintain the eccentric orbit of our hypothetical habitable moon-planet. A moon in a circular orbit would not experience the variations in tidal force needed to heat its interior.

50 R. Greenberg et al., "Habitability of Europa's Crust: The Role of Tidal-tectonic Processes," *Journal of Geophysical Research* 105 (2000): 17551–62.

51 Indeed, many are lost even up to about a 300-day orbital period—just shy of Earth's 365 days. J. W. Barnes and D. P. O'Brien, "Stability of Satellites around Close-in Extrasolar Giant Planets," *Astrophysical Journal* 575 (2002): 1087–1093; V. Dobos et al., "Survival of Exomoons around Exoplanets," *Publications of the Astronomical Society of the Pacific* 133 (2021): id.094401.

52 One could argue that because the Galilean Moons exhibit a systematic decline of volatile content, from the rocky Io to the icy Callisto, so too will Earth-size moons over a range of distances from their host planet. Ganymede, interestingly, is larger but less massive than the planet Mercury. At some intermediate distance, then, there is a chance that a moon will have just the right volatile content for life.

> But even on such a moon, there would be other problems for life. Most of the differences among the Galilean Moons probably resulted from differences in tidal heating: Io experienced the most and Callisto the least. The intense volcanic activity on Io drove away most of its initial endowment of volatiles. This was possible because its surface gravity was not strong enough to retain its volatiles in a liquid or vapor state. An Earth-size moon in the same spot would suffer far less volatile loss from tidal heating than Io did. The abundance of volatiles rises steeply from just inside the water condensation boundary to just outside of it. Thus, a giant planet would have to form at just the right distance from its host star and then migrate inward into the habitable zone to have moons with the right level of volatiles. The precision of this choreography makes this extremely unlikely.

53 Miki Nakajima, et al., "Large Planets May not Form Fractionally Large Moons," *Nature Communications* 13, no. 1 (2022): 568.

54 Perhaps such moon dwellers would get stuck on the notion that everything in the universe revolved about their giant-planet host.

CHAPTER 6: OUR HELPFUL NEIGHBORS

1 Ivars Peterson, *Newton's Clock: Chaos in the Solar System* (New York: W. H. Freeman and Company, 1993), 286.

2 We owe this insight to philosopher of science Robin Collins. Stanley Jaki makes a similar point about the Earth-Moon system in *Maybe Alone in the Universe After All* (Pinckney, MI: Real View Books, 2000), as does Peterson, *Newton's Clock*, 286, 293, regarding the solar system.

3 Kepler's shift from assuming circular planetary orbits to considering the possibility of elliptical orbits may seem like a minor paradigm shift. But it was major. Kepler first tried to fit the orbit of Mars to a circle, but failed. Only after great effort did he finally convince himself that its orbit is an ellipse. He had broken with the long-held view that planets must orbit in perfect circles. The eccentricity of the orbit of Mars is 0.093; Earth's is 0.017.

4 Of course, some of the stars now known to have only one or two planets might one day be found to have other smaller planets. To date only one star, Kepler-90, is known to be a host to as many planets as the Sun.

5 Peterson, *Newton's Clock*, 293.

6 The high eccentricity of Mercury's orbit further complicates matters for potential Mercurians. During perihelion (closest approach to the Sun) Mercury's rapid orbital motion overtakes its slower rotation and, for a few Earth days, the Sun would appear to an observer on Mercury's surface to move backwards in the sky.

7   The hierarchical mass distribution in the solar system also aided the discovery of the laws of planetary motion. Kepler's laws are actually inexact. The planets don't orbit the Sun, but, rather, they orbit the center of mass. Newton showed that Kepler's Third Law, $P^2 = K*a^3$, is slightly different for each planet, given their nonzero masses. In addition, the planets' mutual gravity causes their orbits to depart from simple ellipses. Both these minor deviations from Kepler's simple laws would have been more severe had the planets been more massive, had they been closer in mass to the Sun, severe enough perhaps to prevent discovery of the Third Law. We will return to the topic of hierarchical clustering in the universe in chapter 10.

8   We lack empirical evidence to support our claim that Earth is a better platform than other places in the Solar System for discovering the laws of planetary motion. Perhaps an enterprising planetarium operator can put together an experiment to test it.

9   Possibly one such curiosity is the Titus-Bode relation, first published in 1772 by German astronomer Johann Bode. It relates the relative spacings of the planets to a simple mathematical series. It was used to predict the existence of the dwarf planet Ceres and the planet Uranus, but it has only been partially successful when applied to exoplanets. Also, there appears to be some physical basis for it. See L.-C. Yeh, I.-G. Jiang, and S. Gajendran, "On the Scaling and Spacing of Extra-Solar Multi-Planet Systems," *Astrophysics and Space Science* 365 (2020): 1–11. Interestingly, the following study finds that Earth is by far the largest outlier when fitting Bode's Law to the Solar System: L. Neslušan, "The Significance of the Titius–Bode Law and the Peculiar Location of the Earth's Orbit," *Monthly Notices of the Royal Astronomical Society* 351, no. 1 (2004): 133–136.

10   This limitation doesn't apply to compact exoplanet systems, wherein the planet orbital spacings are small enough that an observer on one of the planets could resolve the other planets without optical aid.

11   A possible correction to this is the evidence for very small amounts of water ice in the regolith on the Moon's polar regions. See A. Colaprete, et al. "Detection of Water in the LCROSS Ejecta Plume." *Science* 330, no. 6003 (2010): 463–68.

12   Crater counting from ground-based observatories is a successful endeavor because the lunar maria are almost exclusively on its near side. Unlike the crater-saturated highlands, a mare presents a relatively smooth surface on which a reliable crater density can be measured. Astronomers use crater density estimates from surfaces of different ages to reconstruct the Moon's cratering history.

13   The Moon has no atmosphere to burn up incoming meteoroids. Consequently, meteorites till the upper surface so frequently and vigorously that they destroy older and more subtle structures, rendering the Moon generally less information-friendly than the gentle depositional processes on Earth's surface. Even without impact tilling, the Moon would still lack the clear sedimentary layering produced in water media on Earth. For a comparison of sedimentation processes on Earth and Moon, see A. Basu and E. Molinaroli, "Sediments of the Moon and Earth as End-Members for Comparative Planetology," *Earth, Moon, and Planets* 85–86 (2001): 25–43.

14   Currently, there are about 190 impact structures known on Earth, with two to three new ones discovered each year.

15   That there was a spike in large impacts on the Moon about 3.9 billion years ago remains controversial. See the review paper: W. F. Bottke, and M. D. Norman, "The Late Heavy Bombardment," *Annual Review of Earth and Planetary Sciences* 45 (2017): 619–47.

16   J. Armstrong, L. Wells, and G. Gonzalez, "Rummaging through Earth's Attic for Remains of Ancient Life," *Icarus* 160 (2002): 183–96; S. H. C. Cabot, and G. Laughlin. "Lunar Exploration as a Probe of Ancient Venus," *The Planetary Science Journal* 1, no. 3 (2020): 66. The Moon collects fragments not only from the early Earth, but also from early Mars and Venus. Since nothing remains on Venus from its early days, the Moon probably provides a unique source of information about this planet.

17   One research group has derived an average terrestrial albedo of 0.297 ± 0.005 from observations obtained at the Big Bear Solar Observatory in California: P. R. Goode et al., "Earthshine Observations of the Earth's Reflectance," *Geophysical Research Letters* 28

(2001): 1671–74. Long term monitoring has revealed that Earth's albedo measurably decreased: P. R. Goode et al., "Earth's Albedo 1998–2017 as Measured from Earthshine," *Geophysical Research Letters* 48, no. 17 (2021): e2021GL094888.

18 See N. J. Woolf et al., "The Spectrum of Earthshine: A Pale Blue Dot Observed from the Ground," *Astrophysical Journal* 574 (2002): 430–33.

19 Jaki, *Maybe Alone in the Universe After All*, 20–26.

20 Consider what an eclipse of a moon would look like from another planet. For example, an eclipse of one of the Galilean Moons would look quite different to inhabitants of Jupiter's cloud tops. Since Jupiter is immensely larger than its moons, the curve of its shadow would be more difficult to discern. The same could be said of the two small moons around Mars. In general, the curve of a planet's shadow is more discernable when its moon is not much smaller than its parent planet.

21 A very informative introduction to distance measurement in astronomy is S. Webb, *Measuring the Universe: The Cosmological Distance Ladder* (Chichester, UK: Praxis Publishing Ltd, 1999). A less technical and more historical treatment is by Kitty Ferguson, *Measuring the Universe: Our Historic Quest to Chart the Horizons of Space and Time* (New York: Walker and Company, 1999).

22 Webb, *Measuring the Universe*, 28–32.

23 Hipparchus also estimated the distance to the Moon using lunar eclipses. See Webb, *Measuring the Universe*, 32–34.

24 See A. H. Batten, "Aristarchos of Samos," *Journal of the Royal Astronomical Society of Canada* 75, no. 1 (1981): 29–35.

25 An observer would have to measure the angle between the Sun and Moon to within eight minutes of arc to get a useful distance. It could also be done by comparing the time intervals between first and third quarters versus third and first quarters, but this would require careful consideration of the Moon's noncircular orbit.

26 Such a precise measurement is theoretically just possible with the naked eye, but in practice the irregular surface of the Moon does not permit it. Telescopic measurements, especially when aided with a camera makes it relatively easy. See the following papers for modern applications of the methods of Aristarchus to determine the distances to the Sun and Moon: F. Momeni et al., "Determination of the Sun's and the Moon's Sizes and Distances: Revisiting Aristarchus' Method," *American Journal of Physics* 85, no. 3 (March 1, 2017): 207–15, K. Krisciunas et al., "The First Three Rungs of the Cosmological Distance Ladder," *American Journal of Physics* 80, no. 5 (May 1, 2012): 429–38.

27 In a series of studies published since 2012, Mario Livio and collaborators explore how a planetary system's architecture affects its asteroid belt and how that relates to planetary habitability. Some examples are R. G. Martin, and M. Livio, "On the Formation and Evolution of Asteroid Belts and their Potential Significance for Life," *Monthly Notices of the Royal Astronomical Society: Letters* 428, no. 1 (2012): L11–L15, R. G. Martin and M. Livio, "Asteroids and Life: How Special Is the Solar System?", *The Astrophysical Journal Letters* 926 (2022): L20, R. G. Martin and M. Livio, "On the Formation and Evolution of Asteroid Belts and their Potential Significance for Life," *Monthly Notices of the Royal Astronomical Society* 428 (2013): L11–L15, J. L. Smallwood et al., "Asteroid Impacts on Terrestrial Planets: The Effects of Super-Earths and the Role of the v_6 Resonance," *Monthly Notices of the Royal Astronomical Society* 473, no. 1 (2018): 295–305. They conclude that the existence of an asteroid belt and the delivery of asteroids to a terrestrial planet are sensitive to the planetary architecture, and in particular, the delivery of asteroids to Earth is sensitive to the orbits of Jupiter and Saturn.

28 Due to Earth's gravity the Moon's effective cross section is actually a bit more than its physical cross section. Thus, the ratio of impacts on the Moon to Earth depends on the Moon's distance, which has been increasing since it formed. There's a second, super-speculative way the other planets may have helped Earthly life. Early on, Earth probably had several nasty encounters with giant asteroids. Some may have vaporized its oceans and sterilized the entire planet. These impacts would have hurled many Earthly fragments into

space. Some of these fragments could have seeded Mars with organisms. Thus, Mars, or even Venus, if it was hospitable to life early on, could have served as a temporary refuge for life while Earth recovered; a later impact on Mars or Venus could then than have reseeded Earth. An isolated planet would lack such temporary refuges for life. N. Sleep, et al., "Annihilation of Ecosystems by Large Asteroid Impacts on the Early Earth," *Nature* 342 (1989): 139–42.

29   Another source of data on large impacts that could have occurred on Earth are the impact craters mapped on Venus by the Magellan probe in 1990. Due to its relatively young surface, only impacts spanning the last 500 to 700 million years are preserved on its surface. In addition, its thick atmosphere does not allow small objects to reach its surface, so the crater count is only a lower limit on the total number of impactors.

30   N. Sleep, et al., "Annihilation of Ecosystems by Large Asteroid Impacts on the Early Earth," *Nature* 342 (1989): 139–42.

31   Impacts would have hurled many Earthly fragments into space. Some of these fragments could have seeded Mars with organisms. Thus, Mars, or even Venus, if it was hospitable to life early on, could have served as a temporary refuge for life while Earth recovered; a later impact on Mars would have reseeded Earth. An isolated planet would lack such temporary refuges for life. For a related idea see L. Wells et al., "Reseeding of Early Earth by Impacts of Returning Ejecta During the Late Heavy Bombardment," *Icarus* 162 (2003): 38–46.

CHAPTER 7: STAR LIGHT, STAR BRIGHT

1   Cited in *The Book of the Cosmos: Imagining the Universe from Heraclitus to Hawking*, ed. Dennis Richard Danielson (Cambridge, MA: Basic Books, 2000), 319.

2   As Comte said, "I still believe that the average temperature of the stars must forever elude us. . . . Such being the case, I do not think I am limiting astronomy too much by assigning to it the discovery of the laws governing the geometrical and mechanical phenomena presented by the celestial bodies, that, and that only." *The Essential Comte: Selected from Cours de Philosophie Positive*, trans. Margaret Clarke (London: Croom Helm, 1974), 76.

3   Astronomers also employ finer divisions within each spectral-type category. Thus, a G0 star is slightly hotter than the Sun, a G2 star.

4   This is because of the diversity of radiative transitions their electrons can partake in (atomic, vibrational, and rotational). The additional atoms present in a molecule allow more degrees of motion than does an isolated atom. The two atoms in a hydrogen molecule, for instance, can rotate or vibrate. The changes in the vibrational or rotational states are quantized. Thus, the electrons around molecules preferentially interact with photons of specific wavelengths. The rotational states have the smallest energy steps, so the rotational spectral lines are bunched close together in the spectrum.

5   The basic reason for this is that one type of molecule, such as titanium oxide, produces tens of thousands of absorption lines over just a few hundred Ångstroms. Therefore, since only a few titanium oxide lines need be measured to determine the sum of the titanium and oxygen abundances, all the other thousands of titanium oxide lines are superfluous and contribute only towards obscuring other lines from other species.

6   Meteorites give us the isotopic abundances of most elements, while they do not give us reliable abundances for the volatiles. The solar spectrum gives us reliable atomic abundances for most elements, including the volatiles, but relatively few isotope abundances.

7   Even if a given star is too far away to detect its tangential motion relative to other stars, we can still measure its radial velocity—its motion along our lines of sight. All that is required to determine a star's radial velocity is to measure the displacement in the wavelengths of its spectrum relative to emission lines produced in a glowing gas-filled tube on Earth's surface. The shift in the spectrum is most reliably measured if it contains sharp features, such as absorption lines. The wavelength shifts are translated into radial velocities via application of the Doppler equation.

8   Observation of stellar Doppler shifts also tells us Earth's orbital velocity around the Sun. Like stellar parallaxes, stellar Doppler shifts are observational evidence of the heliocentric model of the solar system.

9   That the radial motion of a star could be determined from spectra also impressed Huggins: "The foundation of this new method of research, which, transcending the wildest dreams of an earlier time, enabled the astronomer to measure off directly in terrestrial units the invisible motions in the line of sight of the heavenly bodies." Danielson, *The Book of the Cosmos*, 325.

10   Consider an interstellar nebula for contrast. A nebula is spread out over a large volume of space and is typically separated from other nebulae by distances only a few orders of magnitude greater than its size. The atoms in a nebula are relatively weakly bound by gravity, so it doesn't take much to distort them. Close encounters between clouds, close passages by stars, or the asymmetric pull from the large-scale gravitational field of the Milky Way galaxy will do the trick. This distributes light asymmetrically, making it hard to define the nebula's center. Nebulae also have low surface brightness and irregular shapes, making them very poor reference points compared to stars. Modern stellar astrometry has yet to reach a level of measurement precision limited by the angular sizes of stars.

11   For example, the raising of the tides on Earth by the Moon affects the orbital dynamics of the two bodies. Extrapolating the Moon's recession into the distant past or future requires some knowledge of the interior structure of Earth and how it responds to tidal stresses. The same kind of knowledge would be required of stars if they were much larger or much closer together.

12   A technical discussion of the astrophysical application of pulsars' stability is S. M. Kopeikin, "Millisecond and Binary Pulsars as Nature's Frequency Standards—II. The Effects of Low-frequency Timing Noise on Residuals and Measured Parameters," *Monthly Notices of the Royal Astronomical Society* 305 (1999): 563–590. A less technical review of pulsar observations is given by: D. R. Lorimer, "Binary and Millisecond Pulsars at the New Millennium," *Living Reviews in Relativity* 4 (2001): 5, https://doi.org/10.12942/lrr-2001-5.

13   Bodies only a fraction of an Earth mass have been found orbiting one pulsar. See V. Wolszczan, "Confirmation of Earth-Mass Planets Orbiting the Millisecond Pulsar PSR B1257+12," *Science* 264 (1994): 538–542.

14   J. H. Taylor and J. M. Weisberg, "Further Experimental Tests of Relativistic Gravity using the Binary Pulsar PSR 1913+16," *Astrophysical Journal* 345 (1989): 434–450.

15   Taylor and Weisberg, "Further Experimental Tests."

16   D. Castelvecchi, "Monster Gravitational Waves Spotted for First Time," *Nature* 619 (2023): 13–14. They found the expected "Hellings-Downs" curve, predicted on the basis of isotropically distributed gravitational waves.

17   https://www.ligo.caltech.edu/.

18   S. -S. Huang, "Occurrence of Life in the Universe," *American Scientist* 47 (1959): 397–402; J. S. Shklovsky, and C. Sagan, *Intelligent Life in the Universe* (San Francisco: Holden-Day, 1966); M. H. Hart, "Habitable Zones about Main Sequence Stars," *Icarus* 37 (1979): 351–57; J. Kasting, D. P. Whitmire, and R. T. Reynolds, "Habitable Zones around Main Sequence Stars," *Icarus* 101 (1993): 108–28; S. Franck et al., "Habitable Zone for Earth-like Planets in the Solar System," *Planetary and Space Science* 48 (2000): 1099–1105.

19   The oceans do not literally boil away. Rather, water gets into the stratosphere, where it is photodissociated and its hydrogen is lost to space.

20   The location of the outer boundary is harder to determine since it's hard to model the effects of carbon dioxide clouds.

21   S. A. Franck et al., "Reduction of Biosphere Life Span as a Consequence of Geodynamics," *Tellus* 52B (2000): 94–107.

22   Here we are restricting our discussion to the Solar System, but there is likely considerable variation in the properties of asteroid belts around other stars.

23   Much of what we say about asteroids in this section also applies to comets, but their distribution peaks near 1.8 Aus. So, it is Mars that wanders through the most dangerous region of the Solar System as far as comets are concerned.

24 Mars also stirs some up, deflecting asteroids towards the inner planets. Which effect dominates should depend mostly on Mars's mass, but additional research is needed to quantify them. Our guess is that Mars helps more than it hurts us.

25 I. J. Daubar, et al., "The Current Martian Cratering Rate," *Icarus* 225, no. 1 (2013): 506–16.

26 Y. JeongAhn, and M. Renu, "The Current Impact Flux on Mars and Its Seasonal Variation," *Icarus* 262 (2015): 140–53.

27 This is particularly so since the energy implanted by a small body colliding with a planet increases as the square of its impact velocity.

28 A more speculative impact-related factor concerns "life-reseeding events." The fraction of ejecta that escape from a planet following a large impact depends on the velocity of the impactor. If its velocity is small enough, then it is possible that no ejecta will be lost from a planet. Thus, the fraction of impacts that do not produce ejecta is greater for increasing distance from the Sun (given the smaller encounter speeds). However, sterilizing impacts can still occur for the larger impactors, even when they are slow moving. Without microbial life riding on impact-generated ejecta from a sterilizing impact, there is no possibility of reseeding a planet. See L. Wells, J. Armstrong, and G. Gonzalez, "Impact Reseeding During the Late Heavy Bombardment," *Icarus* 162 (2003): 38–46. Including impact sterilization and reseeding into the mix would place additional constraints on the CHZ, though a closer study of our nearest neighbors needs to be made before we can begin to make an educated guess on the importance of these processes.

29 In addition, the total mass of the atmosphere should not be too great. Otherwise, the surface pressure would be too high to allow air breathing land creatures to survive. The amount of energy expended by an animal in breathing would be prohibitive. See Michael Denton, *Nature's Destiny* (New York: The Free Press, 1998), 127–28.

30 Earth receives more sunlight per unit area on its surface than a planet at an average location in the CHZ. On the other hand, a planet in the outer region of the CHZ will have more carbon dioxide in its atmosphere. While the additional carbon dioxide enhances biological productivity of autotrophs, it is harmful to higher life forms. In addition to distance to its host star, a planet's size also has a bearing on the total biological productivity. While the distance to the host star determines the surface energy density available to life, planet size, in part, determines the total biological productivity. The total biological productivity will be proportional to a planet's surface area, which, in turn, is proportional to the square of its size. Thus, even if Mars had been at Earth's location and somehow managed to maintain its atmosphere, its maximum total biological productivity would only be about one-quarter of Earth's. This is important, because the ability of a biosphere to support large mobile organisms depends on the total productivity at the lower levels in the food pyramid.

31 However, the increase in biological productivity with light intensity is not as dramatic as one may think since light is not the limiting factor for regions on a planet near its equator.

32 When compared to Earth, meteorites and their associated parent asteroids provide some evidence for this. In particular, the carbon and water content of the carbonaceous meteorites and their parent asteroids are much greater than that of the bulk Earth. Carbonaceous chondrite meteorites contain up to 20 percent water and 4 percent carbon, while the bulk Earth contains about 0.1 percent water and 0.05 percent carbon (Ward and Brownlee, *Rare Earth*, 46).

33 Ward and Brownlee, *Rare Earth*, 47.

34 It's believed that a large fraction of a terrestrial planet's core consists of material that accreted early in its formation. Later in the formation process, objects contributing to a planet's mass probably come from a larger area in the disk, due to the increasing eccentricities of their orbits.

35 See John S. Lewis, *Worlds Without End: The Exploration of Planets Known and Unknown* (Reading, UK: Helix Books, 1998), 57–59.

36 For a recent discussion of how much potassium (and uranium and thorium) might be in Earth's core, see W. F. McDonough, "Compositional Model for the Earth's Core," *Treatise on Geochemistry* (2014): 559–77.

37 See K-P. Schröder, and R. C. Smith, "Distant Future of the Sun and Earth Revisited," *Monthly Notices of the Royal Astronomical Society* 386, no. 1 (2008): 155–63.

38 Today, astronomers can calculate the location of the CCHZ for a star of arbitrary mass. For more recent reviews of the subject, see M. Lingam, and A. Loeb, "Colloquium: Physical Constraints for the Evolution of Life on Exoplanets," *Reviews of Modern Physics* 91, no. 2 (2019): 021002 and F. de Sousa Mello, and A. C. S. Friaça, "Planetary Geodynamics and Age Constraints on Circumstellar Habitable Zones around Main Sequence Stars," *International Journal of Astrobiology* (2023): 1–45.

39 As we noted in chapter 4, asteroids make their way to Earth via complex orbital dynamics involving giant-planet resonances and the Yarkovsky effect. The Yarkovsky effect changes the mean distances of asteroids from the Sun on timescales of millions of years. It is most pronounced for small asteroids closest to the Sun. Yarkovsky-induced asteroid migration will become more important as the Sun continues to brighten over the next several billion years.

40 A rotationally synchronized planet does offer one significant advantage: the strong tidal forces on the planet from the host star keep its tilt stabilized without the need for a large moon.

41 The following study explores the habitability of planets around M dwarf stars with a range of possible atmospheres: M. J. Heath et al., "Habitability of Planets around Red Dwarf Stars," *Origins of Life and Evolution of the Biosphere* 29 (1999): 405–24. The basic result is that a carbon dioxide level much greater than Earth's is required to maintain equable temperatures and prevent the atmosphere from freezing out. Such an atmosphere would preclude the existence of complex oxygen-breathing life. Even if some significant fraction of the atmosphere were composed of oxygen, say 10 to 20 percent, this would still not allow animallike life. And it is not clear that a planetary atmosphere can have high carbon dioxide and high oxygen levels at the same time. Earth's atmosphere went from being one dominated by methane and carbon dioxide and simple life to one dominated by oxygen and nitrogen and complex life. If an M dwarf planet made the same atmospheric transition, it would quickly cool.

42 A.H. Lobo, et al., "Terminator Habitability: The Case for Limited Water Availability on M-dwarf Planets," *The Astrophysical Journal* 945, no. 2 (2023): 161.

43 Heath et al. argue that a rotationally synchronized planet with enough vigorous and globally-connected ocean circulation can exchange enough heat between the lit and dark sides to keep water liquid over much of its surface. This would still not prevent all the water from freezing out on its dark side, however, if there continental landmasses on the dark side. In that case, water ice would form and stay frozen—like the Antarctica and Greenland ice masses on Earth. Once ice begins forming on land, the ocean level will begin to drop around the planet, exposing more land where stable ice can form, lowering the water level more, exposing more land, and so on until all the water freezes.

In addition, as the water level drops, the oceans will become less efficient at transporting heat between the hot and cold regions, further accelerating the water freeze-out on the cold, dark side. The planet will grow hot and dry on the lit side and cold and dry on the dark side. Some liquid water may remain at the bottom of some of the basins on the dark side, warmed by heat from the planet's interior, but that's about it. Even some elevated land near the terminator (such as mountain ranges) can trigger a runaway planetary freeze as long as ice can form at altitude. Even shallow seas on the dark side would allow for ice to accumulate and the sea level to drop as long as the ice can rest on the sea bottom. In short, surface relief will tend to destabilize a rotationally synchronized planet towards the complete freezing out of its water on its dark side.

At the same time, a waterworld is a poor solution. As we argued elsewhere, a planet with water and little surface relief doesn't allow for a diverse biosphere and may even be sterile. Another major destabilization point is reached when the dropping temperature on the dark side allows carbon dioxide to begin freezing out; this would eventually cause freeze-out of the entire atmosphere. One needs a suspiciously contrived planet, with only continents on the lit side and deep oceans everywhere else, to prevent freeze-out of its water. But as soon as tectonics and volcanic activity form land on its dark side, the entire planet will be on its way to the freeze-out of all its water. Even with an initially thicker carbon dioxide atmosphere, removing carbon dioxide via the deposition of carbonates on the ocean floor would cool the planet.

44 E. K. Pass, et al., "Mid-to-late M Dwarfs Lack Jupiter Analogs," *The Astronomical Journal* 166, no. 1 (2023): 11; I. Ribas, et al., "The CARMENES Search for Exoplanets around M Dwarfs. Guaranteed Time Observations Data Release 1 (2016–2020)," *Astronomy and Astrophysics* 670 (2023): A139.

45 Flares also emit particle radiation that eventually reach Earth and hit the upper atmosphere, sometimes after spiraling along its magnetic field lines. The stronger flares produce more energetic protons, which hit atoms in the upper atmosphere, generating secondary particles. These include muons, which reach the ground and threaten life there. Additionally, the secondary electrons produced in the atmosphere by high-energy ionizing radiation will produce UV photons that can reach the planet's surface.

46 This will be true for flares of all energies, but particularly of the rare energetic ones. Although astronomers have been monitoring solar flares continuously for several decades, they probably have not observed the full intensity range of flares. The strongest flares will occur less often, once per century, once per millennium, etc. If solar flares have posed any degree of threat to Earth life in the past half-billion years or so, then flares on an M dwarf star of the comparable age will pose a substantially greater threat to life on a planet in orbit about it. See M. Lingam, and A. Loeb, "Risks for Life on Habitable Planets from Superflares of their Host Stars," *The Astrophysical Journal* 848, no. 1 (2017): 41. Of course, some simple life might survive beneath the surface of a planet around such a star, where temperature swings would be less pronounced, flare radiation would not pose a threat, and geothermal heating could maintain liquid water. But this would not be an adequate environment for complex or technological life.

47 See J. Scalo, J. C. Wheeler, and P. Williams, "Intermittent Jolts of Galactic UV Radiation: Mutagenetic Effects," in *Frontiers of Life; 12th Rencontres de Blois*, ed. L. M. Celnikier, arxiv .org/abs/astro-ph/0104209(2001); D. S. Smith, J. Scalo, and J. C. Wheeler, "Importance of Biologically Active Aurora-like Ultraviolet Emission: Stochastic Irradiation of Earth and Mars by Flares and Explosions," *Origins of Life and Evolution of the Biosphere* 34 (2004): 513–32.

48 See R. Spinelli, et al., "The Ultraviolet Habitable Zone of Exoplanets," *Monthly Notices of the Royal Astronomical Society* 522, no. 1 (2023): 1411–18. The authors consider the required amount of ultraviolet radiation for building prebiotic molecules. They define a UV habitable zone, which overlaps the conventional CCHZ for a restricted range of stellar spectral types; M dwarfs are excluded from the UV habitable zone. See also L. Do Amaral, et al., "The Contribution of M-dwarf Flares to the Thermal Escape of Potentially Habitable Planet Atmospheres," *The Astrophysical Journal* 928, no. 1 (2022): 12.

49 Simon L. Grimm, et al., "The Nature of the TRAPPIST-1 Exoplanets," *Astronomy & Astrophysics* 613 (2018): A68.

50 S. Zieba, et al., "No Thick Carbon Dioxide Atmosphere on the Rocky Exoplanet TRAPPIST-1 c," *Nature* (2023): 1-4.

51 These planets may be poor in volatiles due either to strong stellar activity during its pre-main-sequence phase and/or to the lack of giant planets. R. Luger and R. Barnes, "Extreme Water Loss and Abiotic O2 Buildup on Planets throughout the Habitable Zones of M Dwarfs," *Astrobiology* 15, no. 2 (2015): 119–43; E. F. Fromont, et al., "Atmospheric Escape from Three Terrestrial Planets in the L 98-59 System," *The Astrophysical Journal*, in press; A. C. Childs, R. G. Martin, and M. Livio, "Life on Exoplanets in the Habitable Zone of M Dwarfs?," *The Astrophysical Journal Letters* 937, no. 2 (2022): L42.

52 See G. Gonzalez, "Is the Sun Anomalous?" *Astronomy & Geophysics* 40, no. 5 (1999): 5.25–5.29, and G. Gonzalez, "Are Stars with Planets Anomalous?," *Monthly Notices of the Royal Astronomical Society* 308 (1999): 447–58. The overall chemical composition of the Sun is anomalous: J. Melendez et al., "The Peculiar Solar Composition and Its Possible Relation to Planet Formation," *The Astrophysical Journal* 704, no. 1 (2009): L66. Additional anomalies

include a low lithium abundance: M. Carlos et al., "The Li-age Correlation: The Sun Is Unusually Li Deficient for its Age," *Monthly Notices of the Royal Astronomical Society* 485, no. 3 (2019): 4052–4059, and a high C/O abundance ratio: M. Bedell et al., "The Chemical Homogeneity of Sun-like Stars in the Solar Neighborhood," *The Astrophysical Journal* 865, no. 1 (2018): 68.

53 A. R. G. Santos, et al., "Temporal Variation of the Photometric Magnetic Activity for the Sun and *Kepler* Solar-like Stars," *Astronomy and Astrophysics* 672 (2023): A56. See also the following studies describing stellar variability for all spectral types: S. J. Adelman, "On the Variability of Stars," *Information Bulletin on Variable Stars* No. 5050 (2001): 1–2; L. Eyer, and M. Grenon, "Photometric Variability in the HR diagram," *ESA SP-402* (July 1997): 467–72. The least variable stars are found to be A0 to K0 spectral types of luminosity classes II through IV. These stars make poor hosts for life for reasons we've already covered, namely relatively rapid evolutionary changes. The main sequence stars (luminosity class V) also tend to be stable, but many of them are young, active stars. The Sun's stability is thought to be due to its age and particular composition.

54 As noted in chapter 2, there is increasing evidence that Earth's climate is strongly linked to the Sun's activity. Such evidence includes the correlation between carbon-14 variations in the atmosphere and auroral activity and the "Medieval Maximum" and the "Little Ice Age." For more on this topic, see L. J. Gray, et al., "Solar Influences on Climate," *Reviews of Geophysics* 48, no. 4 (2010).

55 In those two years Friedrich Bessel, Wilhelm Struve, and Thomas Henderson, reported the parallaxes of three stars, 61 Cygni, Vega, and Alpha Centauri, respectively. A very readable review of the history of parallax measurements (successful and not) is A. W. Hirshfeld, *Parallax: The Race to Measure the Cosmos* (New York: W. H. Freeman and Company, 2001).

56 The closest star, Proxima Centauri, at a distance of 4.2 light-years, exhibits the largest parallax angle, 0.77 seconds of arc. Bessel's target, 61 Cygni, has a parallax of only one-third of a second of arc. It is no wonder progress in stellar parallax work proceeded as slowly as it did. Earth's atmosphere blurs stellar images to sizes near one second of arc, but a few sites sometimes achieve one-half of a second of arc for short periods. Bessel was able to measure a parallax angle smaller than this atmosphere-imposed limit for two reasons. First, it is possible to measure the position of the center of a star's blurred image with a precision about three to five times better than its apparent size. Second, many observations, accumulated over several years, reduce the statistical uncertainty of the final solution. Nevertheless, the early parallax measurements were difficult and just barely possible from Earth's surface.

57 The so-called p-mode oscillations have been detected with the Doppler method from the ground in bright sunlike stars and from space with *Kepler* and *TESS* missions. For a review see J. Jackiewicz, "Solar-like Oscillators in the *Kepler* Era: A Review," *Frontiers in Astronomy and Space Sciences* 7 (2021): 595017.

58 For example, a K5 dwarf star, about 60 percent the mass of our Sun, should have velocity and brightness amplitudes only 30 and 50 percent those in the Sun. See H. Kjeldsen, and T. R. Bedding, "Amplitudes of Stellar Oscillations: The Implications for Asteroseismology," *Astronomy & Astrophysics* 293 (1995): 87–106.

59 The nuclear reactions involved in the synthesis of boron 8 in the Sun's core produce large numbers of high-energy neutrinos every second, many of which pass through Earth. While the chance of any particular atom interacting with a solar neutrino is extremely small, the chance that any one of many such neutrinos will interact with some atom in a large detector is vastly greater. The larger the detector, the better the chances. So, solar neutrino detectors are big. Beginning in 1965 Raymond Davis led a small team of scientists to set up an experiment to detect neutrinos produced in the Sun. His detector consists of one hundred thousand gallons of perchloroethylene in the Homestake Gold Mine in South Dakota. This detector is based on the nuclear reaction that results when a chlorine atom in a perchloroethylene molecule absorbs a neutrino, converting it into an argon atom in the process. Davis's experiment was the only one monitoring solar neutrinos between 1968 and

1988. Even with such a large "neutrino telescope," it detected only one solar neutrino about every three days. This was not only because neutrinos interact very weakly with matter, but also because, in Davis's experiment, the chlorine reaction is produced almost exclusively by the high-energy solar boron 8 neutrinos. Davis's experiment also detected a small fraction of the intermediate-energy solar neutrinos produced by beryllium 7, though none of the abundant proton-proton reaction neutrinos. For his tireless efforts, Davis was awarded a Nobel Prize in physics in 2002, an indication of the value the scientific community places on solar neutrino research. For a review of the history of solar neutrino astronomy, see: J. N. Bahcall, and R. Davis Jr., "The Evolution of Neutrino Astronomy," *Publications of the Astronomical Society of the Pacific* 112 (2000): 429–33.

60   The production of the important boron 8 neutrinos is very temperature sensitive. About 1.5 billion years ago, the Sun's boron 8 neutrino flux was about one-third its present value. See E. Brocato et al.," Stars as Galactic Neutrino Sources," *Astronomy & Astrophysics* 333 (1998): 910–917. The production rate depends approximately on the 25th power of the Sun's central temperature. Had the temperature of the Sun's core been only 5 percent less, it would have produced one-quarter of the boron 8 neutrino flux. In fact, because its core temperature has been increasing since it first began fusing hydrogen in its core, the Sun's boron 8 neutrino flux has also been increasing steadily. A star's core temperature increases during its stay on the main sequence because its mean molecular weight (and hence density) increases as protons combine to form helium nuclei. The Sun also gets bigger, because the layers outside its core expand in response to the increasing luminosity.

61   The fundamental proton-proton (p-p) reaction produces most of the energy (and most of the neutrinos) in the Sun. The temperature-sensitive boron 8 side-reaction only produces one in ten thousand solar neutrinos (though they are more energetic and easier to detect). This means the luminosity of a sunlike star is directly proportional to its p-p neutrino production rate, but the boron 8 neutrino production rate is much more sensitive to it. Thus, because the visible light energy leaving the Sun's surface scales with the p-p neutrino production rate, observers in the CHZ of another sunlike star of different mass will be exposed to the same p-p neutrino flux. This is because both the neutrino flux and light intensity fall off with the square of the distance. This relation does not hold for stars significantly more massive than the Sun, which produce most of their energy from the CNO reactions.

62   See Q. R. Ahmad et al., "Measurement of the Rate of $ne+d \rightarrow p+p+e-$ Interactions Produced by ^8B Solar Neutrinos at the Sudbury Neutrino Observatory," *Physical Review Letters* 87 (2001): 071301.

63   Neutrino telescopes built to study the Sun also led to the serendipitous detection of neutrinos from supernova 1987A in the Large Magellanic Cloud, one of the Milky Way's satellite galaxies. Such events are very rare in the Milky Way and its immediate vicinity. That event gave astronomers a unique look into the inner workings of a supernova.

64   Note that neutrino transmutations are also studied using neutrinos produced in Earth's atmosphere by cosmic rays and also by particle accelerators. See T. Kajita, and T. Yoji, "Observation of Atmospheric Neutrinos," *Reviews of Modern Physics* 73 (2001): 85–118. However, these other neutrino experiments were partly motivated by the early success of the solar neutrino experiments. It was the discrepancy between the observed neutrino flux from the Sun and the expected value that led physicists to propose the neutrino transmutation model in the first place. The solar neutrino experiments have the advantage that the production rate of neutrinos is very sensitive to the state of the Sun's core, which astronomers can study with the neutrinos. On the other hand, if they have enough confidence in the solar models, then astronomers can use the Sun as a kind of calibrated neutrino source.

CHAPTER 8: OUR GALACTIC HABITAT

1   Michael J. Denton, *Nature's Destiny* (New York: The Free Press, 1998), 372.
2   Quoted in Ken Croswell, in *The Alchemy of the Heavens* (Oxford, UK: Oxford University Press, 1995), 13.

3   The specific Hubble type of the Milky Way is SAB(rs)bc, according to G. de Vaucoulers, "Five Crucial Tests of the Cosmic Distance Scale Using the Galaxy as Fundamental Standard," *Proceedings of the Astronomical Society of Australia* 4, no. 4 (1982): 320–27.

4   By "zero-age," we mean very young objects (less than a few million years old) relative to age of the Milky Way. The scale height is the distance one has to go beyond the midplane such that the concentration of objects of a given type is reduced to 37 percent of its value in the midplane.

5   J. J. Matese et al., "Periodic Modulation of the Oort Cloud Comet Flux by the Adiabatically Changing Galactic Tide," *Icarus* 116 (1995): 255–68. Currently, the Solar System is about 55 light-years from the midplane and headed away from it.

6   P. C. Frisch, "The Galactic Environment of the Sun," *American Scientist* 88 (2000): 53–54.

7   For a modern plot of the optical zone of avoidance on the sky, see P. A. Woudt, and R. C. Kraan-Korteweg, "A Catalogue of Galaxies Behind the Southern Milky Way. II. The Crux and Great Attractor Regions (l~289⁰ to 338⁰)," *Astronomy & Astrophysics* 380 (2001): 441–59. The zone of avoidance obscures about 10 percent of the sky in the infrared region of the spectrum.

8   See W. C. Keel and R. E. White III, "Seeing Galaxies through Thick and Thin. IV. The Superimposed Spiral Galaxies of NGC 3314," *Astronomical Journal* 122 (2001): 1369–82.

9   R. Drimmel, A. Cabrera-Lavers, and M. Lopez-Corredoira, "A Three-Dimensional Galactic Extinction Model," *Astronomy & Astrophysics* 409 (2003): 205–15. It should be noted that there is a "dust hole" at the galactic center, though it does have some dust due to stars that continue to form there.

10 It's not an uncommon misconception that we are in a spiral arm. The best evidence indicates we are about halfway between two major spiral arms and close to the inner edge of the Local Arm. See K. Immer and K. L. J. Rygl, "An Updated View of the Milky Way from Maser Astrometry," *Universe* 8, no. 8 (2022): 390.

11 But most of the bulge would not be so enhanced. Besides the standard light fall-off with the inverse square of the distance, the amount of light extinction from interstellar dust makes the light fall with distance even more steeply. Because of the patchiness of dust between us and different parts of the bulge, it is difficult to estimate exactly how much brighter the entire bulge would appear from the inner disk.

12 The only exception might be the relatively rare "blue stragglers." Most globular clusters only have a few. They are believed to result from the merger of two G dwarfs.

13 In *Worlds Without End: The Exploration of Planets Known and Unknown* (Reading, UK: Helix Books, 1998), 167, planetary scientist John S. Lewis describes in detail the view from a globular cluster in the halo.

14 The latest GAIA data release (DR3) includes distances for nearly 1.5 billion stars. This is about one per cent the estimated number of stars in the Milky Way galaxy.

15 Specifically, astronomers have used radio frequencies of carbon monoxide and neutral hydrogen to map a large fraction of our galaxy's spiral structure. At radio frequencies the galaxy is largely transparent.

16 For an introduction to techniques used to map the spiral structure of the Milky Way Galaxy, see D. Mihalas and J. Binney, *Galactic Astronomy: Structure and Kinematics* (New York: W. H. Freeman and Company, 1981), 464–568. Here, we focus on radial velocities and not proper motions (which tell us something about the transverse velocity), because radial velocities can be measured to arbitrary distances, but proper motions become progressively more difficult to measure at increasing distance. Furthermore, proper motions are easiest to measure for stars, not interstellar clouds, which lack a well-defined position.

17 Edwin Hubble first classified galaxies according to their morphology, calling ellipticals early type and spirals late type. Some astronomers once believed this Hubble classification sequence formed an evolutionary sequence. Although astronomers no longer believe this, one can find enough examples of galaxies in the local universe to form a nearly continuous sequence from ellipticals to spirals. Several morphological features do change smoothly

across the sequence. In particular, the central bulge becomes less prominent than the flattened disk as one goes from ellipticals to spirals. Thus, for spirals like the Milky Way, the disk dominates over the bulge, meaning that most stars in our galaxy participate in the differential rotation of the disk (as opposed to the less ordered orbits in the bulge) and are amenable to studies of the type described above.

18   The material in this section is based mostly on the following papers: G. Gonzalez, D. Brownlee, and P. D. Ward, "The Galactic Habitable Zone: Galactic Chemical Evolution," *Icarus* 152 (2001): 185–200; G. Gonzalez, "Habitable Zones in the Universe," *Origins of Life and Evolution of Biospheres* 35 (2005): 555–606; G. Gonzalez, "Setting the Stage for Habitable Planets," *Life* 4, no. 1 (2014): 35–65. A less technical summary is given in: G. Gonzalez, D. Brownlee, and P. D. Ward, "The Galactic Habitable Zone," *Scientific American* (October 2001), 60–67. For a more updated treatment of the Galactic Habitable Zone, see G. Kokaia, and M. B. Davies, "Stellar Encounters with Giant Molecular Clouds," *Monthly Notices of the Royal Astronomical Society* 489, no. 4 (2019): 5165–80.

19   The relative distribution of the masses of stars that form in a given cluster or galactic region is called the initial mass function. There is evidence that the initial mass function depends on metallicity and age. See J. Li et al., "Stellar Initial Mass Function Varies with Metallicity and Time," *Nature* 613, no. 7944 (2023): 460–62.

20   See J. B. Pollack et al., "Formation of the Giant Planets by Concurrent Accretion of Solids and Gas," *Icarus* 124 (1996): 62–85.

21   For reviews of this topic, see: G. Gonzalez, "The Chemical Compositions of Stars with Planets: A Review," *Publications of the Astronomical Society of the Pacific* 118, no. 849 (2006): 1494–1505; V. Adibekyan, "Heavy Metal Rules. I. Exoplanet Incidence and Metallicity," *Geosciences* 9, no. 3 (2019): 105; E. A. Petigura et al., "The California-Kepler Survey. IV. Metal-Rich Stars Host a Greater Diversity of Planets," *The Astronomical Journal* 155, no. 2 (2018): 89. This minimum threshold for metallicity admits of three possible explanations. First, if the core instability accretion model is correct, then building giant planets may require a certain minimum initial metallicity; we'll call this the "primordial" explanation. Second, planets may tend to migrate in a system that is initially richer in metals, thus making a giant planet easier to detect with the Doppler method; we'll call this the "migration" explanation. Third, some planets might fall into their host stars, increasing the metals in the atmospheres of the host stars; we'll call this the "self-enrichment" explanation.

> Planet migration may result in the accretion of planetesimals and planets by their host star, possibly even enhancing its surface metallicity. While there remains some debate on the significance of this self-enrichment effect, it's unlikely it can fully account for the observed high average metallicity of the stars with planets. For instance, some stars with planets have very deep outer convection zones and yet have very high metallicities; it would take the accretion of an unreasonable amount of planetary material to increase the surface metallicities of these stars even by a modest amount. Nevertheless, there is some evidence for self-enrichment among stars with planets. See J. Y. Galarza et al., "Evidence of Rocky Planet Engulfment in the Wide Binary System HIP 71726/ HIP 71737," *The Astrophysical Journal* 922, no. 2 (2021): 129. The following studies give evidences for the migration as the explanation for hot-Jupiters: M. Rice, S. Wang, and G. Laughlin, "Origins of Hot Jupiters from the Stellar Obliquity Distribution," *The Astrophysical Journal Letters* 926, no. 2 (2022): L17; T. O. Hands and R. Helled, "Super Stellar Abundances of Alkali Metals Suggest Significant Migration for Hot Jupiters," *Monthly Notices of the Royal Astronomical Society* 509, no. 1 (2022): 894–902. For a short review of the various mechanisms invoked to account for hot-Jupiters, see R. Malhotra, "Migrating Planets," *Scientific American* (September 1999): 56–63.

22   Kiersten M. Boley, et al., "Searching for Transiting Planets Around Halo Stars. II. Constraining the Occurrence Rate of Hot Jupiters," *The Astronomical Journal* 162, no. 3 (2021): 85.

23   Studies of the dynamics of the known extrasolar planetary systems show that the high eccentricities of their planets do indeed reduce the chance that terrestrial planets would

remain in circular orbits within their habitable zones. See, K. Menou and S. Tabachnik, "Dynamical Habitability of Known Extrasolar Planetary Systems," *Astrophysical Journal* 583 (2003): 473–88; M. Noble, Z. E. Musielak, and M. Cuntz, "Orbital Stability of Terrestrial Planets Inside the Habitable Zones of Extrasolar Planetary Systems," *Astrophysical Journal* 572 (2002): 1024–30.

24 A. V. Shapiro et al., "Metal-rich Stars are less Suitable for the Evolution of Life on their Planets," *Nature Communications* 14, no. 1 (2023): 1893.

25 A. R. G. Santos et al., "Temporal Variation of the Photometric Magnetic Activity for the Sun and Kepler Solar-Like Stars," *Astronomy & Astrophysics* 672 (2023): A56.

26 Planet-mass bodies might be able to form in an initially metal-poor system via the condensation of metal-enriched gas from the ejecta of a supernova. It seems a stellar companion needs to be present before the supernova event. This process has been invoked to account for the planets found around the pulsar PSR B1257+12. See M. Konacki and A. Wolszczan, "Masses and Orbital Inclinations of Planets in the PSR B1257+12 System," *Astrophysical Journal* 591 (2003): L147–L150. It is not clear that these low-mass objects should even be called planets, given their very different nature from the planets in our solar system. In any case, such systems are not habitable. Therefore, we will not consider them further.

27 Petigura et al., "The California-Kepler Survey."

28 Adibekyan, "Heavy Metal Rules."

29 There is evidence that stars hosting Jupiter analogs orbit stars with similar metallicity to the Sun. See L. A. Buchhave et al., "Jupiter Analogs Orbit Stars with an Average Metallicity Close to that of the Sun," *The Astrophysical Journal* 856, no. 1 (2018): 37.

30 Only a small part of this spread is intrinsic to stars formed at a given distance from the galactic center at a given time. Other bodies have deflected the orbits of many stars from their original birthplaces at different distances from the galactic center. Such gravitational deflection is called orbital dynamical diffusion. Stars in the thin disk are born in nearly circular orbits. Over time, gravity from Giant Molecular Clouds, spiral arms, and other stars increasingly distorts their orbits from this initial state and even causes them to migrate in the disk. See the following study for a discussion of the solar system's birth location in the disk and how that might affect its habitability, Junichi Baba, T. R. Saitoh, and T. Tsujimoto, "Exploring the Sun's Birth Radius and the Distribution of Planet Building Blocks in the Milky Way Galaxy: A Multizone Galactic Chemical Evolution Approach," *Monthly Notices of the Royal Astronomical Society* 526, no. 4 (2023): 6088–6102.

31 At the Sun's location the metallicity gradient is about 5 percent per 1,000 light-years (the gradient is approximately linear with radius on a logarithmic abundance scale, about –0.07 "dex" per kiloparsec). Astronomers measure the metallicity gradient in the disk using spectral features in solar type stars, B stars, open-cluster stars, nebulae ionized by O and B stars (H II regions), and planetary nebulae. Studies employing different metallicity indicators have converged to give similar answers only since about 2019. Since the H II regions are very luminous, astronomers have been able to use them to measure the gradient in about two dozen nearby galaxies to date. See E. Trentin et al., "Cepheid Metallicity in the Leavitt Law (C-MetaLL) Survey–II. High-Resolution Spectroscopy of the Most Metal Poor Galactic Cepheids," *Monthly Notices of the Royal Astronomical Society* 519, no. 2 (2023): 2331–48; L. Spina et al., "The GALAH Survey: Tracing the Galactic Disc with Open Clusters," *Monthly Notices of the Royal Astronomical Society* 503, no. 3 (2021): 3279–96. The overlap of the thin and thick disks also yields a declining metallicity with increasing vertical distance from mid-plane. This is the result of diffusion of stars' orbits away from the midplane as they age. One could object that since the content of hydrogen and helium varies by several orders of magnitude among the planets in the solar systm, the variation of initial metallicity by tens of percent in the galactic disk may get "lost in the noise." This objection is not very relevant to this discussion. The planets in the solar system fall into two very distinct classes: terrestrial and gas giant planets. The terrestrial planets inhabit the inner region of the solar system and have very little hydrogen and helium; they were probably formed close to their present

locations. The gas giants that inhabit the outer solar system are mostly hydrogen and helium. There is not a smooth gradation of hydrogen content among the planets. What counts in the present context is the formation of a terrestrial planet of the right mass in the CHZ.

32   That is, its metallicity is increasing at about 8 percent every billion years.

33   As we noted in chapter 7 with respect to the CHZ, the concentration of sulfur and potassium 40 in the Earth's core determines, in part, its geophysics. Changing the ratio of the mass of the nickel-iron core's mass to that of the mantle should profoundly affect a planet's geophysics. Such a ratio will depend on the ratio of iron to silicon plus magnesium. Iron, silicon, and magnesium condense out of cooling gas at similar temperatures. Thus, terrestrial planets forming at different locations in a given system should start with similar relative endowments of these elements. However, a giant collision may alter the core-to-mantle mass ratio. For example, the large core mass of Mercury is thought to be due to a giant collision that removed part of its mantle. Similarly, the giant collision that formed the Moon added to the Earth's core. Such collisions also affect the volatile inventory of a planet. See also Junichi Baba, "Exploring the Sun's Birth Radius."

34   Among the most abundant products of a Type II supernova are oxygen, silicon, magnesium, calcium, and titanium. Also produced are the so-called rapid or "r-process" elements. In the few minutes following the start of the explosion, many free neutrons are liberated, and they are preferentially and quickly absorbed by heavier elements. This leads to the production of isotopes up to thorium and uranium from initial "seeds" as light as iron.

35   Unlike a Type II supernova, the most abundant "ashes" from a Type Ia supernova are the iron-peak elements (iron, nickel, cobalt). Lacking are the abundant free neutrons (and thus r-process elements) produced in a core-collapse event.

36   R. E. Davies and R. H. Koch, "All the Observed Universe Has Contributed to Life," Philosophical Transactions of the Royal Society of London B 334 (1991): 391–403.

37   F. X. Timmes, S. E. Woosley, and T. A. Weaver, "Galactic Chemical Evolution: Hydrogen through Zinc," Astrophysical Journal Supplement 98 (1995): 617–58; L. Potinari, C. Chiosi, and A. Bressan, "Galactic Chemical Enrichment with New Metallicity Dependent Stellar Yields," Astronomy & Astrophysics 334 (1998): 505–39; M. Samland, "Modeling the Evolution of Disk Galaxies. II. Yields of Massive Stars," Astrophysical Journal 496 (1998): 297–306.

38   See C. O'Neill, J. Lowman, and J. Wasiliev, "The Effect of Galactic Chemical Evolution on Terrestrial Exoplanet Composition and Tectonics," Icarus 352 (2020): 114025.

39   Francis Nimmo et al., "Radiogenic Heating and its Influence on Rocky Planet Dynamos and Habitability," The Astrophysical Journal Letters 903, no. 2 (2020): L37; note, that they also qualify their conclusions that there is still much more to learn on this topic. See also C. T. Unterborn et al., "Mantle Degassing Lifetimes through Galactic Time and the Maximum Age Stagnant-lid Rocky Exoplanets can Support Temperate Climates," The Astrophysical Journal Letters 930, no. 1 (2022): L6.

40   H. J. Habing et al., "Disappearance of Stellar Debris Disks around Main-sequence Stars after 400 Million Years," Nature 401 (1999): 456–57.

41   G. J. Melnick, "Discovery of Water Vapour around IRC+10216 as Evidence for Comets Orbiting Another Star," Nature 412 (2001): 160–63. The researchers argue that the amount of water vapor they detect is unexpected for this star. The best explanation they can offer is that the recent increase in the star's luminosity is causing comets around it to vaporize.

42   See J. J. Matese and J. J. Lissauer, "Characteristics and Frequency of Weak Stellar Impulses of the Oort Cloud," Icarus 157 (2002): 228–40. To be more precise, the comets in the Kuiper Belt are in less eccentric orbits, and less affected by the galactic tides; close passing stars are the major perturbers of the Kuiper Belt comets.

43   A. D. Erlykin et al., "Mass Extinctions Over the Last 500 Myr: An Astronomical Cause?," Palaeontology 60, no. 2 (2017): 159–67; M. R. Rampino and A. Prokoph, "Are Impact Craters and Extinction Episodes Periodic? Implications for Planetary Science and Astrobiology," Astrobiology 20, no. 9 (2020): 1097–1108.

44 Astronomers believe that after comets formed in the Sun's protoplanetary disk, many were flung from the giant planet region and into the Oort cloud via close encounters. A planetary system that forms fewer and/or less massive giant planets will place fewer comets into its Oort cloud. Such a system is likely to form from a more metal-poor cloud. Thus, the number of comets in a star's Oort cloud should be highly sensitive to metallicity. Other factors that affect the formation of the giant planets will also affect the formation of an Oort cloud. Comets in the Kuiper Belt are believed to form in situ. See M. Masi, L. Secco, and G. Gonzalez, "Effects of the Planar Galactic Tides and Stellar Mass on Comet Cloud Dynamics," *The Open Astronomy Journal* 2, no. 1 (2009); L. Secco, F. Marco, and F. Marzari. "Habitability on Local, Galactic and Cosmological scales," arXiv preprint arXiv:1912.01569 (2019).

45 Astrophysicists John Scalo and Craig Wheeler of the University of Texas have argued that high levels of radiation (such as those provided by extraterrestrial sources) accelerate evolution through the generation of beneficial mutations. See J. Scalo, J. C. Wheeler, and P. Williams (2003), "Intermittent Jolts of Galactic UV Radiation: Mutageneic Effects," *Frontiers of Life: Proceedings XIIth Rencontres de Blois*, eds. L. M. Celnikier, L. & J. Tran Thanh Van (Hanoi, Vietnam: The Gioi Publishers), 221–28. They base their case on simulations of evolution and supposed laboratory evidence for induced evolution in bacterial cultures. Their primary empirical support for this claim is M. Vulic, R. E. Lenski, and M. Radman, "Mutation, Recombination, and Incipient Speciation of Bacteria in the Laboratory," *Publications of the National Academy of Sciences* 96 (1999): 7348–51. While this study has come closer than any other in claiming to produce a new species of bacteria, it has not actually done so. The researchers produced a genetic barrier between two identical lines, which they admit is "much smaller than the barrier between such clearly distinct species as E. coli and Salmonella enterica." More fundamentally, now new functions or structures emerged. Thus, even with the highly artificial and extreme selective pressures applied by the scientists to rapidly reproducing bacteria in a laboratory setting, they were unable to provide evidence that random genetic mutations yield evolutionary innovations much above the species level. The mutations in the experiment were largely negative mutations. In a few cases mutations may have conferred an isolated survival benefit but at a cost to overall fitness. Add to this the fact that the early Earth experienced much higher radiation levels than it has now, via nearby supernovae, solar flares, and potassium 40 in the oceans, and life hardly did anything interesting for about two billion years.

Even if we granted Scalo and Wheeler's premises about biological evolution, there remains a problem in applying their theory to astrophysical settings. They argue that a higher rate of low-intensity, nonlethal intermittent radiation events will accelerate evolution. However, in every astrophysical setting we can think of, an increase in low-intensity events is accompanied by an increase in high intensity events. (This is true for supernovae, gamma-ray bursts, stellar flares, etc.) Thus, within their model, the conditions necessary for an accelerated rate of evolution will be accompanied by a higher probability of sterilization. Complete sterilization will be less likely for the simplest life but much likely for the most complex life.

46 There are two main scenarios related to damage to the ozone layer. First, damage will let through more ultraviolet radiation to the ground. This can happen if radiation ionizes atoms in the Earth's stratosphere and generates ozone-destroying nitrogen oxides. Second, an increase in the particle-radiation flux in the upper atmosphere leads to a greater flux of secondary particles at the surface. Radiation flux intense enough to make itself felt at the surface will also damage the ozone layer. Nearby supernovae are likely one of the types of causes of the damage. For a discussion of the physical processes involved, see B. C. Thomas and A. M. Yelland, "Terrestrial Effects of Nearby Supernovae: Updated Modeling," *The Astrophysical Journal* 950 (2023): 41. The authors argue that the previously adopted supernova lethality distance should be increased from 10 to 20 parsecs.

47 The exceptions are the long duration gamma-ray bursts, which are more likely to occur in metal-poor environments. See R. Spinelli and G. Ghirlanda, "The Impact of GRBs on

Exoplanetary Habitability," *Universe* 9, no. 2 (2023): 60; R. Spinelli et al., "The Best Place and Time to Live in the Milky Way," *Astronomy & Astrophysics* 647 (2021): A41.

48 J. N. Clarke, "Extraterrestrial Intelligence and Galactic Nuclear Activity," *Icarus* 46 (1981), 94–96; G. Gonzalez, "Habitable Zones in the Universe," *Origins of Life and Evolution of Biospheres* 35 (2005): 555–606; A. Balbi and F. Tombesi, "The Habitability of the Milky Way During the Active Phase of its Central Supermassive Black Hole," *Scientific Reports* 7, no. 1 (2017): 1–6.

49 Only since about the turn of the millennium have we had evidence that we might be close to the corotation circle. It has only become stronger in recent years. See Douglas A. Barros, et al., "Dynamics of the spiral-arm corotation and its observable footprints in the Solar Neighborhood," *Frontiers in Astronomy and Space Sciences* 8 (2021): 644098. See also V. V. Bobylev and A. T. Bajkova, "Determination of the Spiral Pattern Speed in the Galaxy from Three Samples of Stars," *Astronomy Letters* 49, no. 3 (2023): 184–92.

50 The Sun is located between the major arms, Perseus and Sagittarius, but it's near the inner edge of the Local arm. The Local Arm is not one of the major arms; it only extends about 5.5 kilo-parsecs in length. See Y. Xu et al., "The Local Spiral Structure of the Milky Way," *Science Advances* 2, no. 9 (2016): e1600878.

51 Catherine Zucker, et al., "Star formation near the Sun is driven by expansion of the Local Bubble," *Nature* 601, no. 7893 (2022): 334–37. See also Evan Gough, "Nearby Supernovae Exploded Just a few Million Years Ago, Leading to a Wave of Star Formation Around the Sun," *Universe Today* (January 14, 2022), https://www.universetoday.com/154047/nearby-supernovae-exploded-just-a-few-million-years-ago-leading-to-a-wave-of-star-formation-around-the-sun/. One of the coauthors of the paper stated, "But about five million years ago, the Sun's path through the galaxy took it right into the bubble, and now the Sun sits—just by luck—almost right in the bubble's center."

52 S. Ranasinghe and D. Leahy, "Distances, Radial Distribution, and Total Number of Galactic Supernova Remnants," *The Astrophysical Journal* 940 (2022): 63; I. S. M. A. I. L. Yusifov and I. Küçük, "Revisiting the Radial Distribution of Pulsars in the Galaxy," *Astronomy & Astrophysics* 422, no. 2 (2004): 545–53; D. R. Lorimer et al., "The Parkes Multibeam Pulsar Survey–VI. Discovery and Timing of 142 Pulsars and a Galactic Population Analysis," *Monthly Notices of the Royal Astronomical Society* 372, no. 2 (2006): 777–800.

53 R. Spinelli et al., "The Best Place and Time to Live."

54 Many mysteries remain about gamma-ray bursts, such as the precise amount of "beaming" of their radiation and their possible causes, such as merging neutron stars or the explosion of a very massive star.

55 Long duration gamma-ray bursts are more likely to occur in metal-poor star-forming regions. See Spinelli and Ghirlanda, "The Impact of GRBs."

56 According to John Lewis, a dust grain traveling that fast "would have a kinetic energy equal to the explosive power of over one hundred thousand times its weight of TNT." Lewis, *Worlds Without End*, 169.

57 Carbon and oxygen serve as probes of the spiral-arm structure of the galaxy in the form of carbon monoxide, which is a better tracer of relatively dense gas than hydrogen; this includes the gas disks around forming planetary systems.

58 In particular, continuing studies of giant black holes in galactic nuclei, supernovae in the nearby universe, distant gamma-ray bursts, comets in our Solar System and others, and stellar dynamics in our galaxy's disk will help us better understand the myriad threats to the survival of complex life.

59 This is because the average metallicity of a galaxy correlates with its luminosity. Since luminous galaxies contain more stars, galaxies at least as luminous as the Milky Way contain about 23 percent of the stars. When comparing our galactic setting to other galaxies, it is not clear which is the more appropriate statistic. If the total luminosity (or mass) of a galaxy is also relevant (and not just the metallicity of a given star), then the smaller statistic is more appropriate. For observational evidence of the correlation between galaxy luminosity and metallicity, see D. R. Garnett, "The Luminosity-Metallicity Relation, Effective Yields, and Metal Loss in Spiral and Irregular Galaxies," *Astrophysical Journal* 581 (2002): 1019–31.

60 Of course, since a given galaxy will have its own bell curve of stellar metallicities, a galaxy moderately less luminous than the Milky Way galaxy will contain some solar metallicity stars. However, they will tend to be closer to the dangerous nucleus. Thus, there's no sharp dividing line between galaxies with Earth-mass planets and those without. Nevertheless, a galaxy with a mean metallicity less than one-tenth solar is not likely to have any Earth-mass terrestrial planets.

61 Even in an elliptical galaxy a direct physical collision between two stars is rare. However, close encounters between stars can perturb the orbits of their planets.

CHAPTER 9: OUR PLACE IN COSMIC TIME

1 "A Sober Assessment of Cosmology at the New Millennium," *Publications of the Astronomical Society of the Pacific* 113 (2001): 653.

2 For a very readable historical account of Hubble's discoveries and the events leading up to them, see R. Berendzen, R. Hart, and D. Seeley, *Man Discovers the Galaxies* (New York: Columbia University Press, 1984). See also A. Sandage, "Edwin Hubble 1889–1953," *The Journal of the Royal Astronomical Society of Canada* 83, no. 6 (1989): 351–62. In later decades Hubble moved to the newer, and larger, 200-inch telescope at Mt. Palomar, also in California.

3 Astronomer Vesto Slipher had earlier noticed that the spiral nebulae mostly had large redshifts, but it was left to Hubble and Lemaître to discover the connection between distance and redshift. Also, in 1917 astronomer Willem de Sitter found static solutions to Einstein's equations that exhibited redshifts. However, the de Sitter models predicted a quadratic redshift dependence on distance, whereas a simple expanding universe model predicts linear redshift increase with distance. (His solutions also required a universe without matter, suggesting it might not apply to the actual universe, which contains more than a bit of matter.) The first published observations of Hubble in the 1920s were insufficient to exclude the de Sitter models. It was not until 1931 that Hubble's observations of galaxies reached sufficient distance to exclude the de Sitter static models. For more on this, see L. M. Lubin and A. Sandage, "The Tolman Surface Brightness Test for the Reality of the Expansion. IV. A Measurement of the Tolman Signal and the Luminosity Evolution of Early-Type Galaxies," *Astronomical Journal* 122 (2001): 1084–1103.

   Georges Lemaître published what is now called "Hubble's Law" two years prior to Hubble. See J. P. Luminet, "Editorial note to: 'Georges Lemaître, A homogeneous universe of constant mass and increasing radius accounting for the radial velocity of extra-galactic nebulae,'" *General Relativity and Gravitation* 45 (2013): 1619–33.

4 C. F. von Weizsäcker, *The Relevance of Science* (New York: Harper & Row, 1964), 151. Weizsäcker was at one time an assistant to the famous German physicist Werner Heisenberg. Quoted in the introduction to Neil A. Manson, editor, *God and Design: The Teleological Argument and Modern Science* (London: Routledge, 2003), 3.

5 Originally published as "Über die Krummung des Raumes," *Zeitschrift für Physik* 10 (1922): 377–86.

6 *La Revue des Questions Scientifiques*, 4e serie 20 (1931): 391. Lemaître actually published evidence of an expanding universe two years prior to Hubble. It wasn't noticed at the time because he published it in an obscure journal: "Un Univers Homogène de Masse Constante et de Rayon Croissant Rendant Compte de la Vitesse Radiale des Nébuleuses Extra-Galactiques," *Annales de la Société Scientifique de Bruxelles* A47 (1927): 49–59. The title, in English, reads, "A Homogeneous Universe of Constant Mass and Growing Radius Accounting for the Radial Velocity of Extragalactic Nebulae."

7 If one considers the Cepheid P-L relation as a simple linear equation with a constant and a slope, then the constant is the "zero-point." The zero-point is calibrated with observations of Cepheids with known distances. The state-of-the-art calibration of the Cepheid zero-point

is good to about 1 percent error. See M. R. Reyes and R. I. I. Anderson, "A 0.9% Calibration of the Galactic Cepheid Luminosity Scale Based on Gaia DR3 Data of Open Clusters and Cepheids," *Astronomy & Astrophysics* 672 (2023): A85.

8   RR Lyrae variables have typical pulsation periods near half a day. Unlike the Cepheids, all RR Lyrae variables have the same mean luminosity, except for a weak dependence on metallicity. They are found in globular clusters and the halo of the galaxy.

9   Astronomers in the first half of the twentieth century failed to notice that there were actually two classes of cepheids: the Population I (or Classical, or Type I) and Population II (or Type II) Cepheids. The names derive mostly from stellar ages, Population I being younger than Population II. Population II stars are found in elliptical galaxies and the halos of spiral galaxies, including their globular clusters. A Classical Cepheid is about four times brighter than a Type II Cepheid of the same pulsation period. Since Classical Cepheids' typical age is around a few million years, they are only found in galaxies with continuing star formation (in the Milky Way and Andromeda galaxies, Classical Cepheids are more common than Type II Cepheids). The discovery of this difference in the 1950s resulted in a correction to galaxy distances. Classical Cepheids and Type II Cepheids have different P-L relations.

10 Today, astronomers determine distances to galaxies containing Classical Cepheids and Type Ia supernovae to determine the value of the Hubble constant with high precision. See A. G. Riess et al., "A Comprehensive Measurement of the Local Value of the Hubble Constant with 1 km s^{-1} Mpc^{-1} Uncertainty from the Hubble Space Telescope and the SHOES team," *The Astrophysical Journal Letters* 934, no. 1 (2022): L7. The following study first demonstrated the usefulness of Type Ia supernovae as standard candles: A. G. Riess, W. H. Press, and R. P. Kirshner, "A Precise Distance Indicator: Type Ia Supernova Multicolor Light-curve Shapes," *Astrophysical Journal* 473 (1996): 88–109.

> Improvements in both the microwave background measurements and the Cepheid/Type Ia distance ladder has revealed a discrepancy in the derived values of the Hubble constant from these different datasets within the standard cosmological model. Many solutions have been proposed, but most fall into one of two categories: systematic errors in the Cepheid P-L calibration or new physics. We don't yet know the solution.

11 Astronomers also sometimes employ stars that do not vary in brightness as standard candles. As we already noted, astronomers can observe B stars, with masses comparable to Classical Cepheids, throughout much of the Milky Way Galaxy. Also useful are tip-of-the-red-giant branch stars. The problem with non-variable standard candles is that astronomers need more information, photometric or spectroscopic, to be certain of an identification. If the candidate star is in a distant galaxy, it's fairly easy to confuse it with other unresolved stars; what looks like a single star may instead be several stars too close together to resolve. A variable star's light amplitude, though, provides a built-in check on possible contaminating stars. Variable stars of a given type pulsate with a certain light amplitude. If a bright star is positioned so close to a variable star that they cannot be resolved, then its light will combine with the light from the variable star. This will cause its light amplitude to appear smaller. A variable star observed to have an anomalously small light amplitude will be immediately tagged as suspect and removed from further consideration.

12 "The End of the World: from the Standpoint of Mathematical Physics," *Nature* 127 (1931): 447.

13 Because it was detected in the microwave part of the electromagnetic spectrum, it's called the cosmic microwave background radiation or CMBR. Although we tend to associate microwaves with boiling water and burnt popcorn, the background radiation spectrum corresponds to an object 2.7 degrees above absolute zero.

14 The James Webb Space Telescope has made some surprising findings about early galaxy formation and growth that will require astronomers to revise their models. But what Webb has revealed is still broadly consistent with the standard model of cosmology. See Y.-Y. Wang, et al., "Modeling the JWST High-redshift Galaxies with a General Formation Scenario and the Consistency with the ΛCDM Model," *The Astrophysical Journal Letters* 954, no. 2 (2023): L48; G. Sun, et al., "Bursty Star Formation Naturally Explains the Abundance

of Bright Galaxies at Cosmic Dawn," *The Astrophysical Journal Letters* 955 (2023): L35; G. Desprez et al., "ΛCDM Not Dead Yet: Massive High-z Balmer Break Galaxies are Less Common than Previously Reported," *Monthly Notices of the Royal Astronomical Society*, submitted (October 2023).

15 J. D. Barrow, and F. J. Tipler, *The Anthropic Cosmological Principle* (Oxford: Oxford University Press, 1986), 380.

16 The measure of this spectrum is called the "CMBR angular power spectrum." Among other quantities, the CMBR power spectrum can help astronomers estimate the following: the curvature of the universe (flat, open, or closed), the matter energy density, the dark energy density (cosmological constant), the neutrino mass, the reionization redshift, the age of the universe, and the Hubble constant. For a detailed analysis of CMBR data, see M. Tegmark and M. Zaldarriaga, "Current Cosmological Constraints from a 10 Parameter Cosmic Microwave Background Analysis," *Astrophysical Journal* 544 (2000): 30–42. For the results of analysis of the more recent Planck data, see N. Aghanim et al., "Planck 2018 Results-VI. Cosmological Parameters," *Astronomy & Astrophysics* 641 (2020): A6. In short, which cosmological parameters are determined depends on what additional data are combined with the CMBR data (for example, Type Ia supernovae, Baryonic Acoustic Oscillations).

17 Specifically, from the CMBR power spectrum. See Figure 7 of the important study by Perlmutter et al., "Measurements of Omega and Lambda from 42 High Redshift Supernovae," *Astrophysical Journal* 517 (1999), 565–86. An updated version is Figure 5 from N. Suzuki, et al., "The Hubble Space Telescope Cluster Supernova Survey. V. Improving the Dark-Energy Constraints above z > 1 and Building an Early-Type-Hosted Supernova Sample," *The Astrophysical Journal* 746, no. 1 (2012): 85.

18 Note that matter and radiation give us information not only about their immediate local environment but also about global parameters of the universe. For example, the light we observe from supernovae tells us not only about the supernova itself but also about its host galaxy and the bulk properties of the universe.

19 While the abundances of helium-4 and deuterium relative to hydrogen are in good agreement with predictions based on observations of the CMBR, lithium-7 is observed to be a factor of three to four times smaller than predicted. Many solutions have been proposed to this "cosmological lithium problem." Some are based on the nuclear physics of the early universe, while others are based on what happens to the lithium in a star's envelope. Lithium is a relative delicate element, and its abundance is altered in the relatively cool shallow layers of a star. For a recent example of a proposed astrophysical solution, see M. Deal, and C. J. A. P. Martins, "Primordial Nucleosynthesis with Varying Fundamental Constants-Solutions to the Lithium Problem and the Deuterium Discrepancy," *Astronomy & Astrophysics* 653 (2021): A48.

20 There are still discrepancies in some of these tests of the big bang standard model, such as the lithium problem noted earlier and the value of local Hubble constant relative to the CMBR-derived value. While observational error might partly account for them, most likely these discrepancies point either to missing physics or a violation of one of the assumptions of the standard model. Still, taken together, the various successful tests of the standard model provide compelling support for it.

21 This test predicts that the surface brightness of a galaxy should decrease as $(1+z)^4$, where z is the redshift. R. C. Tolman, "On the Estimation of Distances in a Curved Universe with a Non-Static Line Element," *Proceedings of the National Academy of Sciences* 16 (1930): 511–20.

22 A. Sandage, "The Tolman Surface Brightness Test for the Reality of the Expansion. V. Provenance of the Test and a New Representation of the Data for Three Remote Hubble Space Telescope Galaxy Clusters," *The Astronomical Journal* 139, no. 2 (2010): 728. Despite the apparent effectiveness of this test, it does have the weakness that it is sensitive to galaxy evolution over the history of the universe. While Sandage attempted to correct for it, any uncorrected galaxy evolution errors would give systematic errors in the overall test.

23 Specifically, general relativity, which underwrites the big bang model, predicted that time dilation should increase as $(1+z)$. This means that a supernova at a redshift of 1 will take twice as long to go through its brightness variations as a nearby one. The observed trend of

increasing time dilation with redshift also contradicts the tired light theory, which predicts no time dilation effect. See "Errors in Tired Light Cosmology," https://www.astro.ucla .edu/~wright/tiredlit.htm. See also S. Blondin et al., "Time Dilation in Type Ia Supernova Spectra at High Redshift," *The Astrophysical Journal* 682, no. 2 (2008): 724.

24 G. F. Lewis, and B. J. Brewer. "Detection of the Cosmological Time Dilation of High Redshift Quasars," *Nat Astron* (2023): https://doi.org/10.1038/s41550-023-02029-2.

25 The Alcock-Paczynski test compares radial and transverse distances in surveys of distant galaxies or quasars. An incorrect cosmological model will yield a discrepancy between the radial (from redshifts) and transverse (from angular separations) distance measures. X.D. Li, et al., "The Redshift Dependence of the Alcock–Paczynski Effect: Cosmological Constraints from the Current and Next Generation Observations," *The Astrophysical Journal* 875, no. 2 (2019): 92.

26 The following studies report on measurements of the temperature of the CMBR at large redshift values: P. Molaro, et al., "The Cosmic Microwave Background Radiation Temperature at zabs = 3.025 toward QSO 0347-3819," *Astronomy & Astrophysics* 381 (2002): L64–L67 and R. Srianand, P. Petitjean, and C. Ledoux, "The Cosmic Microwave Background Radiation Temperature at a Redshift of 2.34," *Nature* 408 (2000): 931–35. Both studies used "fine-structure transition" absorption lines of neutral and ionized carbon seen against the light of background quasars. Commenting on the work of Srianand et al., John Bahcall remarked, "It is almost as if nature planted an abundance of clues in this anonymous cloud in order to allow some lucky researchers to infer the temperature of the CMB when the Universe was young." In "The Big Bang is Bang On," *Nature* 408 (2000): 916. Bahcall was impressed that the properties of the cloud studied by Srianand et al. would conspire to permit them to measure the temperature of the background and to eliminate other possible explanations of the observations. For more recent measurements, see D. A. Riechers, et al., "Microwave Background Temperature at a Redshift of 6.34 from H_2O Absorption," *Nature* 602, no. 7895 (2022): 58–62.

27 See Lubin and Sandage, 1086.

28 A quick Google search reveals the many competing cosmological models proposed over the years. Having many tests allows us to exclude many competing models. Another test involves the question, "How many neutrino species are there?" Particle physics says there are three; evidence from the cosmic microwave background radiation confirms that.

Yet another test involves the age of the universe derived from observations of stars. The oldest stars with well-determined ages agree with the age of the universe determined from the cosmic microwave background. J. Ying, et al., "The Absolute Age of M92," *The Astronomical Journal* 166 (2023): 18. M92 is a globular star cluster in the Milky Way containing about 330,000 stars. The authors derive an age 13.80 0.75 billion years, and the observations are consistent with the cluster stars forming nearly simultaneously. To date, this is the most accurate determination of age for a group of coeval stars. The age of the universe derived from the CMBR is 13.80 0.06 billion years. Within the quoted error, M92 formed up to 0.75 billion years after the big bang. Other methods for dating stars, such as uranium/thorium abundances and white dwarf cooling ages, generally have larger error bars. But they are consistent with the microwave background value. Discovering stars that are clearly older than 13.8 billion years would provide a straightforward way to disprove the big bang theory.

See also the following study: A. Cimatti and M. Moresco, "Revisiting Oldest Stars as Cosmological Probes: New Constraints on the Hubble Constant," *The Astrophysical Journal* 953 (2023): 149. In Table 2 the authors list age determinations greater than 13.3 billion years. What is remarkable about the table is the consistency among the various estimates, despite the very different methods employed. None of the values are in clear contradiction of the age derived from the CMBR. The current record holder for the oldest individual star is HD 140283. The study cited by Cimatti and Moresco from 2013 (second to last value in Table 2) give it an age of 14.46 0.8 billion years. However, a more recent study using a more accurate *GAIA* parallax confirms its age is near 14 billion years, but the authors don't quote a precise value or errors: M. Spite, et al., "$^{12}C/^{13}C$ Ratio and CNO Abundances in the Classical Very Old Metal-poor Dwarf HD 140283," *Astronomy & Astrophysics* 652 (2021): A97.

29   Many gas clouds also contain significant quantities of elements other than hydrogen and helium, which impress other absorption lines on quasar spectra. This permits astronomers to determine the overall composition of the gas clouds.

30   As we noted earlier, some static models of the universe do result in redshifts, but in de Sitter's static model, the redshifts are quadratic with distance. This would result in the spectra of more distant galaxies being less well-separated in redshift-space.

31   This point was first made by Robert H. Dicke in a very early application of the Anthropic Principle (before it was given that name): "Dirac's Cosmology and Mach's Principle," *Nature* 192 (1961), 440–41. A few others have echoed the idea that we, in some way, select our "now," including Martin Rees, John Barrow, and Paul Davies. For further discussion, see Barrow and Tipler, *The Anthropic Cosmological Principle*, 16–17.

32   Even among disk galaxies similar in mass to the Milky Way, the Milky Way is unusual in that its disk seems to have formed unusually early, and it avoided destructive mergers over the past 10 billion years. These were likely important factors in making the Milky Way a life-friendly galaxy. See V. A. Semenov et al., "Formation of Galactic Disks I: Why did the Milky Way's Disk Form Unusually Early?," *The Astrophysical Journal* 962, no. 1 (2024): 84.. The following study considers the globular cluster population of the Milky Way, and the authors also conclude that it formed anomalously early: G. De Lucia, et al., "On the Origin of Globular Clusters in a Hierarchical Universe," *Monthly Notices of the Royal Astronomical Society* 530, no. 3 (2024): 2760-77.

33   C. H. Lineweaver, "An Estimate of the Age Distribution of Terrestrial Planets in the Universe: Quantifying Metallicity as a Selection Effect," *Icarus* 151 (2001): 307–13; R. Spinelli, et al., "The best place and time to live in the Milky Way," *Astronomy & Astrophysics* 647 (2021): A41.

34   The changing carbon isotope ratio might be another time-dependent factor over billions of years. The Solar System was formed from a cloud with a carbon 12 to 13 isotope ratio near 80 (that is, one atom of carbon 13 for every 80 atoms of carbon 12). This ratio should decline as red giant stars continue to return processed matter to the interstellar medium, typically with carbon isotope ratios between six and twelve. This would not directly influence life, but it would affect the habitability of a planet via the greenhouse effect. Greenhouse gases such as carbon dioxide (CO_2) and methane (CH_4) transmit sunlight in the optical region of the spectrum but absorb outgoing infrared light emitted by the warm surface and atmosphere. The absorption is not continuous throughout the infrared but occurs in bands. The precise locations of the absorption bands of a given molecule depend on the isotopes that make it up. The number of absorption bands in the infrared region of the spectrum is increased if one of the minor isotopes in a greenhouse gas is increased to levels comparable with the primary isotope. Thus, molecules of carbon dioxide and methane in an atmosphere with an increased level of carbon 13 would produce a more efficient greenhouse effect. An Earth twin with relatively more carbon 13 would achieve the same surface temperature with less carbon dioxide and methane. Since the Earth already has a low carbon dioxide content, such a planet might not be able to support most plant life. To compensate, the inner edge of the Circumstellar Habitable Zone would need to be farther out. As we noted in chapter 7, several processes go into defining the Circumstellar Habitable Zone—change one and the net effect may significantly narrow its total extent. But we are not yet prepared to say to what extent habitability will be affected by a changing carbon isotope ratio.

35   https://webb.nasa.gov/.

36   E. H. Gudmundsson and G. Björnsson, "Dark Energy and the Observable Universe," *Astrophysical Journal* 565 (2002): 1–16. Astronomers Fred Adams and Greg Laughlin speculated about the state of the universe 10^{150} years into the future! The discovery of dark

energy quickly made their scenario obsolete. See F. Adams, and G. Laughlin, *The Five Ages of the Universe: Inside the Physics of Eternity* (New York: The Free Press, 1999).

37 Gudmundsson and Björnsson, "Dark Energy and the Observable Universe," 11.

38 Lawrence Krauss and Glenn Starkman give the following description of objects approaching the event horizon: "As the light travels from its source to the observer, its wavelength is stretched in proportion to the growth in a(t) [the scale factor]. Objects therefore appear exponentially redshifted as they approach the horizon. Finally, their apparent brightness declines exponentially, so that the distance of the objects inferred by an observer increases exponentially. While it strictly takes an infinite amount of time for the observer to completely lose causal contact with these receding objects, distant stars, galaxies, and all radiation backgrounds from the big bang will effectively 'blink' out of existence in a finite time; as their signals redshift, the timescale for detecting these signals becomes comparable to the age of the universe." In "Life, the Universe, and Nothing: Life and Death in an Ever-expanding Universe," *Astrophysical Journal* 531 (2000), 23.

39 Michael Rowan Robinson's entire quote is as follows:

> It was pointed out by Hoyle that the energy densities of the microwave background, of cosmic rays, of the magnetic field in our Galaxy, and of starlight in our Galaxy are all of the same order, $\sim 10^{-13}$ W m^{-3}. . . . But the coincidence of these three Galactic energy densities with the energy density of the microwave background, whose spectral shape and isotropy point to a cosmological origin, remains a mystery. Possibly we just have to accept this as a coincidence, as we have to accept the similar apparent sizes of sun and moon.

In *Cosmology* (Oxford: Clarendon Press, 1977), 144–45. Here energy density is simply the amount of energy per unit volume, averaged over a large volume of space.

40 A useful introduction to the various foreground contaminants of the CMBR is M. Lachieze-Rey and E. Gunzig, *The Cosmological Background Radiation* (Cambridge: Cambridge University Press, 1999).

41 Relatively nearby galaxies do not recede from us as quickly as expected from the Hubble Law, because their mutual gravitational attractions are strong enough to counteract the cosmological expansion. For example, the Andromeda Galaxy is actually *approaching* the Milky Way galaxy.

42 Krauss and Starkman, "Life, the Universe, and Nothing," 23.

43 Kalogera *et al.*, "The Coalescence Rate of Double Neutron Star Systems," *The Astrophysical Journal* 556 (2001): 340–56. The typical lifetime of binary neutron stars before they merge is several billion years, so these should continue to be available for would-be astronomers considerably longer, but eventually the implacable hand of time will reduce the frequency of these late blooming phenomena as well.

44 Deuterium, helium 3, and lithium 7 are particularly important, but these isotopes are easily destroyed in stars. (Certain processes also produce lithium, further complicating the reconstruction of its original abundance). The helium 4 abundance will continue to deviate from its original value as stars steadily cycle interstellar gas through their interiors.

45 Cosmologist Phillip J. E. Peebles observes, "A galaxy at redshift z~1 is expected to have an angular size on the order of one arc second, which coincidentally is comparable to the angular resolution, or seeing, permitted by our atmosphere." *Principles of Physical Cosmology* (Princeton: Princeton University Press, 1993), 326. Here, Peebles is discussing specifically the angular sizes of galaxies comparable in physical size to the Milky Way's luminous disk.

46 See L. M. Krauss, and R. J. Scherrer, "The Return of a Static Universe and the End of Cosmology," *General Relativity and Gravitation* 39 (2007): 1545-50. Their abstract states:

> We demonstrate that as we extrapolate the current CDM [cold dark matter] universe forward in time, all evidence of the Hubble expansion will disappear, so that observers in our "island universe" will be fundamentally incapable of determining the true

nature of the universe, including the existence of the highly dominant vacuum energy, the existence of the CMB, and the primordial origin of light elements. With these pillars of the modern Big Bang gone, this epoch will mark the end of cosmology and the return of a static universe. In this sense, the coordinate system appropriate for future observers will perhaps fittingly resemble the static coordinate system in which the de Sitter universe was first presented.

In their conclusion to the paper, the authors state:

> But this coincidence [the present near equality of dark energy and dark matter] endows our current epoch with another special feature, namely that we can actually infer both the existence of the cosmological expansion, and the existence of dark energy. Thus, we live in a very special time in the evolution of the universe: the time at which we can observationally verify that we live in a very special time in the evolution of the universe!

In a similar study, A. Loeb, "Long-term Future of Extragalactic Astronomy," *Physical Review D* 65 (2002): 047301, the concluding paragraph reads, "Within 10^{11} years, we will be able to see only those galaxies that are gravitationally bound to the local group of galaxies, including the Virgo cluster and possibly some parts of the local supercluster (where the global overdensity in a sphere around Virgo is larger than a few). All other sources of light will fade away beyond detection and their fading image will be frozen at a fixed age."

47 Of course, we might survive for billions of years into the future and artificially maintain a habitable environment, but then we're talking about our technological expertise built on the finely tuned habitability and measurability foundation of our own present age.

48 Olbers was actually reframing a similar question asked by Jean Philippe Leys de Cheseaux of Lausanne in 1744, "Why is the sky dark?" Astronomers have been asking this question in some form at least since Kepler.

CHAPTER 10: A UNIVERSE FINE-TUNED FOR LIFE AND DISCOVERY

1 *The Cosmic Blueprint: New Discoveries in Nature's Creative Ability to Order the Universe* (New York: Touchstone Books, 1989), 203.

2 The classic tome on the subject is still John D. Barrow and Frank J. Tipler, *The Anthropic Cosmological Principle* (Oxford: Oxford University Press, 1986). Still, much progress has been made in the intervening years. New examples of fine-tuning have been discovered, and some classic cases have been found faulty. For a recent treatment we suggest, Luke A. Barnes and Geraint F. Lewis, *A Fortunate Universe: Life in a Finely Tuned Cosmos* (Cambridge: Cambridge University Press, 2016).

3 Arguably the most important discussions in the nineteenth century of the fine-tuning of the environment for life were the Bridgewater Treatises. The one that had the greatest influence on Henderson was: W. Whewell, *Astronomy and General Physics, Considered with Reference to Natural Theology* (London, 1833); the ninth edition was published in 1864.

4 For a modern treatment of fine-tuning in chemistry see: R. A. King, et al., "Chemistry as a Function of the Fine-Structure Constant and the Electron-Proton Mass Ratio," *Physical Review A* 81 (2010): 042523. Other elements also seem to be uniquely suited for biology. See Michael Denton's discussion in *Nature's Destiny* (New York: The Free Press, 1998), of magnesium, calcium, iron, copper, and molybdenum: 195–208. See also, J. J. R. Frauso da Silva and R. J. P. Williams, *The Biological Chemistry of the Elements* (Oxford: Oxford University Press, 1991).

5 F. Hoyle, "On Nuclear Reactions Occurring in Very Hot Stars. I. The Synthesis of Elements from Carbon to Nickel," *Astrophysical Journal Supplement* 1 (1954): 121–146. Specifically, Hoyle predicted the resonance at 7.65 million electron volts (MeV) above its ground level. Hoyle's correct prediction is the first practical application of the weak anthropic principle, even though it had not yet been formulated.

6   An alternative to considering fine-tuning of the strong nuclear force is the fine-tuning of the masses of the fundamental particles that make up protons and neutrons, the quarks. Of the six types of quarks and their antimatter pairs, only two (up, and down) make up neutrons and protons. For examples of fine-tuning for these quarks, see S. M. Barr, and A. Khan. "Anthropic Tuning of the Weak Scale and of mu/md in Two-Higgs-doublet Models," *Physical Review D* 76, no. 4 (2007): 045002.

7   This confirmation included self-consistent numerical calculations of the relevant particle physics and stellar models. See T. A. Lähde, U.-G. Meißner, and E. Epelbaum, "An Update on the Fine-Tunings in the Triple Alpha Process," *European Physical Journal A* 56, no. 3 (2020): 89. A recent twist on this story is that the beryllium-8 nucleus becomes bound with a modestly greater change in the light quark mass compared to that required to change the Hoyle resonance energy beyond the life-permitting range. L. Huang, F. C. Adams, and E. Grohs. "Sensitivity of Carbon and Oxygen Yields to the Triple-Alpha Resonance in Massive Stars." *Astroparticle Physics* 105 (2019): 13–24. Lähde et al. find that this stability of the beryllium-8 nucleus does not change the fact of the fine-tuning of the Hoyle resonance in our universe.

8   One could get past the A = 5 bottleneck with lithium-5 or helium-5 with a 10 percent increase in the strong nuclear force. In addition, the resonances in the carbon-12 nucleus are spaced such that a change in the strong force by about 10 percent would bring another resonance into play (Robin Collins, private communication). Of course, there would also have to be another resonance at just the right location in the oxygen nucleus.

9   Barrow and Tipler, *The Anthropic Cosmological Principle*, 326.

10   Note, however, that the elimination of thorium and uranium (with atomic numbers of 90 and 92, respectively) would not necessarily mean that a terrestrial planet would not be able to derive sufficient heating from radioactive decay to drive plate tectonics. They might be replaced by other radioactive elements, which are otherwise stable in our universe. However, they may not have comparable half-lives. There are probably certain ranges of the strong nuclear force that don't yield any isotopes with half-lives between one and ten billion years.

11   Light isotopes with short half-lives would not pose a threat, because they would have decayed away in Earth's crust before humans arrived on the scene. An exception would be something like carbon-14, which is continuously being produced in Earth's atmosphere from cosmic rays. However, it is not produced in sufficient quantities to be harmful.

12   The physicist Maurice Goldhaber famously said that "we feel it in our bones that [the] proton is long lived, for otherwise the radiation from decay would kill you. All you need to know is how much radiation is hazardous for you, and you get a lower limit on proton lifetime, on the order of 10^{18} years." (G. Sanjonivic, "Proton Decay and Grand Unification," arxiv: 0912.5375.) While this is far shorter than the experimentally established lower limit on the proton lifetime (10^{34} years), it is interesting that the anthropic limit is about 10^8 times the current age of the universe. If an isotope of an abundant life-essential element, say oxygen-18, were radioactive, it might pose a threat to life if its lifetime were about 1000 times the current age of the universe. Such a long half-life is not out of the question. The longest experimentally measured half-life belongs to Xenon-124, with a value one trillion times the age of the universe; J. Suhonen, "Dark Matter Detector Observed Exotic Nuclear Decay," *Nature* 568 (2019): 462-63.

13   There are probably more problems in this general neighborhood. Higher relative abundances of some of the stable isotopes of the light elements might also diminish a planet's habitability. As we noted in the previous chapter, significantly increasing the carbon-13 to carbon-12 ratio would make carbon dioxide and methane much more efficient greenhouse gases, which, in turn, would move the Circumstellar Habitable Zone. Significant increases in the abundances of the hydrogen and oxygen isotopes would have similar effects. The particular isotope ratios that went into forming the Solar System depend on the star formation history of the Milky Way galaxy and the details of the nucleosynthesis of the

elements inside stars. Thus, changing any of several constants and forces would significantly change the isotope ratios.

14 In the two examples discussed in this paragraph, we are elaborating on material from M. J. Rees, "Large Numbers and Ratios in Astrophysics and Cosmology," *Philosophical Transactions of the Royal Society of London A* 310 (1983): 101–12.

15 *Just Six Numbers: The Deep Forces that Shape the Universe* (New York: Basic Books, 2001), 30–31.

16 This dividing line depends on the ratio of the electromagnetic to gravitational fine structure constants raised to the 20th power. See B. Carter "Large Number Coincidences and the Anthropic Principle in Cosmology," in *Confrontation of Cosmological Theories with Observation*, ed. M. S. Longair (Reidel, Dordrecht, 1974), 291–98. See also B. J. Carr and M. J. Rees, "The Anthropic Principle and the Structure of the Physical World," *Nature* 278 (1979): 611.

17 For additional discussion of the red-blue dividing line for stars, see chapter 4 of Barnes and Lewis, *A Fortunate Universe*.

18 Rees, *Just Six Numbers*, 30.

19 This is because of its larger surface-area-to-volume ratio. For constant density, a planet half the Earth's diameter would have one-eighth its volume and one-quarter its surface area, resulting in twice its surface area to volume ratio. When self-compression is included, a rocky planet half the size of Earth is about one-tenth the volume and less dense. In a universe with stringer gravity, self-compression is greater for a given mass compared to our universe.

20 See Figure 6.1 of Rees, *Just Six Numbers*, 87.

21 M. Tegmark, et al., "Dimensionless Constants, Cosmology, and other Dark Matters," *Physical Review D* 73, no. 2 (2006): 023505. See also L. A. Barnes, "The Fine-Tuning of the Universe for Intelligent Life," *Publications of the Astronomical Society of Australia* 29 (2012): 529–64.

22 Lewis, and Barnes, *A Fortunate Universe*, 163. In addition to these two problems with trying to explain away the fine-tuning of the cosmological constant, they list an additional six in their book.

23 Lawrence Krauss, "Questions that Plague Physics: A Conversation with Lawrence M. Krauss," *Scientific American* 291, no. 2 (August 2004): 83–84.

24 Fred C. Adams, et al., "Constraints on Vacuum Energy from Structure Formation and Nucleosynthesis," *Journal of Cosmology and Astroparticle Physics* 2017, no. 03 (2017): 021.

25 *The Road to Reality: A Complete Guide to the Laws of the Universe* (New York: Alfred A. Knopf, 2005), 728–30.

26 The first two-parameter fine-tuning plot (strong force/weak force) was produced by P. C. W. Davies in the early 1970s. See Figure 5.7 in *The Anthropic Cosmological Principle*. For an example of the simultaneous variation in the strong and electromagnetic force strengths, see Figure 5 of M. Tegmark, "Is 'The Theory of Everything' Merely the Ultimate Ensemble Theory?," *Annals of Physics* 270 (1998): 1–51. Other examples involving light quark masses, the electron mass, and the weak forces are shown in L. J. Hall, D. Pinner, and J. T. Ruderman, "The Weak Scale from BBN," *Journal of High Energy Physics*, no. 12 (2014): 1–29. See also the diagrams in Barnes and Lewis, *A Fortunate Universe*, chapter 7. For an example of the variation in gravitational and electromagnetic force strengths on stars, see Figure 3 of F. C. Adams, "Constraints on Alternate Universes: Stars and Habitable Planets with Different Fundamental Constants," *Journal of Cosmology and Astroparticle Physics* 2016, no. 02 (2016): 042.

27 Arguably, one of the most impressive clusters of fine-tuning occurs at the level of chemistry. In fact, chemistry appears to be "overdetermined" in the sense that there are not enough free physical parameters to determine the many chemical processes that must be just so (like those discussed in chapter 2). Max Tegmark notes, "Since all of chemistry is essentially determined by only two free parameters, α and β [electromagnetic force constant and electron to proton mass ratio], it might thus appear as though there is a solution to an overdetermined problem with more equations (inequalities) than unknowns. This could be taken as support for a religion-based category 2 TOE [Theory Of Everything], with

the argument that it would be unlikely in all other TOEs." Tegmark, "Is the 'Theory of Everything,'" 27.

28   Another way of thinking about fine-tuning of the forces is to consider variations at more fundamental levels. For example, the fundamental forces are believed to be different manifestations of the same force. The forces become distinguishable at low energies—though, it's not clear that gravity can be made to fit this scheme. Thus, separate variations in strengths of a couple of forces, say weak and electromagnetic, could be described by variations in the strength of a single force at higher energies—say, the electroweak force. For an example of this kind of analysis, see V. Agrawal et al., "Anthropic Considerations in Multiple-domain Theories and the Scale of Electroweak Symmetry Breaking," *Physical Review Letters* 80 (1998): 1822–25. Their analysis suggests that over the many orders of magnitude in the strength of the electroweak force that they considered, life is possible only within a narrow window. They base this conclusion on arguments like the ones we discussed concerning the stability of the deuteron. Note that even if all the forces could be incorporated into a grand unified theory, it would not reduce, much less explain, the coincidence that the value of the grand unified law was just so. As Carr and Rees write, "Even if all apparently anthropic coincidences could be explained in this way, it would still be remarkable that the relationships dictated by physical theory happened also to be those propitious to life," 612.

29   See Virginia Trimble, "Cosmology: Man's Place in the Universe," *American Scientist* 65 (1977): 85. She wrote,

> The changes in these properties required to produce the dire consequences are often several orders of magnitude, but the constraints are still nontrivial, given the very wide range of numbers involved. Efforts to avoid one problem by changing several of the constraints at once generally produce some other problem. Thus, we apparently live in a rather delicately balanced universe, from the point of view of hospitality to chemical life.

John Gribbin and Martin Rees reach a similar conclusion in *Cosmic Coincidences* (New York: Bantam Books, 1989), 269:

> If we modify the value of one of the fundamental constants, something invariably goes wrong, leading to a universe that is inhospitable to life as we know it. When we adjust a second constant in an attempt to fix the problem(s), the result, generally, is to create three new problems for every one that we "solve." The conditions in our universe really do seem to be uniquely suitable for life forms like ourselves, and perhaps even for any form of organic chemistry.

30   E. A. Milne, *Relativity, Gravitation, and World-Structure* (Oxford: Clarendon Press, 1935), 37. Ours is a rephrasing. The exact quote is "This recognition is only possible in virtue of the discreteness of atomic properties. If the world was composed of atoms shading continuously into one another, no such recognition would be possible.

31   N. Liu, et al., "New Multielement Isotopic Compositions of Presolar SiC Grains: Implications for their Stellar Origins," *The Astrophysical Journal Letters* 920, no. 1 (2021): L26.

32   P. Noterdaeme, et al., "The Evolution of the Cosmic Microwave Background Temperature: Measurements of $TCMB$ at High Redshift from Carbon Monoxide Excitation," *Astronomy & Astrophysics* 526 (2011): L7.

33   Paul Davies, *The Mind of God: The Scientific Basis for a Rational World* (New York: Simon & Schuster, 1992).

34   Davies, *The Mind of God*, 159.

35   Harlow Shapley, *Of Stars and Men: Human Response to an Expanding Universe* (Boston: Beacon Press, 1958), 94.

36   Davies, *The Mind of God*, 156–57.

37   George F. R. Ellis has made this point (if only in passing), joining the half dozen or so people who have hinted at the correlation between habitability and measurability. See, "The

Anthropic Principle: Laws and Environments," in *The Anthropic Principle: Proceedings of the Second Venice Conference on Cosmology and Philosophy, November 18–19, 1988*, F. Bertola and U. Curi, eds. (Cambridge: Cambridge University Press, 1993), 31.

38   Within the context of the cooling rate of the expanding universe, Hubert Reeves discusses the minimum requirements for growth of complexity. The key requirement is formation of nonequilibrium structures. Too slow, and the atomic and molecular diversity would be lacking. matter would consist entirely of the final product of equilibrium reactions, iron. Too fast, and large structures, such as planets, stars, and galaxies, could not form. Thus, the cooling rate must be fine-tuned to allow for matter to form complex structures. However, we do not agree with Reeves' suggestion that a "Principle of Complexity" should replace the Anthropic Principle, since we are more than complex structures—we are human observers, who can talk about fine-tuning coincidences. See, "The Growth of Complexity in an Expanding Universe," in *The Anthropic Principle: Proceedings of the Second Venice Conference on Cosmology and Philosophy, November 18–19, 1988*, F. Bertola and U. Curi, eds. (Cambridge: Cambridge University Press, 1993), 67–84.

39   M. S. Safranova, "The Search for Variation of Fundamental Constants with Clocks," *Annalen der Physik* 531, no. 5 (2019): 1800364; J. Alvey, et al., "Improved BBN Constraints on the Variation of the Gravitational Constant," *The European Physical Journal C* 80, no. 2 (2020): 1–6; M. T. Murphy, et al., "Fundamental Physics with ESPRESSO: Precise Limit on Variations in the Fine-Structure Constant Towards the Bright Quasar HE 0515-4414," *Astronomy & Astrophysics* 658 (2022): A123. This last study set an upper limit of 1.3 x 10^{-6} in the relative change of the fine structure constant from cosmological observations. A wide-ranging and lengthy, though now dated, review of this topic is presented by J.-P. Uzan, "Varying Constants, Gravitation and Cosmology," *Living Reviews in Relativity* 14, no. 1 (2011): 1–155. None of these studies have found any evidence of temporal variation in the constants. For a study on placing limits on the speed of light, see Fulvio Melia, "Model-independent confirmation of a constant speed of light over cosmological distances," *Monthly Notices of the Royal Astronomical Society* 527, no. 3 (2024): 7713–18.

40   Penrose, *The Road to Reality*, 762–765.

41   *The Return of the God Hypothesis: Three Scientific Discoveries that Reveal the Mind Behind the Universe* (New York: HarperOne, 2021), 476, note 11.

42   R. Collins, "The Argument from Physical Constants: The Fine-Tuning for Discoverability," in *Two Dozen (or so) Arguments for God*, J. L. Walls and T. Dougherty, eds. (New York: Oxford University Press, 2018), 94.

43   Eugene Wigner, "The Unreasonable Effectiveness of Mathematics in the Natural Sciences," in *Symmetries and Reflections* (Bloomington: Indiana University Press, 1967), 222–37. Philosopher Mark Steiner updated Wigner's musings with much more detail in *The Applicability of Mathematics as a Philosophical Problem* (Cambridge, MA: Harvard University Press, 1998). See also W. L. Craig, "The Argument from the Applicability of Mathematics." *Contemporary Arguments in Natural Theology: God and Rational Belief* (2021): 195.

44   Rees, *Just Six Numbers*, 123.

45   Exceptions include magnetic fields altering the paths of charged particles on the surface of stars, magnetized planets, and the interstellar medium.

46   A. Zee, "The Effectiveness of Mathematics in Fundamental Physics," in *Mathematics and Science*, ed., Ronald E. Mickens (Singapore: World Scientific, 1990.), 20. Thanks goes to Robin Collins for this reference and several examples in his correspondence with us on this issue. Collins refers to this feature of the universe as "separability," that quality of the universe that provides us with bit-sized chunks that often can be accurately analyzed without considering the whole.

47   It's important to distinguish between a law being simple and being easy to discover. Anyone who has tried to follow the movements of the planets across the sky realizes that it's quite tedious. Discovering laws that describe their regularity is even tougher. However, physicists have often used the simplicity of a discovered law as evidence of its truth. To toil away at observing the complicated motions of heavenly bodies for years on end, only to discover a

simple mathematical formulation that describes them, is suggestive—even downright fishy. It provides a clue to the investigator that he has stumbled across something more than a mere generalization or mental construct. This is the sense of simplicity that we speak of here—the simplicity of elegance. It doesn't imply that discovering, say, the inverse square law is so easy that a child could do it.

48   Stuart Clark, *Life on Other Worlds and How to Find It* (Chichester, UK: Springer-Praxis Pub., 2000), 1.

49   Michael Denton, *Nature's Destiny*, 243; *Fire-Maker: How Humans were Designed to Harness Fire and Transform our Planet* (Seattle: Discovery Institute Press, 2016); *The Miracle of Man: The Fine Tuning of Nature for Human Existence* (Seattle: Discovery Institute Press, 2022).

CHAPTER 11: THE REVISIONIST HISTORY OF THE COPERNICAN REVOLUTION

1   Dennis R. Danielson, "The Great Copernican Cliché," *American Journal of Physics* 69, no. 10 (October 2001): 1029.

2   It was also excerpted in *Time* magazine. The article was published one week after the NASA announcement of possible evidence of life in a Martian meteorite on August 7, 1996.

3   The myth that most ancients thought that the earth was flat seems to have been started by Washington Irving in the English-speaking world. For analysis and critique, see Jeffrey Burton Russell, *Inventing the Flat Earth: Columbus and Modern Historians* (New York: Praeger, 1991); Michael Newton Keas, *Unbelievable: 7 Myths About the History and Future of Science and Religion* (Wilmington, DE: ISI Books, 2019).

4   Quoted in John Noble Wilford, "From Distant Galaxies, News of a 'Stop-and-Go Universe,'" *New York Times* (June 3, 2003).

5   Bruce Jakosky, *The Search for Life on Other Planets* (Cambridge, UK: Cambridge University Press, 1998), 299.

6   Stuart Clark, *Life on Other Worlds and How to Find It* (Chichester, UK: Springer-Praxis Pub., 2000), 1. More generally, physicist Steven Weinberg comments, "The more the universe seems comprehensible, the more it also seems pointless." Earlier in the same book, he says: "It is almost irresistible for humans to believe that we have some special relation to the universe, that human life is not just a more-or-less farcical outcome of a chain of accidents reaching back to the first three minutes, but that we were somehow built in from the beginning." Steven Weinberg, *The First Three Minutes* (New York: Basic Books, 1977), 150–55.

7   Bertrand Russell, *Religion and Science* (New York: Oxford University Press, 1961), 222. Quoted in Neil Manson, *Why Cosmic Fine-Tuning Needs to Be Explained*, Dissertation at Syracuse University (December 1998), 146.

8   Stuart Ross Taylor, *Destiny or Chance: Our Solar System and Its Place in the Cosmos* (Cambridge, UK: Cambridge University Press, 1998), 11.

9   Thomas Kuhn, *The Copernican Revolution: Planetary Astronomy in the Development of Western Thought* (Cambridge, MA: Harvard University Press, 1957, renewed 1985), 5.

10 In his *Physics* and *On the Heavens*, excerpted in Dennis Danielson, *The Book of the Cosmos* (Cambridge, MA: Perseus Pub., 2000), 42. Danielson's anthology brings together many of the central Western texts on cosmology. We make generous use of the primary texts included in this volume.

11 Quoted in Danielson, 72. The Christian philosopher Boethius (470–525) who "mediate[d] the transition from Roman to Scholastic thinking," made the same point. See quote and discussion in John Barrow and Frank Tipler, *The Anthropic Cosmological Principle* (Oxford, UK: Oxford University Press, 1986), 45.

12 From chapter 6 of *De Revolutionibus*, quoted in Danielson, 112. Copernicus says earlier in this text, "So far as our senses can tell, the earth is related to the heavens as a point is to a body and as something finite is to something infinite."

13 Kuhn, *The Copernican Revolution*, 112–13.

14 Quoted in Hans Blumenberg, *The Genesis of the Copernican Revolution*, translated by Robert M. Wallace (Cambridge, MA: MIT Press, 1987), xv.

15 For all the talk about the Earth's removal from the center of the cosmos, it is ironic that the claim of both the infinite Newtonian universe and the finite (even if boundless) universe of big bang cosmology is quite different. It's not that we have been displaced from the cosmic center, but that there is no cosmic center.

16 Two nineteenth-century books immortalized the warfare metaphor, *A History of the Warfare of Science with Theology in Christendom* by Andrew Dickson White [1895] (New York: George Braziller, 1955) and John W. Draper, *History of the Conflict between Religion and Science* [1875] (New York: Appleton, 1928). An excellent introduction to the complex relationship between Christianity and modern science is John Hedley Brooke, *Science and Religion: Some Historical Perspectives* (Cambridge, UK: Cambridge University Press, 1991). The warfare metaphor rears its head in the debate over the prevalence of life in the universe. In *Life Everywhere: The Maverick Science of Astrobiology* (New York: Basic Books, 2001), science writer David Darling attacked one of us (GG), attempting to dismiss GG's arguments for life's rarity on the basis of GG's religious views. The book gives the impression that any scientist who holds religious views contrary to Darling's should be disqualified from practicing science. One also gets the impression that Darling doesn't know the difference between an empirical and an *ad hominem* argument.

17 R. Hooykaas, *Religion and the Rise of Modern Science* (Vancouver: Regent College Publishing, 2000; originally published, Edinburgh: Scottish Academic Press, 1972), 162.

18 Blumenberg, 453–87. For detailed discussion, see also Stanley L. Jaki, *The Road of Science and the Ways to God*, The Gifford Lectures, 1974–1975 and 1975–1976 (Chicago: University of Chicago Press, 1978).

19 Hooykaas, 75–96.

20 Christopher Kaiser, *Creation and the History of Science* (Grand Rapids, MI: Eerdmans, 1991), 9.

21 Kaiser, *Creation and the History of Science*, 7–9, 67–69. For a helpful discussion of technological achievements in the Middle Ages, see J. Gimpel, *The Medieval Machine: The Industrial Revolution of the Middle Ages* (New York: Holt, Rinehart & Winston, 1976). The classic source on Christian faith and medieval technology is Lynn White Jr., *Medieval Technology and Social Change* (Oxford, UK: Oxford University Press, 1962) Perhaps we should note, contrary to popular assertion, that nothing in the biblical tradition justifies treating nature as a mere means to an end with no intrinsic value, or as a passive backdrop, with which humans can do as they please.

22 An important study on the centrality of the doctrine of creation in the development of science is Eugene M. Klaaren, *Religious Origins of Modern Science* (Grand Rapids, MI: Eerdmans, 1977).

23 Klaaren, *Religious Origins of Modern Science*, 29–52.

24 We owe this important point to a lecture and personal conversation with nuclear physicist and science historian Peter Hodgson. For background, see Stanley L. Jaki, *Science and Creation* (Edinburgh: Scottish Academic Press, 1986). For detailed discussion, see Thomas F. Torrance, *Divine and Contingent Order* (Oxford: Oxford University Press, 1981). Torrance summarizes his argument in Thomas F. Torrance, "Divine and Contingent Order," in *The Sciences and Theology in the Twentieth Century*, ed. A. R. Peacocke (Notre Dame: University of Notre Dame Press, 1981), 81–97.

25 In his essay "Abstraction in Modern Science," in *Across the Frontiers* (New York: Harper & Row, 1974), 83, 87, the great physicist Werner Heisenberg said:

> This juxtaposition of different intuitive pictures and distinct types of force created a problem that science could not evade, because we are persuaded that nature, in the last resort is uniformly ordered, that all phenomena ultimately take place according to nature's unitary laws.
>
>   Perhaps it is also permissible here to mention yet another comparison, from the field of history. That abstraction arises from continuing to ask questions, from

the striving for unity, can be clearly recognized from one of the most momentous occurrences in the history of religion. The concept of God in the Jewish religion represents a higrhomher stage of abstraction, when compared with the idea of many different nature gods, whose activity in the world can be experienced directly. Only at this higher stage is it possible to recognize the unity of divine activity.

26 Melissa Cain Travis, *Science and the Mind of the Maker: What the Conversation Between Faith and Science Reveals About God* (Eugene, OR: Harvest House Publishers, 2018). Travis makes the case that the beliefs in the comprehensibility of nature and the rationality of the human mind derive from theological principles and are foundational to science.

27 "Faith in the possibility of science, generated antecedently to the development of modern scientific theory, is an unconscious derivative from medieval theology." Alfred North Whitehead, *Science and the Modern World* (New York: Macmillan, 1925), 19.

28 For an introduction to science in ancient, medieval, and Islamic cultures, see David C. Lindberg, *The Beginnings of Western Science: The European Scientific Tradition in Philosophical, Religious, and Institutional Context, 600 B.C. to A.D. 1450* (Chicago: University of Chicago Press, 1992).

29 Blumenberg, 483–84.

30 Of course, Thomas Aquinas did not uncritically adopt Aristotle, as anyone knows who has read his *Summa Theologica*.

31 John Milton reflects this conviction already in *Paradise Lost*, VIII, lines 66–71 (1667):

> For Heav'n
> Is as the Book of God before thee set
> Wherein to read his wondrous works . . ./
> . . . whether Heav'n move or Earth
> Imports not. . . .

Quoted in Owen Gingrich, *The Eye of Heaven* (New York: American Institute of Physics, 1993), 284.

32 Barrow and Tipler call this type of design argument eutaxiological, which requires that order must have a planned cause, which argues from order to a consequent purpose. See *The Anthropic Cosmological Principle*, 29, 44, 88–89, 144.

33 "What did Nicolaus Copernicus do for a living?" Britannica, https://www.britannica.com /question/What-did-Nicolaus-Copernicus-do-for-a-living.

34 As evidence of the large fraction of "scientific stars" who were devout believers in God, see Appendix 2.1 in Rodney Stark, *For the Glory of God: How Monotheism Gave Us Science, Witch Hunts, and the End of Slavery* (Princeton, NJ: Princeton University Press, 2004), 198–99. Stark lists the level of personal piety of 52 scientists living between the critical period 1543 to 1680. Of these, he classifies 32 as devout, 19 as conventional, and 1 as skeptical. Even well into the twentieth century, many leading scientists were devout. Examples include Herschel, Faraday, Maxwell, Kelvin, Millikan, Planck, and Townes.

35 Contrary to the conventional description, the models had not grown appreciably more complicated from the time of Ptolemy to that of Copernicus since there were so few systematic observations to improve on Ptolemy's. The main important additions were *trepidations* added by Islamic astronomers before A.D. 1000 to preserve the accuracy of the Ptolemaic values for the location of stars. Accurate systematic observations did not occur until *after* Copernicus' time. (See Gingrich, 25–26.) Thus, it's an overstatement to claim, as Thomas Kuhn does [in *The Structure of Scientific Revolutions* (Chicago: University of Chicago Press, 1962), 67–8], that a "crisis" and "scandal" precipitated Copernicus' innovation. See Gingrich, 193–204.

36 See Kuhn, *The Structure of Scientific Revolutions*, 6. This contrasts with the common positivist story of Copernicus. In that story, Copernicus was simply following the evidence, and proposed a system that was much simpler than the older one. The story itself is simplistic.

37 Earlier Scholastic critics of Aristotelian cosmology, such as Jean Buridan, Nicole Oresme, and other "nominalists" also played a role. And Copernicus was not the first to postulate heliocentrism. Aristarchus of Samos (c. 310–230 BC) postulated an Earth in a Sun-centered universe with the Earth spinning on its axis.

38 Copernicus's aspirations for natural philosophy, or what we now call science, are admirable. In his preface to *De Revolutionibus*, written to Pope Paul II, he says, "I know that a philosopher's thoughts are beyond the reach of common opinion—for his aim is to search out the truth in all things—so far as human reason, by God's permission, can do that. But I do think that completely false opinions are to be avoided." Quoted in Danielson, *The Book of the Cosmos*, 105.

39 Blumenberg, xiii, 173–85. Copernicus's anthropocentric assumptions give evidence of the influence of the Neo-Stoic strand of the Renaissance.

40 Kuhn, 139.

41 Dennis Danielson in a public lecture, "Copernicus and the Tale of the Pale Blue Dot," to the American Scientific Affiliation, Lakewood, Colorado, July 25–28, 2003.

42 Quoted in Danielson, *The Book of the Cosmos*, 131.

43 Blumenberg, xvii, 230–55.

44 See the discussion of R.G. Collingwood's defense of this point in Eugene M. Klaaren, *Religious Origins of Modern Science*, 11–13.

45 Kuhn, 135.

46 Hans Blumenberg puts it nicely: "Seen in terms of the whole history of his influence, Copernicus triumphed in the end not so much *over* as *through* his opponents," 353.

47 In some ways, Brahe's proposal, although it was a compromise between the ancient and the Copernican models, was a more drastic departure from the Aristotelian/Ptolemaic scheme since it made the nested celestial spheres impossible. His model also simplified the observations, perhaps even more than Copernicus' model. So, the superiority of Copernicus' proposal did not really become clear until Kepler modified it using elliptical orbits.

48 See Gingrich, 39–51; Melissa Cain Travis, *Thinking God's Thoughts: Johannes Kepler and the Miracle of Cosmic Comprehensibility* (Moscow, ID: Romans Roads Press, 2022).

49 Kepler's introduction of elliptical orbits was a real paradigm shift from the old attachment to circular orbits. See D. Boccaletti, "From the Epicycles of the Greeks to Kepler's Ellipse—The Breakdown of the Circle Paradigm," presented at Cosmology through Time—Ancient and Modern Cosmology in the Mediterranean Area (Monte Catone [Rome], Italy) June 18–21, 2001.

50 In Kepler, *Optics* II, 277:21–29, in *Gesammelte Werke* (Munich, Germany: Beck, 1937), translated by Edward Rosen in *Kepler's Conversation with Galileo's Sidereal Messenger* (New York: Johnson Reprint Corporation, 1965), 45. Thanks goes to Michael Crowe for reading portions of our manuscript and alerting us to this remarkable insight of Kepler's.

51 Historians now know that the English mathematician and astronomer Thomas Harriot performed telescopic astronomical observations several months prior to Galileo. He was not a self-promoter like Galileo; the significance of his work was not even recognized until nearly two centuries after his death. See Allan Chapman, "A New Perceived Reality: Thomas Harriot's Moon Maps," *Astronomy & Geophysics* 50 no. 1 (2009): 1.27–1.33. According to Chapman, Harriot's reticence to make his observations public did not come from fear of persecution from the Church of England, but, more likely, from political persecution. Chapman notes, "The Church of England had no official policy of any kind as far as astronomy and cosmology were concerned."

52 Galileo Galilei, *Sidereus Nuncius*, quoted in Danielson, *The Book of the Cosmos*, 150.

53 See Kitty Ferguson's balanced treatment of the subject in her *Measuring the Universe: Our Historic Quest to Chart the Horizons of Space and Time* (New York: Walker and Company, 1999), 94–105.

54 We owe this insight to a conversation with Peter Hodgson, as well as private correspondence and his paper, "Galileo the Scientist."

55 Witch hunts are themselves the subject of much bad history and mythology. For a good corrective, see Stark, *For the Glory of God*, 201–90.

56   See Ron Naylor, "Galileo's Tidal Theory," *Isis* 98, no. 1 (2007): 1–22.

57   Blumenberg, 421–30. For an excellent brief critique of the "Galileo Legend," see also Thomas M. Lessl, "The Galileo Legend as Scientific Folklore," *Quarterly Journal of Speech* 85 (1999): 146–168, and Keas, *Unbelievable*, 75–90.

58   One need not look far for examples. For instance, the author of the feature article of the June 2001 issue of *Sky & Telescope* discusses the future possibility of the first *visual* discovery of an extrasolar planet around Epsilon Eridani, the third-closest naked eye star to the Sun. The story is intrinsically interesting and has nothing to do with Bruno. Nevertheless, it opens with Bruno's famous quote about countless inhabited world and includes a prominent photo of a statue of Bruno, which marks the spot of his execution in Rome. Govert Schilling, "The Race to Epsilon Eridani," 34–41.

59   Hans Blumenberg treats this in detail in a section titled "Not a Martyr for Copernicanism: Giordano Bruno," *The Genesis of the Copernican Revolution*, 353–85. Note that Bruno's claim that the world is eternal was an Aristotelian, not a Copernican, proposition.

60   Herbert Butterfield, *The Origins of Modern Science* (New York: The Free Press, 1957), 13–28. Also important was the "transition" concept of impulse developed by Nicole Oresme and Jean Buridan in the mid-fourteenth century. This is just one example illustrating that the "Middle Ages" were important to the rise of science.

61   This description contradicts the common portrayal of Newton as a semi-deist. For justification of this interpretation, see Edward B. Davis, "Newton's Rejection of the 'Newtonian World View': The Role of Divine Will in Newton's Natural Philosophy," *Science & Christian Belief* 3 (1991): 103–117.

62   In fact, in the General Scholium, within his famous *Principia*, Newton offers just such a design argument. See Isaac Newton, *Mathematical Principles of Natural Philosophy* (Berkeley, CA: University of California Press, 1960), 542–44.

63   Stephen C. Meyer, *The Return of the God Hypothesis* (San Francisco: HarperOne, 2021), 42–50.

64   For discussion of ideas and discoveries that contributed to a form of Epicurean materialism in the Renaissance, see Benjamin Wiker, *Moral Darwinism: How We Became Hedonists* (Downers Grove, IL: InterVarsity Press, 2002), 107–39.

65   This may have been made possible, in part, because of ambivalence in the concept of natural law, which was present in Newton's writings. See John Hedley Brooke, "Natural Law in the Natural Sciences: The Origins of Modern Atheism?" *Science and Christian Belief* 4 (1992): 83–103.

66   Since we're making harsh generalizations about the textbook treatment of the Copernican Revolution, it's only fair to document it. Consider, for instance, *The Universe Revealed*, by Chris Impey and William K. Hartmann (Pacific Grove, CA: Brooks/Cole, 2000). Chapter 3, "The Copernican Revolution," frames the topic by beginning with Galileo's trial. The authors introduce and explain the topic in this way: "The Copernican revolution is one of the most important ideas in history, because it indicates that we are part of a larger cosmic environment, not the masters of nature living in the capital of the universe." No one who understood the history of ideas in the West would speak of the Earth in the earlier cosmology as "the capital of the universe," 53.

      To be fair, astronomy textbooks are less sloppy in making this point than books outside the discipline. As an example of the latter, consider this embarrassing claim by Douglas J. Futuyma, in *Science on Trial* (New York: Pantheon Books, 1983), 195–96: "The geocentric theory of the universe was a theological doctrine developed by St. Clement of Alexandria, Dionysius the Areopagite, the twelfth-century theologian Peter Lombard, and St. Thomas Aquinas. . . .The geocentric theory was unproven; it was untestable; it was held only because it fit Scripture." See also Francisco Ayala's description of Darwinism as the completion of the Copernican Revolution in *Creative Evolution?!*, ed. John H. Campbell and J. William Schopf (Boston: Jones and Bartlett Pub., 1994), 2–5. Such claims would be embarrassing to almost any historian of science. For additional examples, see Keas, *Unbelievable*, 91–107.

67 Dennis R. Danielson, "The Great Copernican Cliché," *American Journal of Physics* 69, no. 10 (October 2001): 1029.

68 From *The Nature of the Gods*, quoted in Danielson, *The Book of the Cosmos*, 56. Daniel Quinn sees this as the fundamental delusion we tell ourselves, namely, "The world was made for man, and man was made to conquer and rule it." He contends that the biblical creation narratives exemplify this more general view. In *Ishmael* (New York: Bantam, 1992), 74. See also Daniel Quinn, *The Story of B* (New York: Bantam, 1996), 129.

> In *The Genesis of the Copernican Revolution*, Hans Blumenberg notes that, unlike the Stoicism of Cicero, for Christian theology the world was not strictly made for man but rather subjugated to him by God, 174. There is a certain irony in the fact that certain Enlightenment figures made "man the measure of all things." In so doing, they exalted man far more than Medieval Christianity ever did. Then, in a startling case of historical amnesia and projection, they began to perpetuate the myth that it was the ancient and medieval world picture that erroneously exalted man above his true status.

69 Blaise Pascal, *Penseés* (New York: Penguin Classics, reprint 1995).

CHAPTER 12: THE COPERNICAN PRINCIPLE

1   Carl Sagan, *Pale Blue Dot* (New York: Ballantine Books, 1994), 7.

2   Quoted in Dennis Overbye, "In the Beginning . . . ." *New York Times* (July 23, 2002), D1.

3   The general theory of relativity neither entails nor implies the cosmological principle. In his highly influential early work in cosmology, E. A. Milne put it this way: "The cosmological principle is simply used [in Milne's work] as a definition, a principle of exclusion, enumerating the class of systems to be considered and excluding all others. It is in no sense used as a 'law of nature', or principle of compulsion, prescribing what is to happen. Whether, when a system is set up satisfying the cosmological principle, it will continue to satisfy it, is a matter always for investigation." *Relativity, Gravitation, and World-Structure* (Oxford, UK: Clarendon Press, 1935), 20.

4   In fact, the early use of the cosmological principle had nothing to do with Copernicus, or with an extension of the Copernican Revolution. Copernicus is rarely if ever mentioned by early twentieth century cosmologists like Milne.

5   Modern writers often simply identify the Copernican principle with the cosmological principle, by defining the former in terms of the homogeneity of the cosmos. As Ernan McMullin describes it, "A homogeneous large-scaled distribution of matter is (for reasons difficult to specify) more likely to be the actual cosmic state of affairs." In "Indifference Principle and Anthropic Principle in Cosmology," *Studies in the History and Philosophy of Science* 24, no. 3 (Feb. 1993): 359.

6   So, J. Richard Gott III argues in "Implications of the Copernican Principle for Our Future Prospects," *Nature* 363, no. 6427 (May 1993): 315. "The Copernican revolution taught us that it was a mistake to assume, without sufficient reason, that we occupy a privileged position in the Universe," he writes. "Darwin showed that, in terms of origin, we are not privileged above other species. Our position around an ordinary star in an ordinary galaxy in an ordinary supercluster continues to look less and less special. . . . The Copernican principle works because, of all the places for intelligent observers to be, there are by definition only a few special places and many nonspecial places, so you are likely to be in a nonspecial place."

7   See discussion in McMullin, "Indifference Principle and Anthropic Principle in Cosmology," 359–67.

8   Of course, the modest version of the Copernican principle could be true and the metaphysical version false. There is, after all, a difference between a worldview and a world picture. But these two can't be safely quarantined from each other. There's bound to be seepage. If evidence overwhelmingly confirmed that our location is utterly unexceptional, we might suspect that we are unexceptional as well. In any case, practitioners very rarely

distinguish between these two senses as we have. In one of the clearest expositions of the subject, Ernan McMullin (in "Indifference Principle") identifies the Copernican principle with the denial of purpose and privilege. "The Copernican principle has to be understood in terms of what it rejects, namely, older teleological beliefs about the uniqueness of human and the likelihood that the human abode would be singled out in some special way, e.g., by being at the cosmic center," he writes. "In practice, the principle reduces simply to asserting that human life is equally likely to be located in any (large) region of cosmic space or (less persuasively) at any point on the cosmic time-line." McMullin, "Indifference Principle and Anthropic Principle in Cosmology," 373.

9   Mario Livio, *The Accelerating Universe: Infinite Expansion, the Cosmological Constant, and the Beauty of the Cosmos* (New York: John Wiley & Sons, 2000), 263.

10   From Percival Lowell, *Mars* (Boston, 1895), excerpted in Dennis Danielson, *The Book of the Cosmos* (Cambridge, MA: Perseus Pub., Helix Books, 2000), 341. Speculation about extraterrestrial life in the Solar System was usually restricted to the other planets, although William Herschel thought the Sun itself might be habitable. See Steven Kawaler and J. Veverka, "The Habitable Sun: One of William Herschel's Stranger Ideas," *Journal of the Royal Astronomical Society of Canada* 75, no. 1 (1981): 46–55. Today, we tend to consider such an idea as quaint, but it follows from the Copernican principle.

11   In John Noble Wilford, *Mars Beckons* (New York: Knopf, 1990), 35, quoted in Danielson, *The Book of the Cosmos*, 341.

12   Quoted in Ken Croswell, *Planet Quest: The Epic Discovery of Alien Solar Systems* (New York: The Free Press, 1998), 185–86. Mayor and Queloz published their discovery in "A Jupiter Mass Companion to a Solar-Type Star," *Nature* 378 (1995): 355–59.

13   L. J. Rosenthal, et al., "The California Legacy Survey. III. On the Shoulders of (Some) Giants: The Relationship Between Inner Small Planets and Outer Massive Planets," *The Astrophysical Journal Supplement Series* 262, no. 1 (2022): 1.

14   The work was first reported in "The Arecibo Message of November, 1974," *Icarus* 26 (1975): 462–66, authored by "the staff at the National Astronomy and Ionosphere Center," which included Carl Sagan and Frank Drake. Apparently, the signal was transmitted as part of the ceremonies to dedicate the upgraded Arecibo radio telescope on November 16, 1974. See discussion in Steven J. Dick, *Life on Other Worlds: The 20th Century Extraterrestrial Life Debate* (Cambridge, UK: Cambridge University Press, 1998), 217.

15   Shapley overestimated the size of the Milky Way and put us 65,000 light years out. Astronomers now estimate that we are approximately 27,000 light years from the galactic center.

16   Ken Croswell, *The Alchemy of the Heavens* (Oxford, UK: Oxford University Press, 1995), 24. Croswell is normally astute at challenging conventional wisdom on such matters, and he does not draw any conclusions in favor of the Copernican principle from Shapley's discovery.

17   S. P. Goodwin, J. Gribbin, and M. A. Hendry, "The Relative Size of the Milky Way," *The Observatory* 118 (Aug. 1998): 201, 207. This paper is shot through with "Copernican" reasoning.

18   P. Villanueva-Domingo, et al. "Weighing the Milky Way and Andromeda Galaxies with Artificial Intelligence," *Physical Review D* 107, no. 10 (2023): 103003.

19   N. Boardman, et al., "Are the Milky Way and Andromeda Unusual? A Comparison with Milky Way and Andromeda Analogues," *Monthly Notices of the Royal Astronomical Society* 498, no. 4 (2020): 4943–54. The authors note, "The MW disc appears to be more compact than the majority of galaxies of MW-like mass . . . and appears to be deficient by around 1σ with respect to the MW's circular velocity." See also, F. Hammer, et al., "The Milky Way, An Exceptionally Quiet Galaxy: Implications for the Formation of Spiral Galaxies," *The Astrophysical Journal* 662, no. 1 (2007): 322. The following study reports that the Milky Way is anomalously *compact*: T. C. Licquia, J. A. Newman, and M. A. Bershady, "Does the Milky Way Obey Spiral Galaxy Scaling Relations?" *The Astrophysical Journal* 833, no. 2 (2016): 220.

CHAPTER 13: THE ANTHROPIC DISCLAIMER

1   Fred Adams and Greg Laughlin, *Five Ages of the Universe: Inside the Physics of Eternity* (New York: The Free Press, 1999), 201.

2   The contemporary philosopher who has done the most to develop this type of argument is William Lane Craig. See especially *The Kalam Cosmological Argument* (Eugene, OR: Wipf & Stock Pub., 1979) and William Lane Craig and Quentin Smith, *Theism, Atheism, and Big Bang Cosmology* (Oxford, UK: Clarendon, 1993).

3   Sir Arthur Eddington, who didn't like the implications of the big bang model, said, "The beginning seems to present insuperable difficulties unless we agree to look on it as frankly supernatural." In *The Expanding Universe* (New York: Macmillan, 1933), 124.

4   Fred Hoyle, "A New Model for an Expanding Universe," *Monthly Notices of the Royal Astronomical Society* 108, no. 5 (1948): 372–82; Hermann Bondi and Thomas Gold, "The Steady-State Theory of the Expanding Universe," *Monthly Notices of the Royal Astronomical Society* 108, no. 3 (1948): 252–70.

5   Denis Sciama, *The Unity of the Universe* (New York: Doubleday, 1961), 70. Quoted in Ernan McMullin, "Indifference Principle and Anthropic Principle in Cosmology," *Studies in the History and Philosophy of Science* 24, no. 3 (February 1993): 368.

6   Stephen C. Meyer, "The Return of the God Hypothesis," *The Journal of Interdisciplinary Studies*, XI, no. 1/2 (1999): 9. The articles substantiating this claim are Alan Guth and Marc Sher, "The Impossibility of a Bouncing Universe," *Nature* 302 (April 7, 1983): 505–7; J. E. Phillip Peebles, *Principles of Physical Cosmology* (Princeton, NJ: Princeton University Press, 1993); and Peter Coles and George Ellis, "The Case for an Open Universe," *Nature* 370 (August 25, 1994): 609–13. Of course, this doesn't mean that cosmologists have given up trying to get rid of an initial singularity, or beginning, given their theism-friendly implications. See, for example, Charles Seife, "Eternal-Universe Idea Comes Full Circle," *Science* 296 (2002): 639.

7   In his widely known book *A Brief History of Time* (New York: Bantam Doubleday Dell, 1998, tenth anniversary edition), Stephen Hawking attempted to accept a finite past while nevertheless avoiding a cosmic beginning (and its presumably distasteful theological implications). He does so with several questionable and highly controversial maneuvers, including the use of imaginary time and Richard Feynman's sum-over-all-histories interpretation of quantum theory. We find his strategy not only unconvincing but desperate in the extreme. For a devastating critique of Hawking's argument, see William Lane Craig in *Theism, Atheism, and Big Bang Cosmology*, 279–300.

> Tellingly, in Hawking's later *The Universe in a Nutshell* (New York: Bantam Books, 2001), he continued to employ similar arguments, showing no awareness of the published critiques of them. He also explicitly endorses the implicit positivism of *Brief History*. This allowed him, albeit inconsistently, to skirt the issue of whether he intended his use of imaginary time to accurately represent reality. If it works mathematically, he claimed, such questions of correspondence with reality are superfluous. He didn't seem to realize that this immunization strategy undercut his desire to do away with a beginning, since we are still left with the question of whether the *actual* universe began to exist. Just because a theoretical physicist concocted a mathematical justification for avoiding the question doesn't mean anyone is obliged to follow him, especially when he refused to argue that his conjecture represents reality. Moreover, once again, he seemed unaware of the trenchant critiques of positivism that have led philosophers of science to abandon it.

8   This point is often misunderstood. See, for instance, Anthony Aguirre, "Cold Big-Bang Cosmology as a Counterexample to Several Anthropic Arguments," *Physical Review* D 64, no. 8 (Sept 2001): https://doi.org/10.1103/PhysRevD.64.083508. The argument does not require that there be only one set of physical constants compatible with complex life. It

only requires that among the range of alternative universes considered in the "universe neighborhood," life-permitting universes are quite rare.

9   Paul Davies, *The Cosmic Blueprint* (New York: Simon & Schuster, 1988), 203.

10 Fred Hoyle, "The Universe: Past and Present Reflections," *Annual Review of Astronomy and Astrophysics* 20 (1982): 16.

11 Stephen Hawking, *A Brief History of Time* (New York: Bantam Doubleday Dell, 1998, tenth anniversary edition).

12 For an excellent presentation of this argument, see Timothy McGrew, Lydia McGrew, and Eric Vestrup, "Probabilities and the Fine-tuning Argument: A Skeptical View," *Mind* 110 (October 2001). Philosopher Robin Collins has responded to their argument in "Fine-Tuning Arguments and the Problem of the Comparison Range," *Philosophia Christi* 7, no. 2 (2005): 385–404; https://doi.org/10.5840/pc20057233.

13 In the new edition *The Design Inference* (Seattle: Discovery Institute Press, 2023), 458–63, William A. Dembski and Winston Ewert handle this "uniformizability" problem with analytic rigor. Thanks to Bill Dembski for his advice on this discussion. See also the treatment by Luke A. Barnes, "A Reasonable Little Question: A Formulation of the Fine-Tuning Argument," *Ergo* 6, no. 2 (2020): 1220–57.

14 To be precise, it's not quite correct to say only one set of constants (in the example, one numeric combination) is compatible with a habitable universe. After all, even the force of gravity relative to the electromagnetic force is "only" fine-tuned out to the fortieth decimal place. So, the illustration requires a *range* of combinations, rather than a single number. Imagine in this story that the Q have discovered that they can add any number to the right of the sequence, that is, past the eighth decimal place, without affecting the outcome. But any other change to the combination above that level of resolution spells disaster. Here's the idea. The Q determined that this number, 91215225.79141425, was required by the Universe Creating Machine to generate a habitable universe, but that the combination tolerated variations in the number past the eighth decimal point. So, combinations such as 91215225.791414251, 91215225.791414252 and so on still permit a habitable universe, but none that they can find without the initial number sequence 91215225.79141425. This would suggest that every number *between* 91215225.79141425 and 91215225.79141426 would permit a habitable universe. And of course, there is a potential infinity of numbers between any two numbers. Nevertheless, even though the Q had found a potentially infinite range of combinations at that level of resolution that generated a habitable universe, they would still not doubt that the original combination had been set. One reason is because a single specific piece of knowledge, which profoundly restricts the range of possibilities, would be necessary to find the "set" of combinations that would work. One might even say the combination to the Universe Creating Machine is still singular, since it can be precisely specified: "The code for generating a habitable universe is any natural number from 91215225.79141425 to 91215225.79141426, excluding 91215225.79141426." No one in such a case would buy the argument that because there are a potentially infinite number of combinations that can produce a habitable universe, we should remain agnostic about whether the combination had been initially set.

15 Brandon Carter initiated the contemporary discussion of the anthropic principle with a formulation similar to this: "What we can expect to observe must be restricted by the conditions necessary for our presence as observers." In "Large Number Coincidences and the Anthropic Principle in Cosmology," in M. S. Longair, ed., *Confrontation of Cosmological Theory with Astronomical Data* (Dordrecht, The Netherlands: Reidel, 1974), 291–98. Reprinted in John Leslie, ed., *Physical Cosmology and Philosophy* (New York: Macmillan, 1990). In addition to the weak anthropic principle, Carter proposed a "Strong Anthropic Principle" (SAP). There is an exegetical debate surrounding the meaning of the Strong Anthropic Principle and, in particular, the meaning of the word "must" in Carter's formulation. Leslie interprets Carter in this way: The Weak sense refers to our location in time and place. The Strong sense refers to the universe as a whole. SAP means, "any universe with observers in it must be observer-permitting." See Leslie's introduction in *Modern*

Cosmology and Philosophy, ed. John Leslie (Amherst, NY: Prometheus, 1998), 2. Whether or not this is a correct interpretation, the Strong Anthropic Principle has been taken in all sorts of ways, and often does not conform to Carter's initial formulation. We are following these definitions of weak anthropic principle and strong anthropic, without judging whether they are consistent with Carter's initial formulation.

16 Carter seems to have understood it in just this way. In "Large Number Coincidences" (reprinted in *Physical Cosmology and Philosophy*, 131), he says, "It consists basically of a reaction against exaggerated subservience to the 'Copernican principle.' Copernicus taught us the very sound lesson that we must not assume gratuitously that we occupy a privileged *central* position in the Universe. Unfortunately, there has been a strong (not always subconscious) tendency to extend this to a most questionable dogma to the effect that our situation cannot be privileged in any sense."

17 Konrad Rudnicki, "The Anthropic Principle as a Cosmological Principle," *Astronomy Quarterly* 7, no. 2 (1990), 117-127, https://doi.org/10.1016/0364-9229(90)90015-S.

18 We thank Paul Nelson for this story.

19 As we explain in endnote 15, there is a confusing variety of different definitions of the Strong Anthropic Principle in the literature. Often it is defined as meaning that the laws and constants somehow had to be what they are, or that they are necessarily what they are. Others define the weak anthropic principle as a selection effect and the strong anthropic principle as implying that the fine-tuning is the result of intentional design. This diversity is confusing, so we've decided to go with the definitions that are easiest to grasp. Nothing crucial in our argument depends on this way of defining the weak anthropic principle and the strong anthropic principle.

20 This is derived from the popular story told by philosophers Richard Swinburne, John Leslie, and others. See John Leslie, *Universes* (London: Routledge, 1989), 13–15, 108.

21 Alan H. Guth, *The Inflationary Universe* (Reading, MA: Perseus Books, 1997). Still more speculative proposals are Andrei Linde, "The Universe: Inflation Out of Chaos," in *Modern Cosmology and Philosophy*, and "The Self-Reproducing Inflationary Universe," *Scientific American* 271 (November 1994), 48–55.

22 For one highly speculative proposal along these lines, see Lee Smolin, *The Life of the Cosmos* (New York: Oxford University Press, 1997). We sympathize with those who object to using the term "universe" in this way, since the term implies singularity. Nevertheless, cosmologists now frequently talk about multiple worlds or universes, so we have decided to follow convention on this point.

23 In *The Anthropic Cosmological Principle* (Oxford, UK: Oxford University Press, 1986), 4, John D. Barrow and Frank J. Tipler observe, "In a sense, the Weak Anthropic Principle may be regarded as the culmination of the Copernican Principle, because the former shows how to separate those features of the Universe whose appearance depends on anthropocentric selection, from those features which are genuinely determined by the action of physical laws."

24 For details, see Craig, *The Kalam Cosmological Argument*, and Craig and Smith, *Theism, Atheism, and Big Bang Cosmology*, 3–191. There is a great deal of confusion about actual infinities. For instance, some physicists and cosmologists assume that, if the universe is open and will continue to expand indefinitely, then the universe contains or will contain an infinite number of planets, stars, and galaxies. This doesn't follow. Even if the matter in a continually expanding universe continued to congeal into stars, planets, and galaxies (which we have no reason to assume), it would never reach an actual infinite set of such objects. Indeed, no matter how large the finite number of such objects became, it would always be infinitely far from being infinite. This is a fairly elementary truth of mathematics, and no amount of theoretical modeling can change it. For examples of authors who make this mistake, see Jonathan I. Lunine, "The Frequency of Planetary Systems in the Galaxy," in Ben Zuckerman and Michael H. Hart, eds., *Extraterrestrials: Where are They?* (Cambridge, UK: Cambridge University Press, 1995), 223–24; and G. F. R. Ellis and G. B. Brundrit, "Life in an Infinite Universe," *Quarterly Journal of the Royal Astronomical Society* 20 (1979): 128–35.

25   As astronomer and science historian Steven Dick explains, "Such a claim [that many universes exist] has emerged over the last few years after pondering the fact that our universe seems to be finely tuned for life." Steven J. Dick, "Extraterrestrial Life and our World View at the End of the Millennium," *Dibner Library Lecture* (Washington, DC: Smithsonian Institution Series, 2000), 29.

26   Adams and Laughlin, *Five Ages of the Universe*, 199, 201. Emphasis in original. Not surprisingly, Adams and Laughlin do not mention even the possibility of design. Similarly, in *Our Cosmic Future: Humanity's Fate in the Universe* (Cambridge, UK: Cambridge University Press, 2000), 239, Nikos Prantzos notes that, despite "the impossibility of any physical contact among the bubble universes," "chaotic inflation is one of the most attractive from a philosophical point of view." One wonders what point of view he has in mind. Martin Rees uses similar reasoning. The Prologue to his book *Our Cosmic Habitat* (Princeton, NJ: Princeton University Press, 2001) is titled: "Could God Have Made the World Any Differently?" He answers "yes," and proposes that what we call laws of nature are "no more than local bylaws—the outcome of historical accidents during the initial instants after our own particular Big Bang" (xvii).

> Although he answers "yes" to the question, he doesn't really mean it. In fact, the question seems to be a red herring, since he appeals to multiple worlds: "In this book I argue that the multiverse concept is already part of empirical science: we may already have intimations of other universes, and we could even draw inferences about them and about the recipes that led to them. In an infinite ensemble, the existence of some universes that are seemingly fine-tuned to harbor life would occasion no surprise; our own cosmic habitat would plainly belong to this unusual subset. Our entire universe is a fertile oasis within the multiverse." Clearly Rees is posing a false dilemma. He hasn't ruled out the possibility that our universe looks fine-tuned because it *is* fine-tuned. He's just failed to consider it.

27   An infinite World Ensemble combines the virtues of chance and necessity. The reason we find ourselves in a habitable universe, according to this explanation, is not due to any purpose. In that sense, it's a result of chance. However, unless the existence of that ensemble is contingent, which the Many Worlds theorist is likely to deny, then the fact that there are habitable universes is due to necessity since the Ensemble is *infinite*. Reality in this scheme is like a cosmic lottery. Individual universes are the tickets. Since every possible universe exists, the chance of our universe existing is exactly one.

28   Stephen C. Meyer, *The Return of the God Hypothesis* (New York: HarperOne, 2021).

29   As philosopher of science Robin Collins has argued in "The Fine-Tuning Design Argument: A Scientific Argument for the Existence of God," *Reason for the Hope Within*, ed. Michael Murray (Grand Rapids, MI: Eerdmans, 1999), 61.

30   A great deal more remains to be said. But a detailed critique of various Many Worlds Hypotheses would take us too far afield. Stephen's Meyer's *The Return of the God of the Hypothesis* offers a detailed recent discussion. See also the delightful treatment in Geraint Lewis and Luke Barnes, *A Fortunate Universe: Life in a Finely Tuned Cosmos* (Cambridge, UK: Cambridge University Press, 2016). A popular general treatment of Many Worlds Hypotheses is John Leslie's *Universes*. For other arguments against Many Worlds Hypotheses, see Roger White, "Fine-tuning and Multiple Universes," *Nous* 34, no. 2 (2000): 260–276; Robin Collins, "The Fine-Tuning Design Argument"; William Lane Craig, "Barrow and Tipler on the Anthropic Principle v. Divine Design," *British Journal for the Philosophy of Science* 38 (1988): 389–395; Jay W. Richards, "Many Worlds Hypotheses: A Naturalistic Alternative to Design," *Perspectives on Science and Christian Faith* 49, no. 4 (1997): 218–227; William Lane Craig, "Cosmos and Creator," *Origins & Design* 20, no. 2 (Spring 1996): 18–28; Richard Swinburne, "Argument from the Fine Tuning of the Universe," in *Modern Cosmology and Philosophy*, ed. John Leslie, 160–179.

31   Nikos Prantzos (op. cit. note 26*)*, 205, claims the opposite.

32   This is his definition of the "cosmological principle," which is a synonym for the Copernican principle. Stuart Clark, *Stars and Atoms: From the Big Bang to the Solar System* (New

York: Oxford University Press, 1995), 23. Konrad Rudnicki also defines the "Generalized Copernican Principle" as the view that "the Universe looks roughly the same for any observer located on any other planet." "The Anthropic Principle as a Cosmological Principle," *The Astronomical Quarterly* 7 (1990): 121. Physicist Richard Feynman calls the principle "the grand hypothesis that nearly every cosmologist makes." He describes in terms of the uniformity of the appearance of the universe from every location. In *Feynman Lectures on Gravitation* (Reading, MA: Addison-Wesley, 1995), 166. He also refers to it as "a completely arbitrary hypothesis, as far as I understand it."

33   A quark is about 10^{-16} m in diameter. The adult human being is roughly 2 meters. The Earth is 10^5 m. The distance between Earth and the nearest stars is about 10^{17} m. The observable universe is about 10^{26} m. See "Scale of the Universe" in Clark, *Stars and Atoms*, 76–77.

CHAPTER 14: SETI, THE DRAKE EQUATION, AND THE UNRAVELING OF THE COPERNICAN PRINCIPLE

1   For a good survey, see Steven J. Dick, *Plurality of Worlds: The Origins of the Extraterrestrial Life Debate from Democritus to Kant* (Cambridge, UK: Cambridge University Press, 1982). For a detailed study of the growth of the idea from Kant to the twentieth century, see Michael J. Crowe, *The Extraterrestrial Life Debate 1750–1900: The Idea of a Plurality of Worlds from Kant to Lowell* (Cambridge, UK: Cambridge University Press, 1986).

2   Steven J. Dick, *Life on Other Worlds: The 20th-Century Extraterrestrial Life Debate* (Cambridge, UK: Cambridge University Press, 1998), 54. This book is an abridgement of a much longer book defending the same thesis, *The Biological Universe* (Cambridge, UK: Cambridge University Press, 1996).

3   Douglas Adams, *The Hitchhiker's Guide to the Galaxy* (New York: Pocket Books, 1979). 76.

4   Courtney Kennedy and Arnold Lau, "Most Americans Believe in Intelligent Life beyond Earth; Few See UFOs as a Major National Security Threat," Pew Research Center (June 20, 2021), https://www.pewresearch.org/short-reads/2021/06/30/most-americans-believe-in-intelligent -life-beyond-earth-few-see-ufos-as-a-major-national-security-threat/.

5   See Dick, *Life on Other Worlds*, 164–65.

6   *Life on Other Worlds*, 3, 261–66.

7   See Dick, *Life on Other Worlds*, 218–21.

8   Frank Tipler, "A Brief History of the Extraterrestrial Intelligence Concept," *Quarterly Journal of the Royal Astronomical Society* 22 (1981): 133–45; Frank Tipler, "Additional Remarks on Extraterrestrial Intelligence," *Quarterly Journal of the Royal Astronomical Society* 22 (1981): 271–91.

9   But we do consider Fermi's Paradox a strong argument. See G. Gonzalez, "Extraterrestrials: A Modern View," *Society* 35, no. 5 (July/August 1998): 14–20. For an interesting survey of the debate, see the essays in B. Zuckerman and M. H. Hart, ed., *Extraterrestrials: Where Are They?* (Cambridge, UK: Cambridge University Press, second edition, 1995). Most of the skeptics are biologists. The British journal *Astronomy and Geophysics* is one of the few scientific journals that has remained balanced in their treatment of the subject, publishing articles from both advocates and critics. (Too many others have become partisan campaigners.) See I. A. Crawford, "How Common are Technological Civilizations?" *Astronomy and Geophysics* 38, no. 4 (1997): 24–26; T. Joseph, W. Lazio and James M. Cordes, "The Number of Civilizations in our Galaxy," *Astronomy and Geophysics* 38, no. 6 (1997): 16–18; I. A. Crawford, "Galactic Civilizations: A Reply," *Astronomy and Geophysics* 38, no. 6 (1997): 19.

10 This response is unlikely to solve Fermi's paradox given the evidence we present in this book. If habitable planets like Earth are as rare as we argue they are, then Earth would have been near the top of any colonizers' target lists. In addition, any spacefaring civilization would have the technology to survey nearby stars for the presence of giant and terrestrial planets. And, since Earth has been inhabited by relatively simple life over most of its history, there

would be little motivation for quarantining it. Thus, it's unlikely the Earth was passed by either due to oversight or by intention in a wave of galactic colonization.

11 For an overview of such responses, see Stephen Webb, *If the Universe Is Teeming with Aliens . . . WHERE IS EVERYBODY?: Seventy-Five Solutions to the Fermi Paradox and the Problem of Extraterrestrial Life, Second Edition* (Heidelberg: Springer International Publishing, 2015). See also Ian Crawford, "Where are They?: Maybe We are Alone in the Galaxy After All," *Scientific American* (July 2000): 29–33.

12 Few have recognized this point explicitly, but planetary scientist Christopher McKay hints at it. "It is clear that the search for life beyond the Earth is based on the assumption that our experience of life on Earth is typical for the cosmos—the Copernican principle," he writes. "The Copernican principle is well established only for stars and elements. . . . [It] is not established with respect to biology, culture, or ethics. In this context, the question 'Are we alone?' is a deep philosophical question." Christopher P. McKay, "Astrobiology: The Search for Life Beyond the Earth," *Many Worlds: The New Universe, Extraterrestrial Life & the Theological Implications*, ed. Steven J. Dick (Philadelphia: Templeton Foundation Press, 2000), 5.

13 Tom Clarke, "SETI Sees the Light: Alien Hunters Look for Light Signals from Other Worlds," *Nature News* (July 25, 2001).

14 Public funding for SETI research waxes and wanes. The federal government backed SETI research for years through NASA, but such funding ended in 1992. Presently, most direct SETI work, such as that done by the SETI Institute, is supported from private sources. Nevertheless, NASA continues to support astrobiology programs at several institutions, programs that bear on the general question of extraterrestrial life. And in 2003, the SETI Institute received a major grant from the federally funded NASA Astrobiology Institute.

15 The meeting took place at the National Radio Astronomy Observatory (NRAO) in Green Bank, West Virginia. For discussion of the equation, and Drake's more recent reflections on the topic, see Frank Drake and D. Sobel, *Is Anybody Out There?: The Scientific Search for Extraterrestrial Intelligence* (New York: Delacorte Press, 1992). Drake had already performed an early search at NRAO called Project OZMA. Independently, Cornell physicists Philip Morrison and Giuseppe Cocconi published a seminal paper, "Searching for Interstellar Communications," *Nature* 184 (1959): 844–46, making the same calculations as Drake. The literature on the subject is quite large. For a brief, sympathetic description of the historical details, see Dick, *Life on Other Worlds*, 200–235. For a description of the variety of estimates that different astronomers have assigned to the Drake Equation, see Dick, *Life on Other Worlds*, 217, and Dick, *The Biological Universe*, 441.

16 The history of the equation is somewhat complicated. Drake's original equation was slightly different from the one here. He used R instead of $N*$ and L instead of fL. R stood for the rate of star formation in the galaxy and L stood for the average number of years that a communicating civilization survives.

17 Although SETI researchers seek empirical evidence, most of them are certain that ETIs exist. They're just trying to prove it. Frank Drake, for instance, saw it as a certainty, as do most of the SETI enthusiasts. See Drake's own discussion of the optimism of the original Green Bank meeting participants in Drake and Sobel, *Is Anybody Out There?*, xi, 45–64. See also David Koerner and Simon LeVay in *Here Be Dragons: The Scientific Quest for Extraterrestrial Life* (Oxford: Oxford University Press, 2000), 162; D. Schwartzman and L. J. Rickard, "Being Optimistic about SETI," *Bioastronomy: The Next Steps*, ed. George Marx (Dordrecht, The Netherlands: Kluwer Academic Publishers, 1988), 305. For an unrealistically optimistic evaluation of the Drake equation, see "The Drake Equation," *Astronomy* (September 2002): 47.

18 Quoted in Dick, *Life on Other Worlds*, 217. Bruce Jakosky says similarly that the "Drake equation is just a mathematical way of saying, who knows?" In *The Search for Life on Other Planets* (Cambridge, UK: Cambridge University Press, 1998), 285.

19 Sagan makes clear his debt to the Copernican principle in his book, coauthored with Soviet astrophysicist I. S. Shklovskii, *Intelligent Life in the Universe* (San Francisco: Holden-Day, 1966). In chapter 25, "The assumption of mediocrity," 357, they say, "The idea that we are not unique has proved to be one of the most fruitful of modern science. . . . Let us assume

that the Earth is an average planet, and that the Sun is an average star. Then the diameter of the Earth, its distance from the Sun, and its albedo, or reflectivity, should be characteristic of planets in general."

20 Dick, *Life on Other Worlds*, 102. Extrasolar planets have also long been a favorite for disciples of the Copernican principle.

21 Quoted by Fred Heeren in "Home Alone in the Universe?" *First Things* 120 (March 2002): 38–46.

22 J. J. Lissauer et al., "Exoplanet Science from Kepler," *Protostars and Planets VII*, in press. The exoplanet statistics quoted in this paragraph are from this study.

23 From the planet's orbital period and the star's mass, which is easy to determine, we can calculate the planet's distance from its host star using Kepler's Third Law of planetary motion.

24 They were defined by R. K. Kopparapu et al., "Habitable Zones Around Main-Sequence Stars: New Estimates," *The Astrophysical Journal* 765, no. 2 (2013): 131. These ranges are scaled from the Solar System to other stars by requiring the same amount of starlight hits the planet.

25 *Kepler* did not detect *any* planets smaller than 1.5 times the size of the Earth with orbital period greater than 128 days. See Figure 2 of D. C. Hsu et al., "Occurrence rates of Planets Orbiting FGK Stars: Combining Kepler DR25, Gaia DR2, and Bayesian Inference," *The Astronomical Journal* 158, no. 3 (2019): 109.

26 See S. Bryson et al., "The Occurrence of Rocky Habitable-zone Planets around Solar-like Stars from Kepler Data," *The Astronomical Journal* 161, no. 36 (2021): 36. The ranges correspond to different assumptions in making the required extrapolations from the observations for each case (conservative and optimistic).

27 See M. L. Hill, et al., "A Catalogue of Habitable Zone Exoplanets," *The Astronomical Journal* 165 (2023): 34.

28 Ulf Houe, "The Protoplasmic Imagination: Ernst Haeckel and H.P. Lovecraft," *Configurations* 30, no. 1 (Winter 2022): https://muse.jhu.edu/article/844466.

29 On the importance of information in biology, see Hubert Yockey, *Information Theory and Molecular Biology* (Cambridge, UK: Cambridge University Press, 1992); Bernd-Olaf Küppers, *Information and the Origin of Life* (Cambridge: MIT Press, 1990); Bernd-Olaf Küppers, *Molecular Theory of Evolution* (Heidelberg: Springer, 1983); W. Loewenstein, *The Touchstone of Life* (New York: Oxford University Press, 1998). On the difference between biological information and the chemical structures that carry that information, see Michael Polanyi, "Life's Irreducible Structure," *Science* 160 (1968): 1308, and Michael Polanyi, "Life Transcending Physics and Chemistry," *Chemical and Engineering News* (Aug. 21, 1967): 54–66.

30 "Life arose here almost as soon as it theoretically could," they write. "Unless this occurred utterly by chance, the implication is that nascent life itself forms—is synthesized from nonliving matter—rather easily. Perhaps life may originate on any planet as soon as temperatures cool to the point where amino acids and proteins can form and adhere to one another through stable chemical bonds. Life at this level may not be rare at all," Peter D. Ward and Donald Brownlee, *Rare Earth: Why Complex Life is Uncommon in the Universe* (New York: Copernicus, 2000), xix. See also 1–14. Stuart Ross Taylor, also a skeptic about the pervasiveness of intelligent life, similarly argues that the origin of life is basically a matter of getting the right chemical conditions. *Destiny or Chance: Our Solar System and its Place in the Cosmos* (Cambridge, UK: Cambridge University Press, 1998), 187.

31 See, for instance, molecular biologist Jacques Monod's *Chance and Necessity* (New York: Knopf, 1971). Astrophysicist Fred Hoyle famously joined biologists in this assessment. See his *The Intelligent Universe* (London: Michael Joseph, 1983).

32 See, for instance, Stuart Kauffman, *At Home in the Universe: The Search for the Laws of Self-Organization and Complexity* (New York: Oxford University Press, 1995) and Christian De Duve, *Vital Dust: A Cosmic Imperative* (New York: Basic Books, 1996).

33 Since we have not dealt with such issues in this book, we will not discuss them in detail in this chapter. But we are deeply skeptical of the standard chemical origin-of-life scenarios.

For details, see Stephen C. Meyer, *Signature in the Cell* (San Francisco: HarperOne, 2009), and Charles Thaxton et al., *The Mystery of Life's Origin: The Continuing Controversy* (Seattle: Discovery Institute Press, 2020). The longer the origin-of-life experiments continue without producing the basic informational molecules of life in plausible settings without undue experimenter intervention, the smaller the chances for the merely chemical origin of life. There's a similar argument for SETI: the longer they go without detection, the smaller the chances for ETI in the Milky Way.

34　See Gould's *Wonderful Life: The Burgess Shale and the Nature of History* (New York: W.W. Norton & Co., 1989).

35　See discussion in Koerner and LeVay, *Here Be Dragons*, 140–53. The title of chapter 6 is "What Happens in Evolution?: Chance and Necessity in the Origin of Biological Complexity." Gould elsewhere criticizes what he calls the "anthropic principle," by which he means the idea that our environment has been designed by a "Supermind." See his "Mind and Supermind," in *Modern Cosmology and Philosophy*, ed. John Leslie (Amherst, NY: Prometheus, 1998). Harvard seems to breed commitment to chance in evolution. Harvard's George Gaylord Simpson argued similarly several decades ago in "The Nonprevalence of Humanoids," *Science* 143 (1964), 769, as did Harvard zoologist Ernst Mayr. See, for instance, Mayr's "The Probability of Extraterrestrial Life;" in *Extraterrestrials: Science and Alien Intelligence*, ed. Edward Regis Jr. (Cambridge, UK: Cambridge University Press, 1985), 23–30.

36　See Simon Conway Morris, *Life's Solution: Inevitable Human's in a Lonely Universe* (Cambridge, UK: Cambridge University Press, 2003). Although many SETI researchers and astrobiologists are encouraged by his argument, Morris himself is not committed to the Copernican principle.

37　This is a quote from Simon Conway Morris in Koerner and LeVay, *Here Be Dragons*, 146.

38　Dick, *Life on Other Worlds*, 169, 195. Stuart Clark, following Paul Davies, argues that astronomers are here assuming the Copernican principle and the principle of plenitude, which means, basically, that whatever can happen, will. We are less confident the principle of plenitude qualifies as a common methodological principle in the way the Copernican principle does. Stuart Clark, *Life on Other Worlds and How to Find It* (Chichester, UK: Springer-Praxis Pub., 2000), 2–6.

39　Ward and Brownlee, *Rare Earth*, 83–242.

40　"My feeling is that the ultimate vindication of the Copernican Principle would be if Earth, as a planet, were not even privileged by the emergence of life. If Copernican ideas can be extended to biology, then there is no reason to assume that life on Earth is a cosmic one-off." Clark, *Life on Other Worlds*, 3. Some SETI advocates manage to blend all these elements into an unstable amalgam. For instance, in *The Search for Life on Other Planets*, 301, Bruce Jakosky says: "Finding non-terrestrial life would be the final act in the change in our view of how life on Earth fits into the larger perspective of the universe [notice the stereotypical history of the Copernican Revolution]. We would have to realize that life on Earth was not a special occurrence, that the universe and all of the events within it were natural consequences of physical and chemical laws, and that humans are the result of a long series of random events."

41　One final prediction of the Copernican principle is: *There's nothing special about water and carbon-based life forms. In other environments, life probably evolves according to different chemistries and energy needs.* Although this belief is hard to falsify, the evidence for carbon and water's unique fitness for life is so strong that leading SETI investigators, Copernican principle devotees to a man, generally refuse to follow the principle on this point. As noted in chapter 2, there's a great deal of evidence that carbon's diverse capacity for chemical bonding makes it especially fitted for the job of bearing biological information. Silicon is a distant second, and no other elements are plausible competitors. And as we've discussed, water is similarly peerless in its role as partner to carbon, thanks to a cluster of life-essential but anomalous properties.

42 Brandon Carter first suggested this idea. Philosopher John Leslie developed it in *The End of the World: The Ethics and Science of Human Extinction* (London: Routledge, 1996). For detailed analysis, see Nick Bostrom, *Anthropic Bias: Observation Selection Effects in Science and Philosophy* (London: Routledge, 2002). J. Richard Gott III defends this type of reasoning in "Implications of the Copernican Principle for Our Future Prospects," *Nature* 363 (May 1993): 315–19. Their various arguments render different judgments and timescales, which we do not discuss here. For a critique of Gott's argument, see Carlton M. Caves, "Predicting Future Duration from Present Age: A Critical Assessment," *Contemporary Physics* 41, no. 33 (2000): 143–53. The Doomsday argument is often combined with additional suggestions that advanced civilizations may have a propensity to destroy themselves, either through weapons of mass destruction or environmental degradation. It may be these politically useful aspects of the argument, rather than its intrinsic plausibility, lead intellectuals to take it seriously.

43 All such temporal applications of the Copernican principle assume that the future is *finite*. If, in contrast, the future is (potentially) infinite, the argument fails since observers will always find themselves near the "beginning" of cosmic time.

44 Amir D. Aczel, *Probability 1: Why There Must Be Intelligent Life in the Universe* (New York: Harcourt Brace, 1998), 208–214.

45 See Steven J. Dick, "Cosmotheology: Theological Implications of the New Universe," *Many Worlds*, 202. D.W. Pasulka explores the spiritual and religious themes of UFOs and reported encounters with non-human intelligences in *American Cosmic* (Oxford, UK: Oxford University Press, 2017).

46 David Darling's defense of SETI optimism, *Life Everywhere: The Maverick Science of Astrobiology* (New York: Basic Books, 2001), is a prime example of the anti-religious tendency of many SETI researchers. We know of one exception. See Michael D. Papagiannis, "May There Be an Ultimate Goal to the Cosmic Evolution?" *Bioastronomy: The Search for Extraterrestrial Life—The Exploration Broadens*, eds. J. Heidmann and M. J. Klein (Berlin: Springer-Verlag, 1990), 366–67.

47 In 2001, Seth Shostak of the SETI Institute gave a public lecture as part of a Templeton Foundation lecture series titled "The Search for Extraterrestrials and the End of Traditional Religion." In a similar vein, see Jill Cornell Tarter (also of the SETI Institute), "SETI and the Religions of the Universe," *Many Worlds*, 143–49.

48 Koerner and LeVay in *Here Be Dragons*, 173.

49 Our own views about the existence of ETIs have changed over the years, but for empirical rather than theological reasons. The possible existence of ETIs raises interesting if ambiguous theological issues. Some scientists, in fact, argue that the existence of ETIs would enhance the prospects for design. Physicist Paul Davies, for instance, argues, "If life is widespread in the universe, it gives us more, not less, reason to believe in cosmic design." "Biological Determinism, Information Theory, and the Origin of Life," *Many Worlds*, 15. In *The Face That Is in the Orb of the Moon*, Plutarch argued similarly for life on the Moon on the basis of intelligent design. See Lewis White Beck, "Extraterrestrial Intelligent Life," in Regis, *Extraterrestrials: Science and Alien Intelligence*, 3–6. Of course, we're not arguing that the existence of ETI would have no theological consequences. It obviously would. For some discussion, see Ernan McMullin, "Life and Intelligence Far from Earth: Formulating Theological Issues," in *Many Worlds*, 151–75. Historically, theological views have influenced astronomical research in a diversity of ways. See Michael J. Crowe, "Astronomy and Religion: Some Historical Interactions Regarding Belief in Extraterrestrial Intelligent Life," *Osiris* 16 (2001): 209–26.

50 We like the approach of C. S. Lewis. He made no theological pronouncements about ETIs, though he included extraterrestrials in his fictional works, most prominently in *Out of the Silent Planet* (1938) and *Perelandra* (1943). Speaking as a Christian, he argued, decades before the discovery of an extrasolar planet, that the existence of ETI is, and should remain, an open question:

> We know that God has visited and redeemed His people, and that tells us just as much about the general character of the creation as a dose given to one sick hen on a big farm tells us about the general character of farming in England. . . . It is, of course, the essence of Christianity that God loves man and for his sake became man and died. But that does not prove that man is the sole end of nature. In the parable, it was one lost sheep that the shepherd went in search of: it was not the only sheep in the flock, and we are not told that it was the most valuable—save insofar as the most desperately in need has, while the need lasts, a peculiar value in the eyes of Love. The doctrine of the Incarnation would conflict with what we know of this vast universe only if we knew also there were other rational species in it who had, like us, fallen, and who needed redemption in the same mode, and they had not been vouchsafed it. But we know of none of these things.

> Lewis dryly commented on atheists' attempts to use both sides of the ETI debate as a weapon against Christianity:

> If we discover other bodies, they must be habitable or uninhabitable: and the odd thing is that both these hypotheses are used as grounds for rejecting Christianity. If the universe is teeming with life, this, we are told, reduces to absurdity the Christian claim—or what is thought to be the Christian claim—that man is unique, and the Christian doctrine that to this one planet God came down and was incarnate for us men and our salvation. If, on the other hand, the earth is really unique, then that proves that life is only an accidental by-product in the universe, and so again disproves our religion. Really, we are hard to please.

> In C. S. Lewis, "Dogma and the Universe," [1943] in *The Grand Miracle and Other Essays on Theology and Ethics from 'God in the Dock,'* ed. by W. Hooper (New York: Ballantine Books, 1990), 14.

51 In Steven J. Dick, "Cosmotheology: Theological Implications of the New Universe," *Many Worlds*, 200–205. In "Extraterrestrial Life and Our World View at the End of the Millennium," *Dibner Library Lecture* (Washington, DC: Smithsonian Institution Series, 2000), 14–25, Dick argues similarly: "I propose that . . . there are two, and only two, 'chief world systems,' in the sense of overarching cosmologies that affect, or should affect, our world view at a lower level of the hierarchy such as philosophical or religious. . . . These two mutually exclusive world views, stemming from the concept of cosmic evolution, are as follows: that cosmic evolution commonly ends in planets, stars, and galaxies, or that it commonly ends in life, mind, and intelligence." Obviously, he's committing one of the oldest and best-known logical fallacies—the false dilemma.

CHAPTER 15: A UNIVERSE DESIGNED FOR DISCOVERY

1   We thank Gershon Robinson and Mordechai Steinman for popularizing this example. For an intriguing argument that *2001: A Space Odyssey* makes a somewhat unconscious argument for design, see their *The Obvious Proof* (CIS Publishers, 1993) as well as their interesting website, "The 2001 Principle," http://www.2001principle.com/.

2   We are alluding here to a common method of reasoning in the historical and origins sciences, often called abductive reasoning or "inference to the best explanation." For discussion, see Stephen C. Meyer, "The Scientific Status of Intelligent Design: The Methodological Equivalence of Naturalistic and Non-Naturalistic Origins Theories," *Science and Evidence for Design in the Universe* (San Francisco: Ignatius Press, 2000), 176–192, and in the same anthology, William A. Dembski and Stephen C. Meyer, "Fruitful Interchange or Polite Chitchat? The Dialogue between Science and Theology," 213–28. See also Peter Lipton, *Inference to the Best Explanation* (London: Routledge, 1991). For a collection of articles about the modern design argument, see Neil A. Manson, editor, *God and Design: The Teleological Argument and Modern Science* (London: Routledge, 2003).

3   See Bell's personal account of the episode in S. Jocelyn Bell Burnell, "Little Green Men, White Dwarfs or Pulsars?" *Annals of the New York Academy of Science* 302 (December 1977): 685–89.

4   Of course, if the universe were deterministic, then, strictly speaking, nothing within it would be the result of chance. Even among those who allow chance explanations, many restrict true chance to the realm of quantum indeterminacy. Protons might decay by chance but falling Scrabble letters do not. On such a view, one might insist that, if we had sufficient knowledge of the initial conditions, we could trace a determining cause for every falling Scrabble piece. We don't intend to take a position on such controversies. Here, we're simply describing the way we normally make causal explanations in everyday life. In this example, our central point is that gravity alone does not determine the specific configuration of the Scrabble letters.

5   Peter Ward and Don Brownlee, *Rare Earth: Why Complex Life is Uncommon in the Universe* (New York: Copernicus, 2000).

6   He first developed and articulated this concept in William A. Dembski, *The Design Inference* (Cambridge, UK: Cambridge University Press, 1998). He developed it for the general reader in *Intelligent Design: The Bridge between Science and Theology* (Downers Grove: InterVarsity Press, 1999). He continued to refine the concept in *No Free Lunch: Why Specified Complexity Cannot Be Purchased without Intelligence* (Lanham, MD: Rowman & Littlefield, 2001). With co-author Winston Ewert, he further refined the arguments in a revised and expanded second edition of *The Design Inference* (Seattle: Discovery Institute Press, 2023). There are many complicated philosophical issues surrounding Dembski's argument, which we have avoided here.

7   Dembski and Ewert, *The Design Inference* (2023), 131-2. We're just hinting at the detailed explanation of specification in this book. For more, see 130-168.

8   Del Ratzsch, *Nature, Design, and Science* (New York: SUNY Press, 2001), 72–73.

9   This phrase is from a public speech by Sir John Polkinghorne at a conference titled Cosmos and Creator, held in Seattle, Washington, April 27, 2001.

10 See John Leslie, *Universes* (London: Routledge, 1989). This type of argument is somewhat subjective if considered in isolation. However, as a supplement to the previous analysis, it seems to capture one of the reasons we often infer design.

11 On both these points, see the excellent brief article by Neil Manson, "Cosmic Fine-Tuning, 'Many Universe' Theories and the Goodness of Life," in *Is Nature Ever Evil?*, edited by William Drees (London: Routledge, 2003), 139–146.

12 Hans Blumenberg, *The Genesis of the Copernican Revolution* translated by Robert M. Wallace (Cambridge, MA: MIT Press, 1987), 3, originally published as *Die Genesis der kopernikanischen Welt* in 1975.

13 Michael J. Denton, *Nature's Destiny: How the Laws of Biology Reveal Purpose in the Universe* (New York: The Free Press, 1998), 50–61.

14 *Nature's Destiny*, 262.

15 Denton has returned to and further mined this terrain in five books published from 2016 through 2022, *The Privileged Species* series. See https://privilegedspecies.com/books/.

16 John D. Barrow and Frank J. Tipler, *The Anthropic Cosmological Principle* (Oxford, UK: Oxford University Press, 1986), 258–76.

17 S. L. Jaki, *Maybe Alone in the Universe After All* (Pinckney, MI: Real View Books, 2000).

18 We both attended, and Guillermo presented, an early version of our thesis at a symposium at Yale on November 2-4, 2000, titled, "Science and Evidence for Design in the Universe." Robin Collins gave a presentation arguing from a more theoretical perspective that the laws of physics are setup for scientific discovery. We were unaware of each other's work on the topic prior to the meeting.

19 Robin Collins, "The Argument from Physical Constants: The Fine-Tuning for Discoverability," in *Two Dozen (or so) Arguments for God*, J. L. Walls and T. Dougherty, eds. (New York: Oxford University Press, 2018), 94. Robin received a grant from the Templeton Foundation to

research this topic. He is expecting to publish an extensive analysis about the time that our book goes to press.

20 In "Resonances," *Physics Today* 27, no. 2 (1974): 73.

CHAPTER 16: FIELDING THE BEST OBJECTIONS TO THE PRIVILEGED PLANET

1   Charles Darwin, *The Origin of Species*, edited with an introduction by Gillian Beer (Oxford, UK: Oxford University Press, 1996), 4 (originally published in 1859).
2   Guillermo Gonzalez, "The Solar System: Favored for Space Travel," *BIO-Complexity* Vol 2020.
3   Robin Collins, "The Argument from Physical Constants: The Fine-Tuning for Discoverability", in Jerry L. Walls, and Trent Dougherty (eds), *Two Dozen (or so) Arguments for God: The Plantinga Project* (New York: Oxford Academic, 2018), 90–108.
4   Indeed, skeptics of our argument are welcome to do just this. Every clever and credible observational technique they invent that would only work in a seemingly less observer-friendly environment would chip away at our argument. One such idea would not destroy our argument because it would take many, many such ideas to show that Earth was less than optimal over a range of observational situations than some uninhabitable place. Such a slew of credible insights is possible, though, and is one more example of how the correlation is falsifiable.
5   Andrea Montovani, "Known knowns, known unknowns, unknown unknowns & Leadership," Medium (April 28, 2020), https://medium.com/@andreamantovani/known-knowns-known-unknowns-unknown-unknowns-leadership-367f346b0953.
6   We made this prediction shortly after we began writing this book, in February or March 2001. Some discussions of the possibility of using gamma ray bursts as standard candles had already appeared in press. Astronomers were still debating the issue when the original edition of this book was published in March 2004.
7   Maria Giovanna Dainotti, et al., "The Gamma-ray Bursts Fundamental Plane Correlation as a Cosmological Tool," *Monthly Notices of the Royal Astronomical Society* 518, no. 2 (2023): 2201–40.
8   Maria Giovanna Dainotti, et al., "Quasars: Standard Candles up to z= 7.5 with the Precision of Supernovae Ia," *The Astrophysical Journal* 950, no. 1 (2023): 45.
9   See J. Armstrong, L. Wells, and G. Gonzalez, "Rummaging through Earth's Attic for Remains of Ancient Life," *Icarus* 160 (2002), 183–96. See also the News and Views piece on the study: C. R. Chapman, "Earth's lunar attic," *Nature* 419 (2002), 791–94.
10 Robin Collins, "The Argument from Physical Constants." Robin has received a Templeton Grant to develop his thesis. He is expecting to publish more extensively on his findings probably about the time our book goes to press. See https://www.templeton.org/grant/the-fine-tuning-for-scientific-discovery.
11 *Just Six Numbers: The Deep Forces that Shape the Universe* (New York: Basic Books), 123.
12 In *The Cosmological Background Radiation*, 7, M. Lachieze-Rey and E. Gunzig mention four such assumptions:

- Homogeneity of space applies. Thus it is assumed that all points of space are equivalent and the properties associated with each point are the same—that is, the laws of physics are the same everywhere. This homogeneity is taken to apply to spatial scales greater than those of galaxies, clusters of galaxies, and even superclusters, i.e. greater than a few hundred Mpc.
- Isotropy of space applies. This means there is no privileged direction in space. (Again this refers to large scales.) These two assumptions make up what is called the cosmological principle.
- That the matter in the universe can be described very simply in terms of what is called a perfect fluid. In this case its properties are completely given by its density rho [Greek symbol] and its pressure p [it].
- That the laws of physics are the same everywhere.

Notice that they repeat the assumption about the laws of physics in the first and fourth item. But homogeneity of space is clearly a separate issue from the uniformity and universality of the laws of physics.

13 See S. Nadis, "Size Matters," *Astronomy* 30 (March 2002), 28–32.

14 The cosmologist P. J. E. Peebles notes that inflation theory violates Einstein's cosmological principle, but the boundary of the universe we can see is distant from the wider chaotic domains. *Principles of Physical Cosmology* (Princeton, NJ: Princeton University Press, 1993), 15.

15 A generalized Copernican principle would subvert our justification for doing cosmology. John Barrow makes a related comment in his *Impossibility: The Limits of Science and the Science of Limits* (Oxford, UK: Oxford University Press, 1998), 169–70:

> Before the possibility of inflation was discovered, it was generally assumed that the Universe should look pretty much the same beyond our horizon as it does inside it. To assume otherwise would have been tantamount to assuming that we occupied a special place in the Universe—a temptation that Copernicus taught us to resist. . . . The general character of inflationary universes reveals that we must expect the Universe to be far more exotically structured in both space and time than we had previously expected.

Barrow seems to imply that chaotic inflationary models contradict the Copernican principle. Actually, these theories, which posit multiple universes that differ in their constants and initial conditions, seem to us an example of the remarkable plasticity—and essentially metaphysical character—of the Copernican principle. Formerly we were told that the Copernican principle required us to assume the uniformity of laws and constants throughout the universe. When such an assumption led to a universe that seemed disturbingly fine-tuned for the existence of complex life, it suddenly and perversely required that we posit a multiverse with varying properties between individual universes.

16 "The Anthropic Principle: Laws and Environments," in *The Anthropic Principle: Proceedings of the Second Venice Conference on Cosmology and Philosophy, November 18_19, 1988* F. Bertola and U. Curi, eds. (Cambridge, UK: Cambridge University Press, 1993), 29.

17 Isaac Newton, *Mathematical Principles of Natural Philosophy* (Berkeley: University of California Press, 1960), 542–44.

18 For a careful treatment of Newton's design argument and why it was not a God-of-the-Gaps argument, see Stephen C. Meyer, *The Return of the God Hypothesis: Three Scientific Discoveries that Reveal the Mind Behind the Universe* (New York: HarperOne, 2021), 412–14, 427–29.

19 In the first edition of *The Design Inference*, William Dembski argued that we can decisively rule out chance for a specified event if, given all the "probabilistic resources" available, the chance for the event is still 1 in 10^{150} (10^{-150}) or less. He arrived at this number by combining the number of elementary particles in the observable universe, the duration of the observable universe until heat death plus the Planck time, which is the smallest unit of time for a physical event. The idea is to treat every particle in the universe as part of giant computer in search of the solution—obviously a wildly generous assumption and purposely so to eliminate any chance of a false positive (detecting design where no design existed). 1 in 10^{158} would obviously exceed this "universal probability bound."

> Of course, we often securely infer design even when the odds of arriving at something by chance are much better than Dembski's very stringent universal probability bound. Moreover, in many cases, we know little about the actual probability of producing a structure. How would we make such a determination, for instance, for Mount Rushmore? See Dembski, *The Design Inference* (Cambridge: Cambridge University Press, 1998), 203–214.

20 Thomas Nagel, *Mind and Cosmos: Why the Materialist Neo-Darwinian Conception Is Almost Surely False* (Oxford, UK: Oxford University Press, 2012).

21 As we mentioned in chapter 13, there is also a powerful philosophical argument against the possibility of a universe with an infinite past. See, for example, William Lane Craig, *The Kalam Cosmological Argument* (Eugene, OR: Wipf & Stock Pub., 1979), and William Lane Craig and Quentin Smith, *Theism, Atheism, and Big Bang Cosmology* (Oxford, UK: Clarendon, 1993). Craig also defends this argument in "In Defense of the *Kalam* Cosmological Argument," *Faith and Philosophy* 14 (1997): 236–247.
22 We owe this illustration to astronomer Kyler Kuehn.
23 Jason Vuic, *The Yugo: The Rise and Fall of the Worst Car in History* (New York, NY: Hill & Wang, 2010).

CONCLUSION: READING THE BOOK OF NATURE

1 T.S. Eliot, from "Little Gidding," *Four Quartets* (Eastbourne, UK: Gardners Books, 2001). Originally published in 1943.
2 Robert Zimmerman, *Genesis: Apollo 8: The First Manned Flight to Another World* (New York: Dell Publishing, 1998), 242, 293.
3 The most complete articulation of Polanyi's view is his *Personal Knowledge: Towards a Post-Critical Philosophy* (Chicago: University of Chicago Press, 1958).

APPENDIX A: THE REVISED DRAKE EQUATION

1 *Nicomachean Ethics*, 1 1094a24–1095a.
2 Peter D. Ward and Donald Brownlee, *Rare Earth: Why Complex Life is Uncommon in the Universe* (New York: Copernicus, 2000).
3 In their version of the Drake equation, Ward and Brownlee give n_g, the number of stars in the galactic habitable zone, but it should be f_g, as we present it in the text.
4 R. A. Wittenmyer et al., "Cool Jupiters Greatly Outnumber Their Toasty Siblings: Occurrence Rates from the Anglo-Australian Planet Search," *Monthly Notices of the Royal Astronomical Society* 492, no. 1 (2020): 377–83.
5 For a remarkably cogent analysis and critique of the very "pursuitworthiness" of SETI, see André Kukla, "SETI: On the Prospects and Pursuitworthiness of the Search for Extraterrestrial Intelligence," *Studies in the History and Philosophy of Science* 32, no. 1 (2001), 31–67.

APPENDIX B: WHAT ABOUT PANSPERMIA?

1 Francis Crick, and Leslie Orgel, "Directed Panspermia," *Icarus* 19 (1973): 341–46.
2 Francis Crick, *Life Itself: Its Origin and Nature* (New York: Simon & Schuster, 1981), 88.
3 W. T. B. Kelvin, *Popular Lectures and Addresses; Constitution of Matter. Volume 2: Geology and General Physics* (London: Macmillan, 1894).
4 Svante Arrhenius, *Worlds in the Making: The Evolution of the Universe* (New York: Harper & Row, 1908).
5 Two other types of panspermia discussed in the literature are pseudo-panspermia and necropanspermia. Pseudo-panspermia is the transport of the building blocks of life, such as amino acids. Necropanspermia is the transport of dead organisms. We don't consider either of these as effective ways to infect another world.
6 H. J. Melosh, "Exchange of Meteoritic Material Between Stellar Systems," *Lunar and Planetary Science XXXII* (2001): 2022.
7 P. S. Wesson, "Panspermia Past and Present: Astrophysical and Biophysical Conditions for the Dissemination of Life in Space," *Space Science Reviews* 156 (2010): 239–52.
8 V. Duda, et al., "Ultramicrobacteria: Formation of the Concept and Contribution of Ultramicrobacteria to Biology," *Microbiology* 81, no. 4 (2012): 379–90.
9 R. J. Worth, S. Sigurdsson, C. H. House, "Seeding Life on the Moons of the Outer Planets via Lithopanspermia," *Astrobiology* 13, no. 12 (2014): 1155–65.

10 L. E. Wells, J. C. Armstrong, and G. Gonzalez, "Reseeding of Early Earth by Impacts of Returning Ejecta during the Late Heavy Bombardment," *Icarus* 162 (2003): 38–46.
11 Worth et al., "Seeding Life"; H. J. Melosh, "Exchange of Meteorites (and Life?) Between Stellar Systems," *Astrobiology* 3, no. 1 (2003): 207–215.
12 Worth et al., "Seeding Life"; Melosh, "Exchange of Meteorites."
13 Idan Ginsburg, Manasvi Lingam, Abraham Loeb, "Galactic Panspermia," *Astrophysical Journal Letters* 868 (2018): L12.
14 Ginsburg et al., "Galactic Panspermia." As an illustration, consider two spacefaring rocks, each with the same initial number of organisms. One rock (A) contains a species with a typical survival time in space of 1 year. The other (B) contains a different species that can survive 2 years. After some years have passed, the number viable organisms in the rock B will be 16 times greater than those in rock A.
15 E. Willerslev, A. J. Hansen, R. Ronn, et al., "Long–term Persistence of Bacterial DNA," *Current Biology* 14, no. 1 (2004): R9–R10.
16 A. V. Shatilovich, et al., "Viable Nematodes from Late Pleistocene Permafrost of the Kolyma River Lowland," *Doklady Biological Sciences* 480 (2018): 100–102.
17 T. K. Lowenstein, B. A. Schubert, M. N. Timofeeff, "Microbial Communities in Fluid Inclusions and Long–term Survival in Halite," *GSA Today* 21(1) (January 2011): 4–9.
18 Lowenstein et al., "Microbial Communities in Fluid Inclusions," 7.
19 This appendix is a condensed and edited version of: G. Gonzalez, "Can Panspermia Explain the Origin of Life?" *The Comprehensive Guide to Science and Faith*, W. A. Dembski, C. Luskin, J. M. Holden, eds. (Eugene, OR: Harvest House Publishers, 2021), 449–56.

INDEX